Tropospheric Ozone

Regional and Global Scale Interactions

NATO ASI Series

Advanced Science Institutes Series

A Series presenting the results of activities sponsored by the NATO Science Committee, which aims at the dissemination of advanced scientific and technological knowledge, with a view to strengthening links between scientific communities.

The series is published by an international board of publishers in conjunction with the NATO Scientific Affairs Division

A Life Sciences B Physics	Plenum Publishing Corporation London and New York
C Mathematical and Physical Sciences	D. Reidel Publishing Company Dordrecht, Boston, Lancaster and Tokyo
D Behavioural and Social Sciences E Applied Sciences	Martinus Nijhoff Publishers Dordrecht, Boston and Lancaster
F Computer and Systems Sciences G Ecological Sciences H Cell Biology	Springer-Verlag Berlin, Heidelberg, New York, London, Paris, and Tokyo

Series C: Mathematical and Physical Sciences Vol. 227

Tropospheric Ozone

Regional and Global Scale Interactions

edited by

I. S. A. Isaksen

Institute of Geophysics,
University of Oslo, Oslo, Norway

D. Reidel Publishing Company

Dordrecht / Boston / Lancaster / Tokyo

Published in cooperation with NATO Scientific Affairs Division

Proceedings of the NATO Advanced Workshop on
Regional and Global Ozone Interaction and its
Environmental Consequences
Lillehammer, Norway
June 1-5, 1987

ISBN-13: 978-94-010-7811-5 **e-ISBN-13: 978-94-009-2913-5**
DOI:10.1007/ 978-94-009-2913-5

Published by D. Reidel Publishing Company
P.O. Box 17, 3300 AA Dordrecht, Holland

Sold and distributed in the U.S.A. and Canada
by Kluwer Academic Publishers,
101 Philip Drive, Norwell, MA 02061, U.S.A.

In all other countries, sold and distributed
by Kluwer Academic Publishers Group,
P.O. Box 322, 3300 AH Dordrecht, Holland

D. Reidel Publishing Company is a member of the Kluwer Academic Publishers Group

Softcover reprint of the hardcover 1st edition 1988

This book contains the proceedings of a NATO Advanced Research Workshop held within the programme of activities of the NATO Special Programme on Global Transport Mechanisms in the Geo-Sciences running from 1983 to 1988 as part of the activities of the NATO Science Committee.

Other books previously published as a result of the activities of the Special Programme are as follows:

BUAT-MENARD, P. (Ed.) – *The Role of Air-Sea Exchange in Geochemical Cycling* (C185) 1986

CAZENAVE, A. (Ed.) – *Earth Rotation: Solved and Unsolved Problems* (C187) 1986

WILLEBRAND, J. and ANDERSON, D. L. T. (Eds.) – *Large-Scale Transport Processes in Oceans and Atmosphere* (C190) 1986

NICOLIS, C. and NICOLIS, G. (Eds.) –*Irreversible Phenomena and Dynamical Systems Analysis in Geosciences* (C192) 1986

PARSONS, I. (Ed.) – *Origins of Igneous Layering* (C196) 1987

LOPER, E. (Ed.) – *Structure and Dynamics of Partially Solidified Systems* (E125) 1987

VAUGHAN, R. A. (Ed.) – *Remote Sensing Applications in Meteorology and Climatology* (C201) 1987

BERGER, W. H. and LABEYRIE, L. D. (Eds.) – *Abrupt Climatic Change - Evidence and Implications* (C216) 1987

VISCONTI, G. and GARCIA, R. (Eds.) – *Transport Processes in the Middle Atmosphere* (C213) 1987

HELGESON, H. C. (Ed.) – *Chemical Transport in Metasomatic Processes* (C218) 1987

SIMMERS, I. (Ed.) – *Estimation of Natural Recharge of Groundwater* (C222) 1987

CUSTODIO, E., GURGUI, A. and LOBO FERREIRA, J. P. (Eds.) – *Groundwater Flow and Quality Modelling* (C224) 1987

TABLE OF CONTENTS

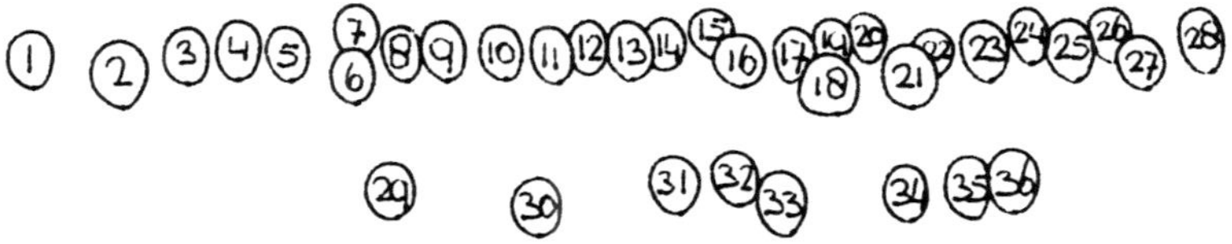

1. R. A. Cox
2. B. Rognerud
3. D. Derwent
4. R. Guicherit
5. R. Cicerone
6. J. Logan
7. I. S. A. Isaksen
8. H. Van Dop
9. S. A. Penkett
10. D. Kley
11. D. Ehhalt
12. W. Seiler
13. S. C. Liu
14. F. Fehsenfeld
15. Ø. Hov
16 S. Joffre
17. H. Kelder
18. H. Hakola
19. J. Jerre
20. G. Ancellet
21. P. Crutzen
22. G. Vaughan
23. A. Volz
24. H. Levy II
25. P. Taalas
26. P. Grennfeldt
27. M. Legrand
28. G. Megie
29. A. Marenco
30. G. Brasseur
31. R. Bojkov
32. J. E. Jonson
33. B. Prinz
34. B. Chatfield
35. M. Isaksen
36. J. Fishman

PREFACE

The main objective of the workshop was to increase our knowledge of ozone formation and distribution in the troposphere, its relation to precursor (NO_x and HC species) distribution, how it is affected by transport processes in the troposphere, and to show how the increasing levels of ozone can cause environmental problem.

The focus was on the interaction of ozone on regional and global scales.

There is mounting evidence that such interactions occur and that the ozone levels are increasing in most of the Northern Hemisphere troposphere. A likely source of ozone increase is human activity. As result of this, tropospheric climate may change significantly within a few decades, either through direct effects by ozone itself or indirectly through its effect on other radiatively active trace species. Furthermore, ozone may have adverse effects on vegetation over large continental areas due to enhanced levels which have been measured to take place. As it is well known that ozone plays a key role in the oxidation of a large number of chemical species in the troposphere, natural as well as man-made, the atmospheric distribution of important trace species like sulfur dioxide, nitrogen oxides and hydrocarbons could be markedly changed as a result of ozone changes.

The rapidly increasing interest in tropospheric ozone, and the key role ozone plays in several atmospheric areas as well the obvious increase in the tropospheric concentration of ozone made ozone a natural choice as a topic for the workshop.

The workshop was divided into two parts. During the first half of the meeting, scientific papers of our present knowledge were presented. During the second half, the participants were divided into four working groups. Each working group was given a specific topic related to ozone.

Several papers were presented which discussed the scientific background of tropospheric ozone. Particular emphasis was given to the exchange of ozone and its precursors (NO_x and HC species) between polluted regional areas and the less polluted background troposphere. It was shown that there are observational evidences that ozone levels are increasing in the troposphere on a regional and also on a global scale, and that present day ozone levels are substantially higher than they were several decades ago when the release of pollutants was small. This change in ozone will have significant consequences for the chemistry of the atmosphere, as ozone is a key species in the chemical turnover in the troposphere. It was also demonstrated that climate effects by ozone itself of by species affected by ozone (e.g.CH_4), and effects on vegetation should be expected with changes in ozone.

The task of the four working groups were to identify areas where gaps in our knowledge exist and make recommendations for work that is needed in the following areas of research relevant to tropospheric ozone: I) Atmospheric transport, II) NO_x in the troposphere, III) Regional ozone and IV) Background ozone.

Based on the discussions in these working groups four group summary reports were prepared. The main suggestions were as follows:

a) The development of sufficiently detailed numerical models (preferably 3-D models) to study the complex interaction between transport and chemistry on different scales in the troposphere in a realistic way.

b) Establish an ozone measuring network with good global coverage for long term trend measurements.

c) Establish emission inventories for NO_x and HC species.

d) Measure the distribution of ozone precursors (particularly NO_x/HC chemistry) the formation of ozone under different atmospheric conditions.

e) Study the effects of ozone and related species on vegetation.

The workshop which was held at Lillehammer, Norway, offered an excellent opportunity to present recent development in the rapidly expanding field of tropospheric ozone research. Several large international projects which are under way (e.g.the EUROTRAC project TOR, the AROCHE project) or planned in the future were presented and discussed at the meeting. Through these projects a considerable amount of new data will be collected over the next few years.

The research workshop was mainly sponsored NATO, and we want to express our gratitude to the NATO Science Committee for the important assistance and cooperation we received. We also like to thank the Norwegian Research Council for Science and Technology (NTNF) and the Department of Environment (MD) for their support of the meeting.

It is hoped that the content of this book will contribute to the understanding of the complex interaction of chemistry and transport which determined ozone distribution and changes in the troposphere. With the rapid increasing interest in the topic of tropospheric ozone and the large amount of data being collected it is important that this workshop is followed by a similar workshop within a few years.

Oslo, October 1987 Ivar S.A. Isaksen

LIST OF AUTHORS AND PARTICIPANTS

Authors

Dr. G. Ancellet
Centre National de la Recherche Scientifique
Service d'Aèronomie
B.P. No. 3
F-91371 Verrieres-Buisson
FRANCE

Dr. Rumen D.Bojkov
Atmospheric Environment Service
4905 Dufferin Street
Downsview, Ontario
CANADA M3H 5T4

Dr.Bob Chatfield
National Center for Atmospheric Research
P.O.3000
Boulder, Colorado 80307
USA

Dr. R.A.Cox
EMS Division Bldg 551
AERE Harwell
Didcot, Oxon OX11 ORA
UNITED KINGDOM

Dr. Paul Crutzen
Max-Planck-Institut für Chemie
Box 3060
D-6500 MAINZ
FEDERAL REPUBLIC OF GERMANY

Dr. Han van Dop
Royal Netherlands Meteorological Institute
Division of Physical Meteorology
P.O. box 201
3730 AE de bilt
THE NETHERLANDS

Dr. Dieter H. Ehhalt
Institut für Chemie
Kernforschungsanlage Jülich
Institut 3
Atmosphärische Chemie
Postfach 1913
D-517 Jülich
FEDERAL REPUBLIC OF GERMANY

Dr.Fred Fehsenfeld
National Oceanic and Atmospheric Administration
Aeronomy Laboratory/ERL
325 Broadway
Boulder
Colorado 80303
USA

Dr. Jack Fishman
NASA Langley Research Center
Atmospheric Sciences Division
Hampton
Virginia 23665-5225
USA

Dr. Robert Guicherit
T N O
Division of Technology for Society
Department of Environmental Chemistry
P.O. Box 217
NL-2600 AE Delft
THE NETHERLAND

Dr.Øystein Hov
Norwegian Institute for Air Research
P.O.Box 64
N-2001 Lillestrøm
NORWAY

Dr. Sylvain Joffre
Finnish Meteorological Institute
Air Quality Department
Sahaajankatu 22E
SF-00810 Helsinki
FINLAND

Dr.Dieter Kley
Institut für Chemie
Kernforschungsanlage Jülich
Chemie der Belasteten Atmosphäre
Postfach 1913
D-5170 Jülich
FEDERAL REPUBLIC OF GERMANY

Dr.Hiram Levy, II
U.S. Department of Commerce
National Oceanic and Atmospheric Administration
Geophysical Fluid Dynamics Laboratory
Princeton University,P.O. Box 308
Princeton,New Jersey 08542
USA

Dr.Shaw C. Liu
National Oceanic and Atmospheric Administration
Environmental Research Laboratories
325 Broadway
Boulder, Colorado 80303
USA

Dr.Jennifer Logan
Harvard University
Department of Earth and Planetary Sciences
Pierce Hall
29 Oxford st.
Cambridge, MA 02138
USA

Dr.Alain Marenco
Laboratoire d'Aèrologie
Universitè Paul Sabatier
31062 - Toulouse Cedex
FRANCE

Dr.Stuart A.Penkett
University of East Anglia
School of the Environmental Sciences
Norwich NR4 7TJ
UNITED KINGDOM

Dr. Bernhard Prinz
Landesanstalt für Immissionsschutz
Nordrheim - Westfalen
Wallneyerstr.6
D-4300 Essen
FEDERAL REPUBLIC OF GERMANY

Dr. Geraint Vaughan
Department of Physics
University College of Wales
Aberystwyth SY23 3BZ
Wales
UNITED KINGDOM

Dr. Andreas Volz
Kernforschungsanlage Jülich
Institut für Chemie 2
Postfach 1913
D-5170 Jülich
FEDERAL REPUBLIC OF GERMANY

Dr. W.C.Wang
Atmospheric Environmental Research Inc.
840 Memorial Drive
Cambridge,MA 02139
USA

Participants

Dr.Guy Brasseur
Institut d'Aeronomie
Spatiale de Belgique
Ave. Circulaire 3
Brussels Uccle B 18
BELGIUM

Dr. Ralph Cicerone
Atmospheric Chemistry Division
National Center for Atmospheric Research
Boulder, Colorado 80307,
USA

Dr.Dick Derwent
EMS Division Bldg 551
AERE Harwell
Didcot, Oxon OX11 ORA
GREAT BRITAIN

Dr. Anver Ghazi
EEC Comission DGX II
Rue de la Loi 200
B-1049 Brussels
BELGIUM

Dr.Peringe Grennfeldt
IVL Box 5207
40224 Gøteborg
SWEDEN

Mrs. Hannele Hakola
Finnish Meteorological Institute
Air Quality Department
Sahaajankatu 22E
SF-00810 Helsinki
FINLAND

Dr.Ivar S.A.Isaksen
University of Oslo
Institute of Geophysics
P.O.Box 1022
0315 Blindern, Oslo 3
NORWAY

Mr. Jon Jerre
Fraunhofer-Institute for Atm.Umweltforschung
Hindenburgstrasse 43
D-8100 Garmish Partenkirchen
FEDERAL REPUBLIC OF GERMANY

Mr. Jan Eiof Jonson
University of Oslo
Institute of Geophysics
P.O.Box 1022
0315 Blindern, Oslo 3
NORWAY

Dr.H.Kelder
KNMI
p.o.box 201
3730 AE de Bilt
THE NETHERLANDS

Dr. Michel Legrand
Laboratoire de Glaciologie
et Geophysique
de l'Environment
B.P. 96
F-38402 St.Martin
D'Heres Cedex
FRANCE

Dr. Gerard Megie
Centre National de la Recherche
Scientifique
Service d`Aeronomie
B.P.No.3
F-91371 Verrieres - Le Buisson Cedex
FRANCE

Mrs. Bjørg Rognerud
University of Oslo
Institute of Geophysics
P.O.Box 1022
0315 Blindern, Oslo 3
NORWAY

Dr. Wolfgang Seiler
Fraunhofer-Institut für Atmosphärische
Umweltforschung
Kreuzeckbahnstrasse 19
D-8100 Garmisch-Partenkirchen
FEDERAL REPUBLIC OF GERMANY

Dr.Frode Stordal
University of Oslo
Institute of Geophysics
P.O.Box 1022
0315 Blindern, Oslo 3
NORWAY

Mr. Petteri Taalas
Finnish Meteorological Institute
Air Quality Department
Sahaajankatu 22E
SF-00810 Helsinki
FINLAND

Part I

PRESENTATIONS

TROPOSPHERIC OZONE: AN OVERVIEW

Paul J. Crutzen
Max-Planck-Institute for Chemistry
P.O. Box 3060
D-6500 Mainz
West Germany

ABSTRACT. Although only about 10% of all atmospheric ozone is located in the troposphere, it is the main driver of the photochemical processes which lead to the recycling of most of the gases that are emitted into the atmosphere by natural processes and anthropogenic activities.

Substantial changes are taking place in the tropospheric ozone abundance. In this paper, an overview is given of the chemical processes that lead to ozone formation and destruction in the troposphere, emphasizing human impact on the global scale. In the northern hemisphere, anthropogenic emissions of NO_x clearly lead to high ozone concentrations during photochemical pollution episodes. However, also in the background free troposphere, pronounced ozone concentration increases have occurred, which are here confirmed by a review of measurements that were made in the 1930's - 50's in Europe.

The potential for tropospheric ozone production from carbon monoxide and hydrocarbon oxidation, with NO_x acting as catalyst, is enormous. At this stage, only at most 10% of this does actually occur, due to insufficient NO_x in the background troposphere. Future human developments that lead to increasing NO emissions may, therefore, lead to further ozone increases, especially in the tropics.

1. INTRODUCTION

"Among the many interesting bodies which the researches of modern chemists have brought to light, few are more remarkable than the substance to which the name of ozone has been given". With this sentence starts an article written in 1855 by Thomas Andrews, M.D. and Professor of Chemistry in Queen's College, Belfast, one of the early pioneers of ozone research. By that time the exact chemical composition of ozone was yet not established. One proposal was that the gas

I. S. A. Isaksen (ed.), Tropospheric Ozone, 3–32.

was the teroxide of hydrogen HO_3, based on the fact it is also formed when pure water is electrolyzed, a discovery which was made in 1840 by the German chemist Schönbein, working in Basel, Switzerland. Soret (1863, 1865) and Andrews (1874) argued that ozone is a pure oxygen allotrope, probably consisting of three oxygen atoms. The question about the exact chemical composition of ozone seems firmly to have been settled first by the beginning of this century. A comprehensive overview of early ozone research is given by Fonrobert (1916).

By the latter quarter of the 19th century, it had been clearly demonstrated by Andrews and others that ozone was a normal constituent of atmospheric air. This had already been proposed in 1845 by Schönbein, who also introduced an "ozonometer", consisting of iodized starch paper, which was extensively used at many sites around the world. The interest in ozone measurements was especially stimulated since Schönbein discussed the possible role of ozone as a disinfectant against epidemic disease and it is claimed by Fonrobert (1916) that a million ozone measurements were made with the Schönbein method. It is unfortunate that the many data on tropospheric ozone that were taken with the Schönbein test paper are only of limited value, because the method is unspecific and even dependent on meteorological factors such as wind speed and atmospheric humidity (Bojkov, 1986).

By 1874 some basic discoveries about surface ozone had nevertheless already been made by some careful researchers. We quote from Andrews (1874): "Ozone is rarely found in the air of large towns, unless in a suburb when the air is blowing from the country. ... It is, in fact, rapidly destroyed by smoke and other impurities which are present in the air of localities where large bodies of man have fixed their habitations, and I have often observed this destructive action extending to a distance of one or two miles from a manufacturing town, even in fine and bright weather. Ozone is rarely, if ever, absent in fine weather from the air of the country; and it is more abundant, on the whole, in the air of the mountain than of the plain".

Unfortunately, we probably only have one long series of measurements which may be taken representative for ground level ozone concentrations during the past century, namely those made by A. Levy and coworkers at the Montsouris observatory in Paris between 1876 and 1907 (Levy, 1907). The quality of those observations has recently been ascertained by Volz and Kley (1987; also this volume). They indicate 3-4 times lower average concentrations of tropospheric ozone than are presently measured at comparable sites. Because at current levels ozone is phytotoxic (Prinz, this volume; Skarby and Sellden, 1984), this is an extremely important observation. Other measurements of tropospheric ozone which are in agreement with these findings will be reviewed in this paper.

Interest in the study of tropospheric ozone strongly declined during the first decades of this century and one may speculate whether this was not largely due to the low quality of the measurements which precluded the discovery of any temporal or spatial patterns in the observations. We cite again from Andrews (1874): "The amount of ozone in the air is greater, according to some observers, in winter than in summer, in spring than in autumn; according to others, it is greater in spring and summer than in autumn and winter. As regards the influence of day and night, the observations do not tell the same tale. Ozone has usually been found more abundantly in the air at night than by day, but some careful observers have found the reverse of this statement to be true". Even after taking great care with the Schönbein measurements, Friesenhof in 1904 could only summarize the following results from a major observational program in Austria: " ... and thus I have already noted that the occurrence of thunderstorms in the vicinity are accompanied by abnormally low ozone concentrations" and further: "The second result was a diurnal period in ozone concentrations with a maximum during the hours prior to midnight and a minimum in the hours before noon" (transl. by author).

More serious research in tropospheric ozone had to await the development of more reliable instrumentation, in the first place optical techniques. With these Strutt (1918) and Fabry and Buisson (1921) deduced that most ozone must be located in the stratosphere with a total abundance of about 3mm STP. The ozone maximum was first thought to be located near 50 km altitude (Cabannes et Dufay, 1927), but this was later corrected to about 27 km (Götz *et al*, 1934). The optical techniques were subsequently also applied to tropospheric ozone measurements (e.g. Fabry and Buisson, 1931; Götz and Ladenburg, 1931). However, because the spectrographic technique was too cumbersome to use for long term observations, by the end of the 1930's improved chemical techniques, mostly based on the conversion of potassium iodide (KI) by ozone to iodine (I_2), were again increasingly used (Cauer, 1935; Paneth and Edgar, 1938; Regener, 1938). With this method many measurements were made during and after worldwar II at several health and vacation resorts in Germany and Switzerland. Most of these data are probably of sufficient quality to reveal any substantial differences between present tropospheric ozone concentrations and those measured 30-50 years ago. Before giving a brief review of past ozone measurements, we will, however, first show why the study of tropospheric ozone is of fundamental importance for an understanding of the photo chemistry of the atmosphere.

2. THE ROLE OF OZONE IN TROPOSPHERIC CHEMISTRY

After it became clear that most ozone is located in the stratosphere, first by ground-based measurements (Strutt, 1918; Götz et al, 1934) and later by in-situ balloon-borne observations (Regener and Regener, 1934; Regener, 1938), it was for a long time generally believed that tropospheric ozone originated from the stratosphere and that most of it was destroyed at the earth's surface, although some loss admittedly could also occur in the troposphere (e.g. Regener, 1943). This idea was supported by the observed strong correlations between total ozone and tropospheric meteorological patterns (Dobson et al, 1929). Furthermore, since Chapman (1930), it had become clear that ozone was produced by the photodissociation of O_2, which can only occur at wavelengths shorter than 240 nm, precluding production in the troposphere. From the mid-1940's, however, it became obvious that ozone could also be produced in the troposphere. After heavy injury to vegetable crops occurred repeatedly in the Los Angeles area, it was shown by Haagen-Smit (1952) that the plant damage could be reproduced by ozone, which had been shown to be present in such high concentrations in the Los-Angeles "photochemical smog" that a stratospheric origin could be excluded (Haagen-Smit and Fox, 1956). The overall reaction mechanism was identified as:

$$NMHC + NO_x + h\nu \longrightarrow O_3 + \text{other pollutants}$$

, where NMHC denotes various reactive non-methane-hydrocarbons, NO_x is $NO + NO_2$, and $h\nu$ is solar radiations of wavelengths less than about 400 nm (Finlayson-Pitts and Pitts, 1985). Ozone formation is possible, because solar radiation between about 300 and 400 nm can reach the earth's surface and dissociate NO_2 into NO and O. The recombination of O with O_2 produces O_3. For about 20 years it was thought that ozone formation through this mechanism could only take place in atmospheric environments that were heavily polluted with automobile exhausts and strongly illuminated with sunlight. It was only in the 1970's that photochemical smog was also discovered in Europe (e.g. Guicherit and van Dop, 1977; Derwent and Stewart, 1973). At about the same time the first proposals were made that significant in-situ photochemical production and destruction could also take place on global scales (Crutzen, 1973, 1974; Chameides and Walker, 1973).

Already in 1971 and 1972 Levy had proposed that ozone photolysis at wavelengths shorter than about 310 nm leads to the production of hydroxyl (OH)

$$O_3 + h\nu \longrightarrow O(^1D) + O_2 \quad (\lambda \leq 310\ nm) \quad (1)$$
$$O(^1D) + H_2O \longrightarrow 2\ OH \quad (2)$$

and that hydroxyl plays an exceedingly important role in the removal of many gases that are produced naturally or by human activities. Levy (1971) also proposed that a chain reaction leading to increased hydroxyl production could occur via the formation of CH_2O in the methane oxidation cycle. McConnell _et al_ (1971) and Wofsy _et al_ (1972) postulated that reaction of hydroxyl with methane could be the most important atmospheric source of carbon monoxide. In earlier years it was either believed that the atmospheric CO budget was totally dominated by natural processes or by anthropogenic activities. McConnell _et al_ (1971) and Wofsy _et al_ (1972) also proposed that the atmospheric CO distribution was mainly governed by the reactions

$$CH_4 + OH \longrightarrow \ldots\ldots \longrightarrow CO \quad (3)$$
$$CO + OH \longrightarrow H + CO_2 \quad (4)$$

However, because carbon monoxide concentrations are much more variable than those of methane, it is clear that a variety of natural and anthropogenic processes must make additional important contributions to the atmospheric CO cycle, such as fossil fuel burning (Seiler, 1974), biomass burning in the tropics (Crutzen _et al_, 1979; Crutzen _et al_, 1985) and oxidation of hydrocarbons emitted by forests.

Crutzen (1973, 1974) pointed out that ozone could both be produced and destroyed by the methane and carbon monoxide oxidation cycles, depending on the concentrations of nitric oxide (NO). For instance, the carbon monoxide oxidation cycles can either proceed via:

$$CO + OH \longrightarrow CO_2 + H \quad (4)$$
$$H + O_2 + M \longrightarrow HO_2 + M \quad (5)$$
$$HO_2 + NO \longrightarrow OH + NO_2 \quad (6)$$
$$NO_2 + h\nu \longrightarrow NO + O \ (\lambda \leq 400\ nm) \quad (7)$$
$$O + O_2 + M \longrightarrow O_3 + M \quad (8)$$
$$\text{net: } CO + 2O_2 \longrightarrow CO_2 + O_3$$

or by

$$CO + OH \longrightarrow CO_2 + H \quad (4)$$
$$H + O_2 + M \longrightarrow HO_2 + M \quad (5)$$
$$HO_2 + O_3 \longrightarrow OH + 2O_2 \quad (9)$$
$$\text{net: } CO + O_3 \longrightarrow CO_2 + O_2$$

Because the rate coefficient for reaction (6) is about 4000 times faster than that for reaction (9), the ozone producing carbon monoxide oxidation branch is more important than the ozone destruction branch for NO to O_3 concentration ratios exceeding 1:4000, i.e. for NO volume mixing ratios larger than about $5\text{-}10\times10^{-12}$ (5-10 pptv) in the lower troposphere.

Similar, but more complex reaction cycles do also take place during the oxidation of hydrocarbon gases, with methane being the most important example in extensive portions of the atmosphere, especially in the remote marine environments. The oxidation of methane plays a large role in the tropospheric ozone and hydroxyl balance. Again the availability of NO plays a decisive role. For example, in NO-rich environments, the oxidation of CH_4 to CH_2O occurs mostly via the reaction steps

$$CH_4 + OH \longrightarrow CH_3 + H_2O \qquad (3)$$
$$CH_3 + O_2 + M \longrightarrow CH_3O_2 + M \qquad (10)$$
$$CH_3O_2 + NO \longrightarrow CH_3O + NO_2 \qquad (11)$$
$$CH_3O + O_2 \longrightarrow CH_2O + HO_2 \qquad (12)$$
$$HO_2 + NO \longrightarrow OH + NO_2 \qquad (6)$$
$$2\text{x } (NO_2 + h\nu \longrightarrow NO + O);\ \lambda \leq 400 \text{ nm} \qquad (7)$$
$$2\text{x } (O + O_2 + M \longrightarrow O_3 + M) \qquad (8)$$
$$\text{net: } CH_4 + 4\ O_2 \longrightarrow CH_2O + H_2O + 2\ O_3$$

In NO-poor environments, other reaction paths, leading to loss of odd hydrogen ($OH + HO_2$) radicals are important:

$$CH_4 + OH \longrightarrow CH_3 + H_2O \qquad (3)$$
$$CH_3 + O_2 + M \longrightarrow CH_3O_2 + M \qquad (10)$$
$$CH_3O_2 + HO_2 \longrightarrow CH_3O_2H + O_2 \qquad (13)$$
$$CH_3O_2H + OH \longrightarrow CH_2O + OH + H_2O \qquad (14a)$$
$$\text{net: } CH_4 + OH + HO_2 \longrightarrow CH_2O + 2\ H_2O$$

and a catalytic subcycle:

$$CH_3O_2 + HO_2 \longrightarrow CH_3O_2H + O_2 \qquad (10)$$
$$CH_3O_2H + OH \longrightarrow CH_3O_2 + H_2O \qquad (14b)$$
$$\text{net: } OH + HO_2 \longrightarrow H_2O + O_2$$

No net effect on either O_3 or odd hydrogen has the cycle:

$$CH_4 + OH \longrightarrow CH_3 + H_2O \qquad (3)$$
$$CH_3 + O_2 + M \longrightarrow CH_3O_2 + M \qquad (10)$$
$$CH_3O_2 + HO_2 \longrightarrow CH_3O_2H + O_2 \qquad (13)$$
$$CH_3O_2H + h\nu \longrightarrow CH_3O + OH \qquad (15)$$
$$CH_3O + O_2 \longrightarrow CH_2O + HO_2 \qquad (12)$$
$$\text{net: } CH_4 + O_2 \longrightarrow CH_2O + H_2O$$

Finally, CH_2O is decomposed to CO either by the reaction

$$CH_2O + h\nu \longrightarrow CO + H_2,\ \leq 350 \text{ nm} \qquad (16a)$$

, or in NO-rich environments by

$$CH_2O + h\nu \longrightarrow CHO + H, \quad \lambda \leq 350 \text{ nm} \qquad (16b)$$
$$CHO + O_2 \longrightarrow CO + HO_2 \qquad (17)$$
$$H + O_2 + M \longrightarrow HO_2 + M \qquad (5)$$
$$2x\ (HO_2 + NO \longrightarrow OH + NO_2) \qquad (6)$$
$$2x\ (NO_2 + h\nu \longrightarrow NO + O) \qquad (7)$$
$$2x\ (O + O_2 + M \longrightarrow O_3 + M) \qquad (8)$$
$$\text{net: } CH_2O + 4O_2 \longrightarrow CO + 2\ OH + 2\ O_3$$

$$CH_2O + OH \longrightarrow CHO + H_2O \qquad (18)$$
$$CHO + O_2 \longrightarrow CO + HO_2 \qquad (17)$$
$$HO_2 + NO \longrightarrow OH + NO_2 \qquad (6)$$
$$NO_2 + h\nu \longrightarrow NO + O \qquad (7)$$
$$O + O_2 + M \longrightarrow O_3 + M \qquad (8)$$
$$\text{net: } CH_2O + 2O_2 \longrightarrow CO + H_2O + O_3$$

In NO-poor environments, the significant CH_2O reactions are:

$$CH_2O + h\nu \longrightarrow CHO + H \qquad (16b)$$
$$CHO + O_2 \longrightarrow HO_2 + CO \qquad (17)$$
$$H + O_2 + M \longrightarrow HO_2 + M \qquad (5)$$
$$2x\ (HO_2 + O_3 \longrightarrow OH + 2\ O_2) \qquad (9)$$
$$\text{net: } CH_2O + 2\ O_3 \longrightarrow CO + 2\ O_2 + 2OH$$

$$CH_2O + OH \longrightarrow CHO + H_2O \qquad (18)$$
$$CHO + O_2 \longrightarrow CO + HO_2 \qquad (17)$$
$$HO_2 + O_3 \longrightarrow OH + 2\ O_2 \qquad (9)$$
$$\text{net: } CH_2O + O_3 \longrightarrow CO + H_2O + O_2$$

Altogether, the oxidation of one methane molecule to carbon monoxide and then to carbon dioxide may yield on the average the following, rather astonishing, net results (Crutzen, 1986):

(a) In NO-poor environments: a net loss of 3.5 odd hydrogen and 1.7 ozone molecules
(b) in NO-rich environments: a net gain of 0.5 odd hydrogen and 3.7 ozone molecules

Heterogeneous processes on aerosol particles and water droplets may modify these results through the removal of intermediate products, especially the peroxides CH_3O_2H and H_2O_2. Hydrogen peroxide is formed by

$$HO_2 + HO_2 \longrightarrow H_2O_2 + O_2 \qquad (19)$$

These results are very important for the photochemistry of the "background" troposphere, because CH_4 and CO are the main reaction partners of OH, and because CH_4 concentrations have been increasing by about 1% per year during the past decades (Rasmussen and Khalil, 1981, 1984; Blake et al, 1982; Blake and Rowland, 1986; Fraser et al, 1981;

Seiler, 1984; Rinsland et al, 1985). Analysis of spectra taken on the Jungfraujoch in Switzerland at about 3.5 km altitude (Rinsland and Levine, 1985) indicates that CO mixing ratios may also have been increasing, on the average by 0.5-4% per year between 1950 and 1977. However, Dianov-Klokov and Yurganov (1981) infer by comparing their own total vertical column CO observations in the 70's to those by Shaw (1958) in the U.S. two decades earlier that CO increased by about 2% per year only during winter time, remaining unchanged during summer.

Natural sources of atmospheric NO_x include lightning, about $3\text{-}8 \times 10^{12}$g N/yr (Noxon, 1978; Chameides et al, 1987b) and emanations from soils, about $5\text{-}15 \times 10^{12}$g N/yr (Galbally and Roy, 1980). More NO is, however, emitted to the atmosphere by anthropogenic activities due to fossil fuel burning at mid-latitudes in the northern hemisphere ($\approx 20 \times 10^{12}$g N/yr) and biomass burning ($5\text{-}10 \times 10^{12}$g N/yr) during the dry season in the tropics (Ehhalt and Drummond, 1982 and this volume; Logan, 1983; Crutzen et al, 1979, 1985). However, because the average lifetimes of NO_x in the atmosphere is of the order of a day, except for some complications due to long-range transport of peroxy-acetyl nitrate (Crutzen, 1979; Singh and Salas, 1983), it is clear that the NO-rich atmospheric environments should mainly be limited to the temperate zone of the northern hemisphere and the boundary layer of the tropics during the dry season. The NO-poor environments are mainly the marine areas, the free troposphere of the tropics and most of the southern hemisphere. Consequently, due to the photochemical reaction chains presented before, we suspect that ozone and hydroxyl concentrations are decreasing in clean atmospheric environments and increasing at mid-latitudes in the northern hemisphere. This gradual shift in the oxidative power of the atmosphere seems to be born out by ozone measurements at some "background" stations (Oltmans, 1985), which show the following average surface ozone concentration trends: Barrow, Alaska (0.78 ± 0.52%/yr), Mauna Loa Observatory, Hawaii (1.20 ± 0.52%/yr) and Samoa (-0.70 ± 0.80%/yr).

Because the sources of atmospheric methane are mainly influenced by anthropogenic activities (see Table I), these trends in "background" ozone concentrations will likely continue.

As a global average, hydroxyl concentrations are estimated at about $5 \pm 2 \times 10^5/cm^3$ (e.g. Crutzen and Gidel, 1983; Volz et al, 1981), with largest concentrations in the tropics (Figure 1). As shown in Figures 2 and 3, this implies that for several important gases, e.g. methane and carbon monoxide, the largest atmospheric sources and sinks are also likely located in the tropics (Crutzen and Gidel, 1983). It is interesting to note that during the dry season, especially in the savanna regions, large ozone (and carbon

	Methane
Sinks	
Tropospheric reactions with OH	290-350
Stratospheric reactions with OH	25-35
Uptake on aerobic soils	10-30
Annual growth (≈1.1%/yr)	50-60
Total required sources (sum of above)	375-475
Sources	
Domestic animals	70-80
Natural gas leaks	≤35
Coal mining	35
Landfills	30-70
Biomass burning	30-100
Wild ruminants	2-6
Other fauna (e.g., insects)	<30
Decay of animals wastes	?
Rice fields and natural wetlands	>44-228

Table I: Estimates of global sources and sinks of methane (units 10^{12}g per year), based on information in Seiler *et al* (1984), Keller *et al* (1983), Harriss *et al* (1982), Rasmussen and Khalil (1984), Khalil and Rasmussen (1986), Blake and Rowland (1986), Crutzen *et al* (1986), Crutzen (1986), Bingemer and Crutzen (1987), Crutzen *et al* (1985).

monoxide) concentrations do indeed build up, as expected from photochemical reactions in the presence of oxides of nitrogen resulting from biomass burning (Crutzen *et al*, 1985; Delany *et al*, 1985; Andreae *et al*, 1987). Elsewhere, in clean environments, and during the wet season, ozone concentrations are substantially lower (Logan and Kirchhoff, 1986).

3. TRENDS IN OZONE CONCENTRATIONS

Because we have reasons to expect that "background" ozone concentrations have increased substantially in the mid-latitude zone of the northern hemisphere, it is important to look at long-term trends in tropospheric ozone concentra-

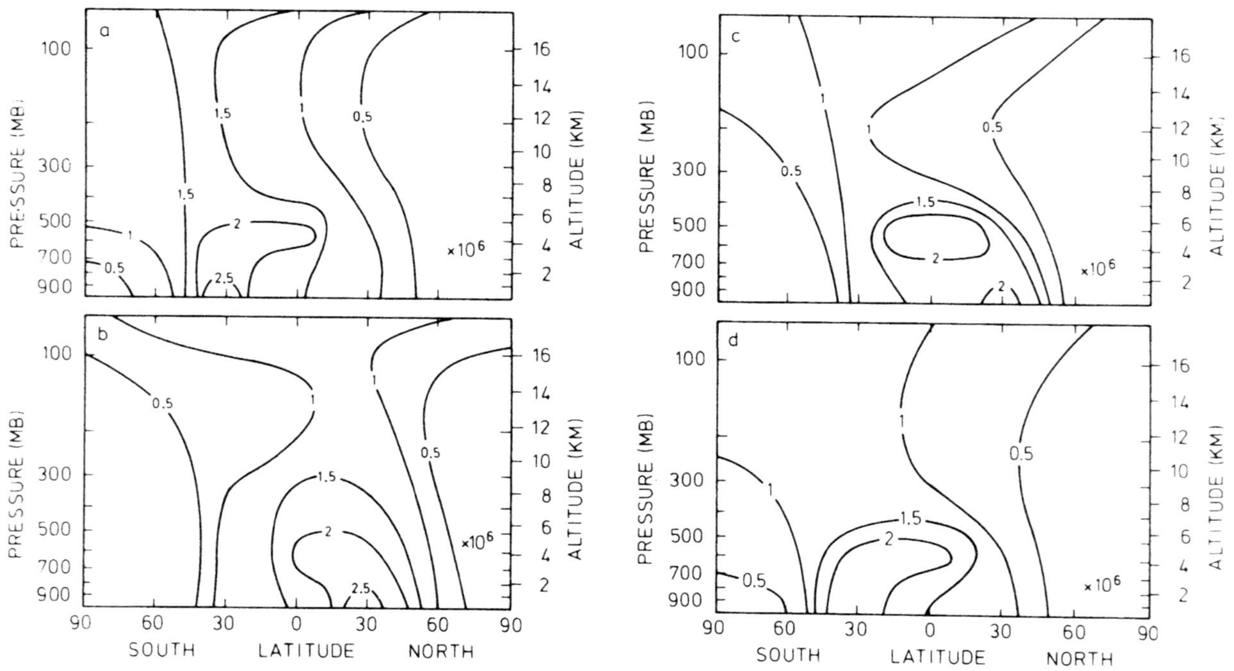

Figure 1: Calculated daytime average meridional distributions of OH (in units of 10^6 molecules cm^{-3}) for (a) January, (b) July, (c) April, and (d) October, (reprinted from Crutzen and Gidel, 1983).

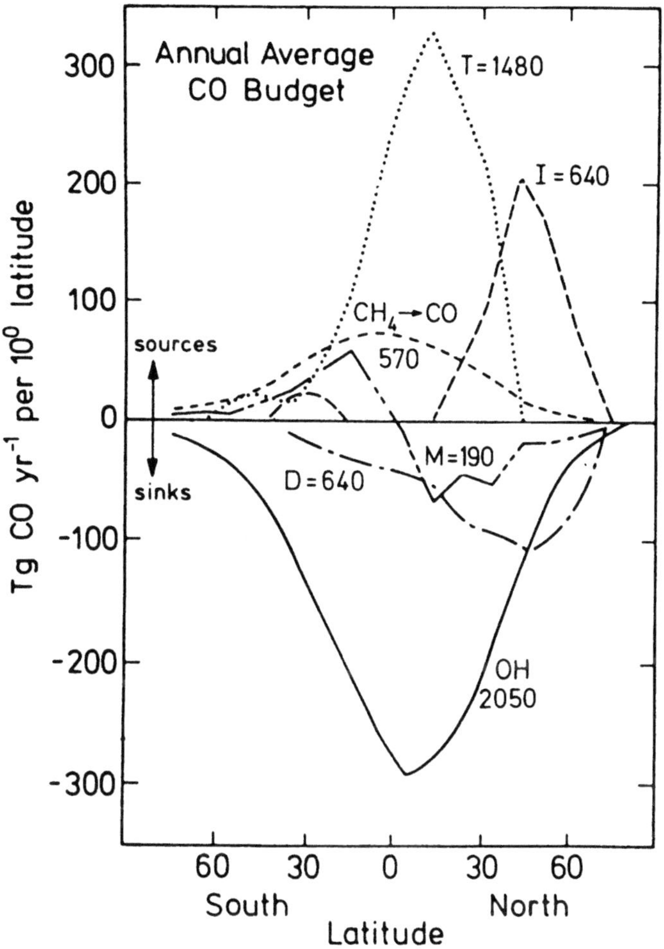

Figure 2: Annual CO budget estimated by Crutzen and Gidel (1983) in units of 10^{12}g/yr; OH: destruction by reaction with OH; D: destruction at the earth's surface (Seiler, 1974); M: transport from northern to southern hemisphere; CH_4: production by oxidation of CH_4; I: fossil fuel burning (Seiler, 1974); T: production by a variety of other sources, mainly biomass burning and oxidation of hydrocarbons in the tropics.

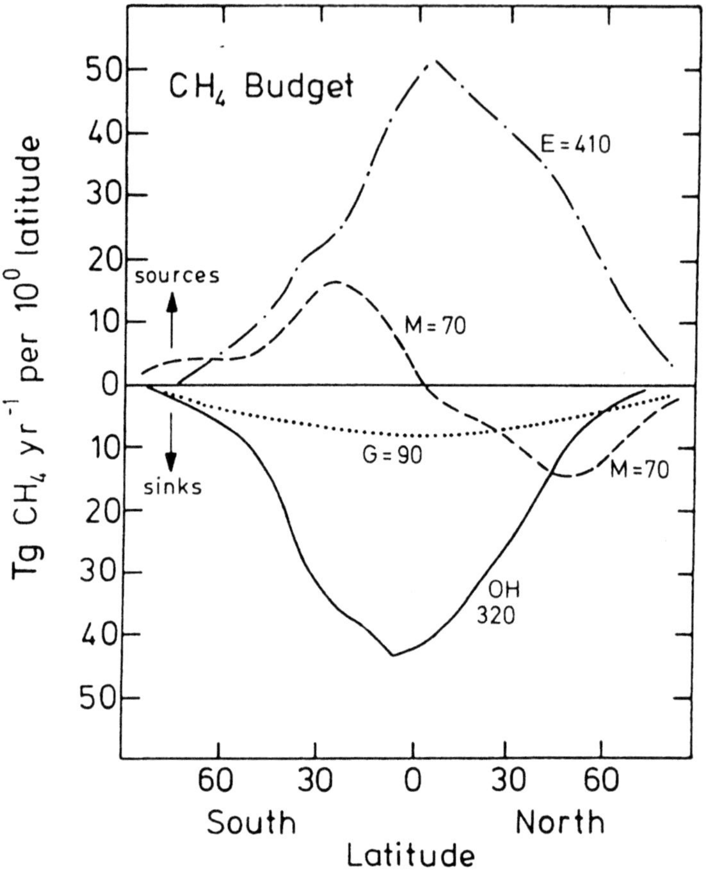

Figure 3: Annual CH_4 budget, according to Crutzen and Gidel (1983) in units of 10^{12}g/yr. OH: destruction by reaction with OH; M: transport from nothern to southern hemisphere; E. variety of sources (see Table 1).

tions. A thorough analysis of this kind, mostly based on measurements in the 60's and 70's by Logan (1985) indeed established that ozone concentrations in rural areas at mid-latitudes in the northern hemisphere have increased by 20-100%. Also in the free troposphere annual ozone concentration increases by 1-2% between 30°N and 75°N were observed. However, the rapid growth in NO_x and hydrocarbon emissions started much earlier (see Figure 4). Consequently, it is important to look at even earlier ozone data. Most

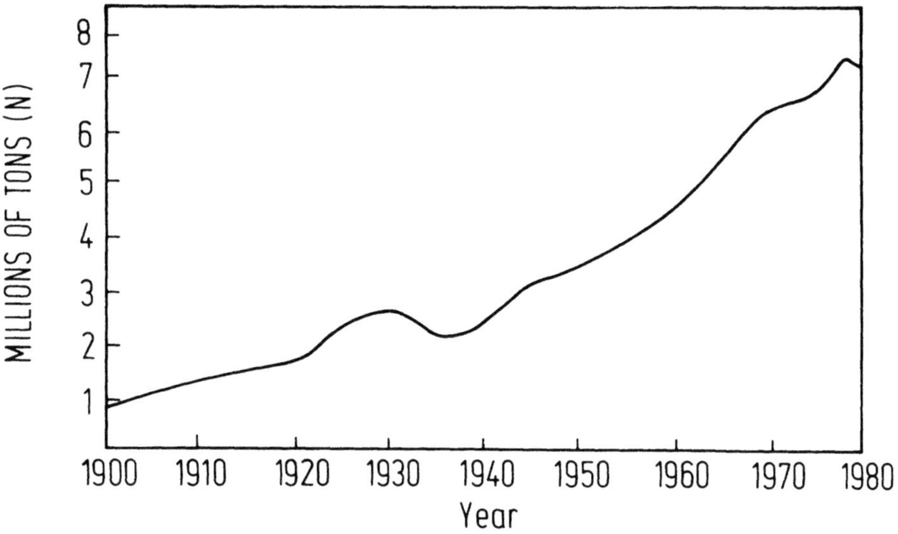

Figure 4: Estimates of NO_x emissions in the U.S. in units of 10^{12}gN per year; adapted from NAPAP (1984).

reliable are those obtained by optical techniques especially during the early 1930's. Unfortunately, only so few measurements were made with this technique that they fit on one-page (see Table II).

It is interesting to note that all ozone observations from low altitude (< 1 km) stations are substantially lower than currently measured with modern chemical techniques at representative stations in similar areas, e.g. Hohenpeißenberg in Southern Germany (Attmannspacher et al, 1984). However, because the exact measurement conditions were often not reported, the comparisons should be supported by more detailed analyses, provided the old field work reports are still available. Most significant are the observations from Arosa and especially those from the Jungfraujoch (alt. 3450 m) in Switzerland, which yield an average volume mixing ratio of about 30 ppbv. In comparison, the summer time ozone volume mixing ratios, that are currently measured above Hohenpeißenberg at a similar altitude as the Jungfraujoch, average about 60 ppbv (Attmannspacher et al, 1984), i.e. about two times more than in 1933. This indicates a substantial large scale increase in tropospheric ozone, confirming the findings by Attmannspacher et al (1984) based on measurements between 1968 and 1984.

Chemical ozone observations, based on the $KI-O_3$ reactions, were improved considerably by Cauer (1935), Paneth and Edgar (1938), Regener (1938), Ehmert and Ehmert

Table II: Compilation of ozone measurements made with optical technique.

Location	Date	Height/m	Mixing ratio/ppbv	Authors
Arosa	03.09.32	1900	8-9	Götz and Maier-Leibnitz (1933)
	28.09.32		19-22	
Chur	29.09.32	600	18-21	
	01.10.32		19-21	
Arosa	28.03.33	1900	31-36	
	31.03.33		25-29	
Jungfrau-joch	21.08.33	3450	27	Challonge and Vassy (1934); Challonge *et al* (1934)
	24.08.33		35	
	25.08.33		25	
	26.08.33		33	
	27.08.33		24	
	28.08.33		35	
	29.08.33		33	
Lauter-brunnen	18.08.33	800	14	
	19.08.33		16	
	21.08.33		11	
	23.08.33		26	
	24.08.33		19	
	25.08.33		20	
	28.08.33		15	
	29.08.33		14	
	30.08.33		18	
	31.08.33		19	
Provence	Oct,1929	300	20-22	Fabry and Buisson (1931)
	March,1930		20-22	
Arosa	04.05.30	2300	22	Götz and Ladenburg (1931), corrected by Götz *et al* (1935)
Arosa	17.,19., 22.03.34	1900	29	Götz and Penndorf (1941)
	22.04.34		25.4	
	24.04.34		27	
	25.04.34		24.5	
	26.04.34		24.7	
	27.04.34		27.7	
	28.04.34		24.3	
	01.05.34		26	
	02.05.34		32	
	22.03.34		32	Götz *et al* (1934)

(1941, 1949) and Ehmert (1949, 1951). With this technique a considerable amount of measurements were made in the 40's and 50's in Europe. The main drawback with the adopted chemical method is a negative interference by SO_2 (Schenkel and Broder, 1982). Observations were, however, taken at vacation and health resorts, while measurement techniques were intercompared, yielding satisfactory results (Regener, 1943). Consequently, it may be safely assumed that several of the stations may have given satisfactory observations. We next review some of the results of the chemical ozone mesurements made from the mid-1930's to the mid-50's.

The first surface ozone measurements by chemical techniques in the 30's were performed by Cauer (1936, 1937) during June - October between 1934 and 1936 in the Tatra mountains (now Poland) in various locations at altitudes between 400 and 1000 m. Only a limited amount (≈ 400) of all measurements were reported. Except during periods of thunderstorm activity, measured daytime ozone volume mixing ratios, with few exceptions, were generally in the range of less than 10 to 25 ppbv. Cauer (1936) also reports some very early measurements on the Jungfraujoch (3450 m alt) in Switzerland made in September, 1932, which show below 10 ppbv of ozone, much less than those determined by optical methods (see Table II). It is clear that these very early measurements with the KI-technique must have been incorrect.

Regener (1938) in August 1938 made 12 measurements at the Jungfraujoch (3600 m altitude). The range of observed ozone volume mixing ratio was 24-43 ppbv with an average of 30 ppbv in very good agreement with the earlier optical observations, discussed above. The average of six ozone measurements near Friedrichshafen at Lake Constance (400 m asl) during daytime gave an average volume mixing ratio of 21 ppbv. An extensive series of measurements near the same site by Auer (1939) between September and December, 1938, from 10 a.m. until 16 p.m. generally gave ozone volume mixing ratios of 10-25 ppbv. During night, ozone concentrations were much lower due to ozone destruction at the ground.

Using the Cauer method, Tichy (1939) in February - April, 1937 and 1938 made a total of 252 ozone measurements during fair weather conditions in Schreiberhau (Silezia, now Poland) at 700 m altitude. The well known diurnal cycle with maximum ozone values in the early afternoon hours, which is caused by turbulent downward transport of ozone from the free troposphere (Regener, 1938), is quite apparent, although Tichy seemed to imply that this is correlated with the measured intensity of ultraviolet radiation. Altogether, early afternoon ozone concentrations are generally in the range 10-15 ppbv, i.e. about 2-3.5 times lower then at Hohenpeißenberg in 1982, but about equal to the measured values at this station in 1968

(Attmannspacher *et al*, 1982).

Extremely interesting are the observations that were made during the period Sept., 15-24, 1940 by Ehmert and Ehmert (1941, 1949) in Friedrichshafen at Lake Constance and on the Pfänder mountain at an altitude of 1064 m on the east side of the same lake. Most ozone volume mixing ratios are between 10 and 15 ppbv and no values are reported that are higher than 20 ppbv. The low altitude Friedrichshafen station clearly shows the well known diurnal behavior of surface ozone concentrations.

Paneth and Edgar (1938) measured ground level ozone mixing ratios between 5 and 23 ppbv in the London area in the spring of 1938. In a late series of measurements during windy conditions in December, 1940, Paneth and Glückauf (1941) near Durham in England measured ozone volume mixing ratios ranging between 15 and 25 ppbv, which values may have been representative for free tropospheric air.

After the second world war, Effenberger (1949) on the island Westerland/Sylt on the German Northsea coast made extensive ozone measurements during May - August, 1947. Earlier, Effenberger (1948) had intercompared the Cauer and Regener chemical ozone measurement devices and had come to satisfactory conclusions. The summer of 1947 must have been particularly bad in West Europe with much windy and rainy weather. Nevertheless, for more than half of the measurement period ozone volume mixing ratios remained below 20 ppbv. Maximum ozone valus (≈ 60 ppbv) were reported in connection with thunderstorm activity. The measurements by Fabian and Pruchniewicz (1977), using automatic sampling systems, at the German stations of Norderney and Westerland during 1971 and 1972 also showed ozone volume mixing ratios of only about 20 ppbv, but generally substantially higher concentrations in 1973 and 1974 (≈ 40 ppbv). Unfortunately, these Northsea island stations, which are located at a distance of only about 150 km from each other, often show large differences in monthly averaged ozone trends, so that the quality of some of the data may be doubtful.

An extensive series of ozone measurement were conducted in Arosa, Switzerland from April, 29, 1954 to October, 18, 1958 (Perl, 1965). Average ozone concentrations show an annual cycle with a minimum of about 12 ppbv in December and a maximum of 30 ppbv in May - June. Maximum measured ozone concentrations are about 20-30% larger than the average. Interestingly, fair weather ozone concentrations are generally somewhat higher than average, indicating the possibility that some ozone buildup by photochemical reactions may already have begun by the latter half of the 1950's. Earlier tropospheric ozone measurements at Arosa by Götz and Volz (1951) from April, 1950 to March, 1951 agree quite well with those by Perl (1965). The measurements made from the middle of March to early May of 1934 by the optical technique (Götz and Penndorf, 1941) gave ozone volume mixing ratio values ranging from 26 ppbv to 35 ppbv, with an average of 32 ppbv. This is about 20% larger than the values measured by Perl (1965) and Götz and Volz (1951), giving credence to the observational methods and the measured ozone concentrations.

A long series of ground based ozone measurements at Arkona on the Baltic coast in the German Democratic Republic have shown a large annual upward trend by 2.8% between 1956 and 1971, followed by a downward trend of -0.74% between 1972 and 1984 (Warmbt, 1979; Feister and Warmbt, 1987). The trend analysis may, however, have been compounded by the introduction of an SO_2 scrubber in 1972.

Altogether, the historical ozone data, which have been reviewed in this chapter, indicate that surface ozone concentrations in the first half of this century were mostly in the range 10-20 ppbv at European clean air stations, somewhat higher than the early measurements by Levy (1907). It is interesting to note that background surface ozone volume mixing ratios in the 10-20 ppbv range are also typical for mid-latitudes in the southern hemisphere. In New Zealand, Farkas (1979) measured ozone volume mixing ratios in the 10-20 ppbv range with some tendency for the higher values to occur during winter and spring, indicating either enhanced downward transport from the stratosphere during winter and spring or photochemical destruction over the oceans during summer. Ozone surface measurements are also made at the baseline station Cape Grim in Tasmania. Volume mixing ratios during 1982-1985 generally ranged from about 15 ppbv in January to 30 ppbv in July (Douglas et al, 1986). Finally, measurements by Winkler (1980) on ship cruises over the Atlantic gave ozone volume mixing ratios between 10 and 15 ppbv in the southern hemisphere, compared to about 30 ppbv at mid-latitudes in the northern hemisphere.

Because ozone is much more efficiently destroyed on land surfaces than on the ocean, the fact that there is now so much more ozone in the northern hemisphere than in the southern hemisphere (Routhier et al, 1980; Fishman, 1985; Logan, 1985; Seiler and Fishman, 1981) is indicative of substantial ozone production in the northern hemisphere (Fishman and Crutzen, 1978; Fishman et al, 1979). The appearance of simultaneous O_3 and CO bands in the northern hemisphere is evidence of this (Fishman et al, 1980, 1987).

According to Becker et al (1985), present day average annual ozone volume mixing ratios at low altitude stations in West Germany are typically in the range 25-35 ppbv, with average summer time (April - September) values of 35-45 ppbv. At more elevated observational sites (alt: 800-1300 m) summer time ozone values are about 5-10 ppbv higher. Maximum volume mixing ratios nowadays frequently exceed 100 ppbv (Becker et al, 1985; Derwent et al, 1987), values which were never reported in the old data sets. Clearly, there have been an enormous increases in background ozone concentrations over the past 30 years, especially in heavily motorized industrial areas. However, also at other more remote background stations substantial ozone increases have taken place.

It is extremely interesting to do a much more thorough analysis of "ancient" ozone data than was possible for this study, e.g. by rebuilding the instruments used, checking them against modern techniques and recalibration (e.g. Volz and Kley, 1987). It may also be so that there is a considerable amount of data that was never published and still stored away in old archives. Finally, it would be very interesting to compare certified old data with modern data taken at the same sites as where the "ancient" data were taken.

4. THE GLOBAL POTENTIAL FOR OZONE FORMATION

As shown before, the oxidation of carbon monoxide and hydrocarbons leads to the production of ozone, provided sufficient concentrations of NO are present in the atmosphere. From knowledge of the mean global distributions of CH_4, CO and hydroxyl, the latter calculated (Crutzen and Gidel, 1983), the atmospheric sinks of CH_4 and CO can be estimated. These are presented in Figures 2 and 3. Altogether about 1.5×10^{15}g CO and 4.1×10^{14}g of CH_4 are emitted in the atmosphere from a variety of sources. In comparison, other alkane emissions are substantially smaller. For instance, the global production of ethane (C_2H_6) is equal to about $10\text{-}16 \times 10^{12}$g per year (Blake and Rowland, 1986). If enough NO were present to produce ozone by the reaction cycles shown before, one ozone for each carbon monoxide and 2.7 ozone molecules for each methane molecule that is oxidized, the maximum global production potential of ozone would be equal to 6×10^{15}g O_3 annually (or 5×10^{11} molecules $cm^{-2}s^{-1}$), about ten times larger than the estimated downward flux of ozone from the stratosphere to the troposphere, and the ozone destruction flux at the earth's surface. To this, the potential ozone production should be added which results from the oxidation of reactive non-methane-hydrocarbons (NMHC) coming from forest emissions. For the U.S., this contribution was previously estimated by Zimmerman <u>et al</u> (1978) to be equal to about 7×10^{13}g C/year, which was extrapolated to a global production of 8.3×10^{14}g C/year. A recent study by Lamb <u>et al</u> (1987) gives an estimated annual emission rate of 3.1×10^{13}g NMHC in the U.S. Contrary to Zimmerman <u>et al</u> (1978), isoprene emissions contributed only about 15% of the total. If we extrapolate this new information by the same method as Zimmerman <u>et al</u> (1978) an annual global NMHC emission flux of about 3.7×10^{14}g is estimated. If we assume that for each C atom in any of these organic molecules, there are produced two ozone molecules under NO-rich conditions (Liu et al, 1987), then an additional potential, global production of about 2×10^{15}g ozone ($\approx 1.5 \times 10^{11}$ molecules $cm^{-2}s^{-1}$) emerges.

Altogether, the total potential for global ozone production may, therefore, add up to about 7×10^{11} molecules $cm^{-2}s^{-1}$, which is more than a factor of ten larger than all other sources and sinks of ozone. Clearly, most hydrocarbon and carbon monoxide are oxidized in the atmosphere without producing ozone, because the necessary concentrations of NO are not present in natural atmospheric environments. Our analysis shows, however, that any future human developments, that lead to NO emissions in the vicinity of forested regions will probably lead to growing ozone concentrations. For the U.S. this has been shown convincingly by Trainer et al (1987).

Compared to the total annual, natural NMHC source of 3.1×10^{13}g in the U.S., the volatilization from various anthropogenic sources of about 2×10^{13}g is not much different (Liu et al, 1987; Derwent et al, 1987). Because much of the latter category of hydrocarbons is emitted together with anthropogenic NO_X, it is likely that a substantial fraction of these contribute to ozone formation.

In a very interesting study, Liu et al (1987) estimated that at least one molecule of ozone is produced for each NMHC that is oxidized. For summer time conditions they estimate an average ozone production rate for the U.S. of about 10^{12} molecules $cm^{-2}s^{-1}$. They relate the O_3 production rates with those of NO_X, such that during summer daily ten ozone molecules are produced for each NO_X molecule. Also for other seasons, they claim a similar ozone production efficiency, because a longer NO_X lifetime compensates for a less efficient daily ozone production rate. Their analysis did, however, neglect reactions of NO_3 and N_2O_5 on wet surfaces during nighttime, so that the latter hypothesis is probably not valid. Extrapolation of this information to other industrialized nations implies that probably of the order of $4\text{-}7 \times 10^{14}$g O_3 ($\approx 5\text{-}10 \times 10^{10}$ molecules $cm^{-2}s^{-1}$) may annually be produced through the oxidation of anthropogenic non-methane hydrocarbons in the northern hemisphere. In fact, this may be a lower limit, because the anthropogenic NMHC emissions may be significantly underestimated (Westberg and Lamb, 1985).

The oxidation of anthropogenic NMHC is clearly an important contribution to the ozone budget of the northern hemisphere. Much of this ozone is produced in the atmospheric boundary layer over limited continental areas, leading to high ozone concentrations. However, a fraction of the hydrocarbons can escape to the free troposphere by convection, especially in the tropics and during summer at mid-latitudes (Gidel, 1983; Chatfield and Crutzen, 1984; Ehhalt et al, 1985). If this happens in industrialized areas, then together with the hydrocarbons also NO_x pollutants will be transported upward, setting up favorable conditions for ozone formation in the upper troposphere with more efficient ozone formation per NMHC molecule emitted (Liu et al, 1987). The convective activity will also be accompanied by NO_x formation through lightning (Chameides et al, 1987 a, b). Because convection is particularly important above tropical forests, a substantial fraction of the large amounts of hydrocarbons that are produced in this ecosystem (Zimmerman et al, 1978; Crutzen et al, 1985; Greenberg and Zimmerman, 1984) will be rapidly transferred to the upper troposphere, where their oxidation may produce significant amounts of ozone.

From tropospheric ozone budget considerations it is quite clear that most CO and organics that are emitted into the atmosphere are oxidized in NO poor atmospheric environments. This is possible because the average atmospheric lifetime of NO_x is only about one day. Effective ozone production does, therefore, probably still only occur in the temperate latitudes in the northern hemisphere and maybe in the upper troposphere of the tropics. Extremely low background NO concentrations ($\leq$ 1 pptv) are indeed observed in clean environments, such as in the lower atmosphere over the Pacific between Hawaii and California (Davis et al, 1987; McFarland et al, 1979; Ridley et al, 1987).

The overall budget of tropospheric ozone remains uncertain (Fishman, 1985). The annual average downward flux of ozone from the stratosphere is in the range $4\text{-}8 \times 10^{10}$ and $3\text{-}5 \times 10^{10}$ molecules $cm^{-2}s^{-1}$ in the northern and southern hemisphere, respectively (Danielsen and Mohnen, 1977; Gidel and Shapiro, 1980). The corresponding ozone destruction rates at the earth's surface are estimated to be about 17×10^{10} and $5\text{-}6 \times 10^{10}$ molecules $cm^{-2}s^{-1}$ (Galbally and Roy, 1980; Lenschow et al, 1985; Fishman, 1985). The average photochemical destruction of ozone in both hemispheres by reactions R1 and R2 are about equal to 7×10^{10} and 5×10^{10} molecules $cm^{-2}s^{-1}$ in the northern and southern hemisphere. From mass balance considerations a net photochemical production of ozone of about 2×10^{11} molecules $cm^{-2}s^{-1}$ must take place in the troposphere of the northern hemisphere. Also in the southern hemisphere some net ozone formation in the "NO-rich" upper troposphere and transport from the northern to the southern hemisphere, adding up to about 5×10^{10} molecules $cm^{-2}s^{-1}$, seems to be required for a satisfactory ozone budget.

5. CONCLUSIONS

In this paper we have shown that anthropogenic activities most likely have caused large increases in ozone concentrations in the northern hemisphere. These changes not only have affected heavily polluted industrial areas, but also the background free troposphere. The processes that produce enhanced ozone concentrations are basically well known, viz. the photochemical oxidation of hydrocarbons in the presence of NO_x.

The maximum possible, global ozone production by oxidation of all hydrocarbons and carbon monoxide which are emitted into the atmosphere is about equal to 10^{16}g of ozone annually ($\approx 10^{12}$ molecules $cm^{-2}s^{-1}$). Only a small fraction ($\leq 10\%$) of this ozone production can actually take place, because in much of the free atmosphere there is very little NO. Future human developments, that are accompanied by enhanced NO_x production, will consequently lead to considerable increased tropospheric ozone concentrations. This may even be of climatic significance (Fishman et al, 1979; Ramanathan and Dickinson, 1979; Wang, this volume).

Because of the potential for large future changes in atmospheric concentrations, and because of the central role in atmospheric chemistry, the study of atmospheric ozone remains a topic of intensive interest. It is remarkable to note that the importance of this gas was already "smelled" more than 100 years ago by researchers, even before its chemical composition was established.

Acknowledgement: Special thanks go to Dr. H. Wende for his assistance in retrieving many ancient ozone papers from libraries, extinct journals and other repositories.

REFERENCES

Andreae, M.O., Browell, E.V., Garstang, M., Gregory, G.L., Harriss, R.C., Hill, G.F., Jacob, D.J., Pereira, M.C., Sachse, G.W., Setzer, A.W., Silva Dias, P.L., Talbot, R.W., Torres, A.L. and Wofsy, S.C. (1987). Biomass burning emissions and associated haze layers over Amazonia, J. Geophys. Res. (in press)

Andrews, T. (1856). On the constitution and properties of ozone, Phil. Trans. Roy. Soc., 6, 1-13.

Andrews, T. (1874). Address on Ozone. J. Scott. Meteorol. Soc., 4, 122-134.

Attmannspacher, W., Hartmannsgruber, R. und Lang, P. (1984). Langzeittendenzen des Ozons der Atmosphäre aufgrund der 1967 begonnenen Ozonmeßreihen am Meteorologischen Observatorium Hohenpeißenberg, Meteorol. Rdsch., 37, 193-199.

Auer R. (1939). Über den täglichen Gang des Ozongehalts in der bodennahen Luft, Gerland Beitr. Geophys., 54, 137-145.

Becker, K.H., Fricke, W., Löbel, J. and Schurath, U. (1985). Formation, transport, and control of photochemical oxidants, in Air Pollution by Photochemical Oxidants, R. Guderian (Editor), Springer, Berlin, p. 11-67.

Bingemer, H.G. and Crutzen, P.J. (1987). The production of methane from solid wastes, J. Geophys. Res., 92 (D2), 2181-2187.

Blake, D.R., Mayer, E.W., Tyler, S.C., Montague, D.C., Makide, Y. and Rowland, F.S. (1982). Global increase in atmospheric methane concentration between 1978 and 1980, Geophys. Res. Lett., 477-480.

Blake, D.R. and Rowland, F.S. (1986). World-wide increase in tropospheric methane, J. Atmos. Chem., 4, 43-62.

Blake, D.R. and Rowland, F.S. (1986). Global atmospheric concentrations and source strength of ethane, Nature, 321, 231-233.

Bojkov, R.D. (1986).Surface ozone during the second half of the nineteenth century, J. Climate and Appl. Meteorol., 25, 342-352.

Cabannes, J. and Dufay, J. (1927). Mésure de l'altitude et de l'épaisseur de la couche d'ozone dans l'atmosphére, J. de Physique, 8, 125-152.

Cauer, H. (1935). Bestimmung des Gesamtoxydationswertes des Nitrits, des Ozons und des Gesamtchlorgehalts roher und vergifteter Luft, Z. anal. Chemie, 103, 385-416.

Cauer, H. (1936). Chemisch-bioklimatische Studien in der hohen Tatra und ihrem Vorland, Der Balneologe, 3, 7-23.

Cauer, H. (1937). Chemisch-bioklimatische Studien im Glatzer Bergland, Der Balneologe, 4, 545-565.

Challonge, D. et Vassy, E. (1934). Recherches sur la transparence de la basse atmosphère et sa teneur en ozone, J. de Physique, 5, 309-319.

Challonge, D., Götz, F.W.P. and Vassy, E. (1934). Simultanmessungen des bodennahen Ozons auf Jungfrauhoch und in Lauterbrunnen, Naturwissenschaften, 22, 297.

Chameides, W.L. and Walker, J.C.G. (1973). A photochemical theory of tropospheric ozone, J. Geophys. Res., 78, 8751-8760.

Chameides, W.L., Davis, D.D., Rodgers, M.O., Bradshaw, J., Sandholm, S. Sachse, G., Hill, G. and Gregory, G., (1987a). Net ozone photochemical production over the Eastern and Central North Pacific as inferred from GTE/CITE 1 observations during fall 1983, J. Geophys. Res., 92 (D2), 2131-2152.

Chameides, W.L., Davis, D.D., Bradshaw, J., Rodgers, M., Sandholm, S. and Bai, D.B. (1987b). An estimate of the NO_X production rate in electrified clouds based on NO observations from the GTE/CITE 1, Fall 1983 field operation, J. Geophys. Res., 92 (D2), 2153-2156.

Chapman, S. (1930). A theory of upper-atmospheric ozone, Mem. Roy. Meteor. Soc., 3, 103-125.

Chatfield, R.B. and Crutzen, P.J. (1984). Sulfur dioxide in remote oceanic air: cloud transport of reactive precursors, J. Geophys. Res., 89, 7111-7132.

Crutzen, P.J. (1973). A discussion of the chemistry of some minor constituents in the stratosphere and troposphere, Pure Appl. Geophys., 106-108, 1385-1399.

Crutzen, P.J. (1974). Photochemical reactions initiated by and influencing ozone in unpolluted tropospheric air, Tellus, 26, 47-57.

Crutzen, P.J. (1979). The role of NO and NO_2 in the chemistry of the troposphere and stratosphere, Ann. Rev. Earth Planet. Sci., 7, 443-472.

Crutzen, P.J. (1986). The role of the tropics in atmospheric chemistry, in Geophysiology of Amazonia (R.E. Dickinson, Editor), Wiley, New York, p. 107-130.

Crutzen, P.J. Heidt, L.E., Krasnec, J.P., Pollock, W.H. and Seiler, W. (1979). Biomass burning as a source of atmospheric gases CO, H_2, N_2O, NO, CH_3Cl and COS, Nature, 282, 253-256.

Crutzen, P.J. and Gidel, L.T. (1983). A two-dimensional photochemical model of the atmosphere. 2: The tropospheric budgets of the anthropogenic chlorocarbons, CO, CH_4, CH_3Cl and the effects of various NO_X sources on tropospheric ozone, J. Geophys. Res., 88 (C11), 6641-6661.

Crutzen, P.J., Delany, A.C., Greenberg, J., Haagenson, P., Heidt, L., Lueb, R., Pollock, W., Seiler, W., Wartburg, A. and Zimmerman, P. (1985). Tropospheric chemical composition measurements in Brazil during the dry season, J. Atmos. Chem., 2, 233-256.

Danielsen, E.F. and Mohnen, V.A. (1977). DUSTORM report: ozone transport and meteorological analyses of tropopause folding, J. Geophys. Res., 82, 5867-5877.

Davis, D.D., Bradshaw, J.D., Rodgers, M.O., Sandholm, S.T. and Kesheng, S. (1987). Free tropospheric and boundary layer measurements of NO over the central and eastern north pacific ocean, J. Geophys. Res., 92 (D2), 2049-2070.

Delany, A.C., Haagenson, P., Walters, S., Wartburg, A.F. and Crutzen, P.J. (1985). Photochemically produced ozone in the emission from large-scale tropical vegetation fires, J. Geophys. Res., 90, 2425-2429.

Derwent, R.G. and Stewart, H.N.M. (1973). Elevated ozone levels in the air of Central London, Nature, 241, 342-343.

Derwent, R.G., et al (1987). Ozone in the United Kingdom, United Kingdom Photochemical Oxidants Review Group, Interim Report, Department of the Environment and Department of Transport, Victoria Road, South Ruislip, Middlessex HA4 ONZ, 112 pp.

Dianov-Klokov, V.I. and Yurganov, L.N. (1981). A spectroscopic study of the global space-time distribution of atmospheric CO, Tellus, 33, 262-273.

Dobson, G.M.B., Harrison, D.N. and Lawrence, J. (1929). Ozon in the Earth's Atmosphere, Proc. Roy. Soc., A 122, 456-486.

Douglas, M.D., Elsworth, C.M. and Galbally, I.E. (1986). Ozone in near surface air, in Baseline Atmospheric Program (Australia), R.J. Francey and B.W. Forgan (Eds), Bureau of Meteorology and CSIRO/Atmospheric Reseach, Melbourne.

Effenberger, E. (1948). Kritische und vergleichende Untersuchungen der luftchemischen Meßmethoden des Ozongehalts und Gesamtoxydationswertes, Meteorol. Rdsch., 1, 488-491.

Effenberger, E. (1949). Messungen des Ozongehaltes und Gesamtoxydationswertes der bodennahen Luftschicht auf Sylt, Meteorol. Rdsch., 2, 94-97.

Ehhalt, D.H. and Drummond, J.W. (1982). The tropospheric cycle of NO_x, in Chemistry of the Unpolluted and Polluted Troposphere (H.W. Georgii and W. Jaeschke, Eds), Reidel, Dordrecht, Holland, 219-252.

Ehhalt, D.H., Rudolph, J., Meixner, F. and Schmidt, U. (1985). Measurements of selected C_2-C_5 hydrocarbons in the background troposphere: vertical and latitudinal variations, J. Atmos. Sci., 3, 29-52.

Ehmert, A., (1949). Über das troposphärische Ozon. Ber. Deutschen Wetterdienstes in der U.S.-Zone, 11, 26-28.

Ehmert, A. und Ehmert, H. (1949). Über den Tagesgang des bodennahen Ozons, Ber. Deutschen Wetterdienstes in der U.S.-Zone, 11, 58-62.

Ehmert, A. und Ehmert, H. (1941, 1949). Über die Bestimmung des Ozongehalts der Luft, Forschungsberichte des Reichswetterdienstes, A13, reprinted in Ber. Deutschen Wetterdienstes in der U.S.-Zone, 11, 67-71.

Ehmert, A. (1951). Ein einfaches Verfahren zur absoluten Messung des Ozongehaltes von Luft, Meteorol. Rdsch., 4, 64-68.

Fabian, P. and Junge, C.E. (1970). Global rate of ozone destruction at the earth's surface, Archiv Meteor. Geoph. Biokl., A19, 161-172.

Fabian, P. and Pruchniewicz, P.G. (1977). Meridional distribution of ozone in the troposphere and its seasonal variation, J. Geophys. Res., 82, 2063-2073.

Fabry, Ch. et Buisson, H. (1921). Etude de l'extrémité ultraviolette du spectre solaire, J. de Physique, Serie VI, 2, 197-226.

Fabry, Ch. et Buisson, H. (1931). Sur l'absorption des radiations dans la basse atmosphère et le dosage de l'ozone, Comptes Rendus, 192, 457-461.

Farkas, E. (1979). Surface ozone variations in the Auckland region, New Zealand J. Sci., 22, 63-76.

Feister, U. and Warmbt, W., 1987. Long-term measurements of surface ozone in the German Democratic Republic, J. Atmos. Chem., 5, 1-22.

Finlayson-Pitts, B.J. and Pitts, J.N. Jr., (1986). Atmospheric Chemistry, Wiley, New York, 1098 pp.

Fishman, J. (1985). Ozone in the Troposphere, Chapter 4 in Ozone in the Free Atmosphere (R.C. Whitten and S.S. Prasad, Editors), Van Nostand, New York, pp. 161-194.

Fishman, J. and Crutzen, P.J. (1978). The origin of ozone in the troposphere, Nature, 274, 855-858.

Fishman, J., Solomon, S. and Crutzen, P.J. (1979). Observational and theoretical evidence in support of a significant in-situ photochemical source of tropospheric ozone, Tellus, 32, 456-463.

Fishman, J., Ramanathan, V., Crutzen, P.J. and Liu, S.C. (1979). Tropospheric ozone and climate, Nature, 282, 818-820.

Fishman, J., Seiler, W. and Haagenson, P. (1980). Simultaneous presence of O_3 and CO bands in the troposphere, Tellus, 32, 456-463.

Fishman, J., Gregory, G.L., Sachse, G.W., Beck, S.M. and Hill, G.F. (1987). Vertical profiles of ozone, carbon monoxide and dew-point temperature obtained during GTE/CITE 1, October - November 1983, J. Geophys. Res., 92 (D2), 2083-2094.

Fonrobert, E. (1916). Das Ozon, Enke Verlag, Stuttgart, 282 pp.

Fraser, P.G., Khalil, M.A.K., Rasmussen, R.A. and Crawford, A.Y. (1981). Trends of atmospheric CH_4 in the southern hemisphere, Geophys. Res. Lett., 9, 1063-1066.

Friesenhof, Frh.G. (1904). Einiges über Ozonbeobachtung, Z. Österr. Ges. Meteor., 21, 380-382.

Galbally, I.E. and Roy, C.E. (1980). Loss of fixed nitrogen from soils by nitric oxide exhalation, Nature, 275, 734-735.

Galbally, I.E. and Roy, C.R. (1980). Destruction of ozone at the earth's surface, Quart. J. Roy. Meteorol. Soc., 106, 599-620.

Gidel, L.T. (1983). Cumulus transport of transient tracers, J. Geophys. Res., 88, 6587-6599.

Gidel, L.T. and Shapiro, M.A. (1980). General circulation model estimates of the net vertical flux of ozone in the lower stratosphere and the implications for the tropospheric ozone budget, J. Geophys. Res., 85, 4049-4058.
Götz, F.W.P. and Ladenburg, R. (1931). Ozongehalt der unteren Atmosphärenschichten, Die Naturwissenschaften, 18, 373-374.
Götz, F.W.P., Meetham, A.R. and Dobson, G,M.B. (1934). The vertical distribution of ozone in the atmosphere, Proc. Roy. Soc. (London), A145, 416-446.
Götz, F.W.P., Schein, M. and Stoll, B. (1934). Atmosphärische Untersuchungen mit dem Lichtzählrohr in Arosa, Helv. Physica Acta, 7, 484-488.
Götz, F.W.P., Schein, M. und Stoll, B. (1935). Messungen des bodennahen Ozons in Zürich, Gerlands Beitr, Geophys, 45, 237-242.
Götz, F.W.P. und H. Maier-Leibnitz (1933). Zur Ultraviolett-absorption bodennaher Luftschichten, Z. Geophys., 9, 253-260.
Götz, F.W.P. und Penndorf (1941). Weitere Frühjahrswerte des bodennahen Ozons in Arosa, Meteorol. Z., 58, 409-415.
Götz, F.W.P. and Volz, F. (1951). Aroser Messungen des Ozongehalts der unteren Troposphäre und sein Jahresgang, Z. Naturforschg, 6a, 634-639.
Greenberg, J.P. and Zimmerman, P.R. (1984). Nonmethane hydrocarbons in remote tropical, continental, and marine atmospheres, J. Geophys. Res., 89, 4767-4778.
Guicherit, R. and van Dop, H. (1977). Photochemical production of ozone in Western Europe (1971-1975) and its relation to meteorology, Atmos. Environ., 11, 145-155.
Haagen-Smit, A.J. (1952). Chemistry and Physiology of Los Angeles smog, Indust. Eng. Chem., 44, 1342.
Haagen-Smit, A.J. and Fox, M.M. (1956). Ozone formation in photochemical oxidation of organic substances, Indust. Eng. Chem., 48, 1484.
Harriss, R.C., Gorham, E., Sebacher, D.I., Bartlett, K.B. and Flebbe, P.A. (1982). Methane flux from northern peatlands, Nature, 315, 652-653.
Keller, M., Goreau, T.J., Wofsy, S.C., Kaplan, W.A. and McElroy, M.B. (1983). Production of nitrous oxide and consumptiun of methane from forest soils, Geophys. Res. Lett., 10, 1156-1159.
Lamb, B., Guenther, A., Gay, D. and Westberg, H. (1987). A national inventory of biogenic hydrocarbon emissions, Atmos. Environ. (submitted).
Lenschow, D.H., Pearson, R. Jr. and Stankov, B.B. (1982). Measurements of ozone vertical flux to ocean and forest, J. Geophys. Res., 87, 8833-8837.

Levy, A. (1907). Analyse de l'air atmosphérique-ozone, Annales d'Observatoire Municipal de Montsouris, 8, 289-291.
Levy, H. (1971). Normal atmosphere: Large radical and formaldehyde concnetrations predicted, Science, 173, 141-143.
Levy, H. (1972). Photochemistry of the lower troposphere, Planet. Space Sci., 20, 919-935.
Liu, S.C., Trainer, M., Fehsenfeld, F.C., Parrish, D.D., Williams, E.J., Fahey, D.W., Hübler, G. and Murphy, P.C. (1987). Ozone production in the rural troposphere and the implications for regional and global ozone budgets, J. Geophys. Res. (in press).
Logan, J.A. (1983). Nitrogen oxides in the troposphere: global and regional budgets, J. Geophys. Res., 99, 10785-10807.
Logan, J.A. (1985). Tropospheric ozone: seasonal behavior, trends and anthropogenic influence, J. Geophys. Res., 90, 10463-10482.
Logan, J.A. and Kirchhoff, V.W.J.H. (1986). Seasonal variations of tropospheric ozone at Natal, Brazil, J. Geophys. Res., 91 (D7), 7875-7881.
McConnell, J.C., McElroy, M.B. and Wofsy, S.C. (1971). Natural sources of atmospheric CO, Nature, 233, 187-188.
McFarland, M., Kley, D., Drummond, J.W., Schmeltekopf, A.L. and Winkler, R.H. (1979). Nitric oxide measurements in the equatorial Pacific region, Geophys. Res. Lett., 6, 605-608.
NAPAP (National Acid Precipitation Assessment Program) (1984). Annual Report to the President and Congress, 1984, Director of Research, 726 Jackson Place, N.W., Washington, D.C. 20503, USA.
Noxon, J.F. (1978). Tropospheric NO_2, J. Geophys. Res., 83, 3051-3057.
Oltmans, S.J. (1985). Tropospheric ozone at four remote observatories, in Atmospheric Ozone, Proceedings Ozone Symposium, Halkidiki, Greece, 3-7 Sept. 1984, C.S. Zerefos and A. Ghazi, Editors, Reidel, Dordrecht, Holland, p. 730-734.
Paneth, F.A. and Edgar, J.L. (1938). Concentration and measurement of atmospheric ozone, Nature, 142, 112-113 (see also p. 571).
Paneth, F.A. and Glückauf, E. (1941). Measurement of atmospheric ozone by a quick electrochemical method, Nature, 147, 614-615.
Perl, G. (1965). Das bodennahe Ozon in Arosa, seine regelmäßigen und unregelmäßigen Schwankungen, Archiv Meteorol., Geophys. Bioklim., A14, 449-458.

Rasmussen, R.A. and Khalil, M.A.K. (1981). Atmospheric methane (CH_4): Trends and seasonal cycles, J. Geophys. Res., 86, 9826-9833.
Rasmussen, R.A. and Khalil, M.A.K. (1984). Atmospheric trends in the recent and ancient atmospheres: concentrations, trends, and interhemispheric gradient, J. Geophys. Res., 89 (D7), 11599-11605.
Regener, V.H. (1938). Neue Messungen der vertikalen Verteilung des Ozons in der Atmosphäre, Z. Phys., 109, 642-670.
Regener, V.H. (1938). Messungen des Ozongehalts in Bodennähe, Meteorol. Z., 55, 459-462.
Regener, E. und Regener, V.H. (1934). Aufnahmen des ultravioletten Sonnenspektrums in der Stratosphäre und vertikale Ozonverteilung, Phys. Z., 35, 788-793.
Ridley, B.A., Carroll, M.A. and Gregory, G.L. (1987). Measurements of Nitric Oxide in the Boundary Layer and Free Troposphere over the Pacific Ocean, J. Geophys. Res., 92 (D2), 2025-2048.
Rinsland, C.P., Levine, J.S. and Miles, T. (1985). Concentration of methane in the troposphere deduced from 1951 infrared solar spectra, Nature, 318, 245-249.
Rinsland, C.P. and Levine, J.S. (1985). Free tropospheric carbon monoxide concentrations in 1950 and 1951 deduced from infrared total column amount measurements, Nature, 318, 250-254.
Routhier, F., Dennett, R., Davis, D.D., Danielsen, E., Wartburg, A., Haagenson, P. and Delany, A.C. (1980). Free tropospheric and boundary layer airborne measurements of ozone over the latitude range of 58°S and 70°N, J. Geophys. Res., 85, 7307-7321.
Schenkel, A. and Broder, B. (1982). Interference of some trace gases with ozone measurements by the KI method, Atmos. Environ., 16, 2187-2190.
Schönbein, Ch.F. (1840). Recherches sur la nature de l'odeur qui se dégage de certaines réactions chimique, Comptes Rendus, 10, 706.
Schönbein, Ch.F. (1845). Einige Bemerkungen über die Anwesenheit des Ozons in der atmosphärischen Luft und die Rolle welche dieser bei langsamen Oxydationen spielen dürfte, Ann. Phys. Chim (Poggendorf), 65, 161-172.
Seiler, W. (1974). The cycle of atmospheric CO, Tellus, 26, 117-135.
Seiler, W. (1984). Contribution of biological processes to the global budget of CH_4 in the atmosphere, in Current Perspectives in Microbial Ecology (M.J. Klug and C.A. Reddy, Eds), American Society for Microbiology, Washington, D.C., 468-477.

Seiler, W. and Fishman, J. (1981). The distribution of carbon monoxide and ozone in the free troposphere, J. Geophys. Res., 86, 7255-7266.

Seiler, W., Conrad, R. and Scharffe, D. (1984). Field studies of methane emission from termite nests into the atmo-sphere and measurements of methane uptake by tropical soils, J. Atmos. Chem., 1, 171-186.

Shaw, J.H. (1958). The abundance of atmospheric carbon monoxide above Columbus, Ohio, Astrophys. J., 128, 428-440.

Singh, H.B. and Salas, L.J. (1983). Peroxy-acetyl nitrate (PAN) in the free troposphere, Nature, 201, 326-327.

Skarby, L. and Sellden, G. (1984). The effects of ozone on crops and forests, Ambio, 13, 68-72.

Soret, J.L. (1863). Sur les relations volumétriques de l'ozone, Comptes rendus, 57, 604-609 and 941-944.

Soret, J.L. (1865). Recherches sur la densité de l'ozone, Comptes rendus, 61, 941-944.

Strutt, R.J. (1918). Ultra-violet transparency of the lower atmosphere and its relative poverty in ozone, Proc. Roy. Soc., A94, 260-268.

Thompson, A.M. and Cicerone, R.J. (1986). Possible perturbations to CO, CH_4 and OH, J. Geophys. Res., 91, (D10), 10853-10864.

Tichy, H. (1939). Gleichzeitige Messungen von Ultraviolett und bodennahem Ozon, Der Balneologe, 6, 125-130.

Trainer, M., Williams, E.J., Parrish, D.D., Buhr, M.P., Allwine, E.J., Westberg, H.H., Fehsenfeld, F.C. and Liu, S.C. (1987). Impact of natural hydrocarbons on rural ozone: modeling and observations (in preparation).

Volz, A., Ehhalt, D.H. and Derwent, R.G. (1981). Seasonal and latitudinal variation of ^{14}CO and the tropospheric concentration of OH radicals, J. Geophys. Res., 86, 5163-5171.

Volz, A. and Kley, D. (1987). Ozone measurements in the 19th century: evaluation of the Montsouris series, Nature (submitted).

Thompson, A.M. and Cicerone, R.J. (1986). Atmospheric CH_4, CO and OH from 1860 to 1985, Nature, 321, 148-150.

Warmbt, W. (1979). Ergebnisse langjähriger Messungen des bodennahen Ozons in der DDR., Z. Meteorologie, 29, 24-31.

Westberg, R.G., and Lamb, B. (1987). Ozone production and transport in the Atlanta, Georgia region (private communication).

Winkler, P. (1980). Meridionalverteilung des bodennahen Ozons über den Atlantik, Ann. Meteorol., 15, 241-242.

Wofsy, S.C. McConnell, J.C. and McElroy, M.B. (1972). Atmospheric CH_4, CO and CO_2. J. Geophys. Res., 77, 4477-4493.

Zimmerman, P.R., Chatfield, R.B., Fishman, J., Crutzen, P.J. and Hanst, P.L. (1978). Estimates on the production of CO and H_2 from the oxidation of hydrocarbon emissions from vegetation, Geopyhs. Res. Lett., 5, 679-682.

TRANSPORT OF POLLUTANTS ON DIFFERENT SCALES IN THE TROPOSPHERE

Dr. H. van Dop
Royal Netherlands Meteorological Institute
P.O. Box 201
3730 AE DE BILT
The Netherlands

ABSTRACT. The mathematical formulation of atmospheric transport models is based on a conservation equation. The solution of this equation requires numerical methods which treat the variables at discrete points in space and time. Atmospheric flow and processes may vary in scale from 10^{-3}-10^{7} m. Many of the smaller scales are relevant for global transport and chemistry. Computer limitations prohibit the application of models which cover the whole range of scales. Therefore, part of the flow structure and a number of physical processes are not resolved in currently applied models, where the grid size may vary from 20-500 km. Unresolved phenomena must be parameterised separately. The mathematical concepts and parameterisations are discussed. Finally some current regional atmospheric boundary layer models are presented.

1. INTRODUCTION

The chemical composition of the troposphere as we know it nowadays has been the result of physical, chemical an biological processes which have been going on for millions of years. In its present composition the average ratio of its major constituents, nitrogen and oxygen seems to be fairly constant. Apart from that, the troposphere consists of a large variety of trace gases. Their concentrations show in general large variations in time and space. Major changes in the concentrations of some of these trace gases, of which carbondioxide, oxides of nitrogen and sulphur, volatile hydrocarbons and ozone are the most important, have occurred in the past age.

The industrial revolution, the population explosion and the increased economic activity have caused drastic changes in this respect. An inevitable question in connection with the production of air pollution is: what are the environmental consequences.

Air pollution transport, transformation and deposition models provide a means to get an answer to this question. The public awareness of adverse effects of air pollution required causal relationships between emissions and its effects. In this way "the art of atmospheric transport modelling" came into being. When put in a historical perspective, the

I. S. A. Isaksen (ed.), Tropospheric Ozone, 33–48.

first efforts in modelling concerned the regional air quality in the direct neighbourhood of cities and industrial plants. In the early sixties the first evidence of air pollution damage due to transport over longer distances was reported by the Scandinavian countries, which claimed that the chemical (and consequently biological) composition of their lakes was affected by emissions of sulphur dioxide in the large industrial centres in West and Central Europe (Eliassen et al., 1978). Not only the airborne concentrations were a matter of concern, but also how and where air pollution returned to the earth surface. A complicating factor was that atmospheric chemical processes should be taken into account in view of the transport times involved in transport on the continental and global scale.

It was recognised that the on-going production of air pollution gradually changes the trace gas composition of the whole troposphere (NAS report, 1985). This could lead to climatic changes in the next century which, according to some, may be hardly measurable, and to others, have an apocalyptic character.

This very brief overview indicates that air pollution transport modelling covers spatial scales from a few kilometers to the global scale, with corresponding time scales of minutes to years. Chemical transformation processes and deposition are necessary ingredients in these models. In this chapter we emphasise the transport only, the other processes being dealt with in other contributions.

Almost all air pollution is emitted at, or near the surface of the earth. Transport and (fast) chemistry of the majority of pollutants is mainly confined to a thin layer of, say 2-3 km, the atmospheric boundary layer (ABL). The understanding of the physics of the ABL, including cloud formation and precipitation is crucial in the development of transport modelling (and also for the mechanisms which transport (a tiny fraction of) the (chemically converted) emissions up into the background troposphere. As a consequence, transport models should include a reasonable amount of properties of the ABL, such as the three dimensional windfield (with adequate time and space resolution), a description of turbulent motion, including the mixed-layer height, radiation, cloud formation and precipitation.

Depending on scale, desired complexity, numerical facilities etc. a model formulation will be chosen. Though a large variety of models exists covering various spatial ranges, two main categories should be distinguished: Eulerian and Lagrangian transport models. Eulerian models describe the dispersion of pollution in a fixed frame of reference (fixed with respect to an observer or to a point on the earth surface). In Lagrangian models the motion of a polluted air parcel is followed from its initial position as it moves along its trajectories. The reason why one decides to use a Lagrangian or a Eulerian formulation lies in the application. This will be discussed in the next section, which is devoted to the main concepts in long range transport modelling.

A sometimes underestimated aspect of long range transport modelling is the preparation of the meteorological variables which describe circulation and eddy (transport) This will be discussed below.

We have approached a point where two modelling activities are about to meet: on the one hand the ABL air pollution models which have been formulated on increasing scales, now covering large parts of a continent. Often the output of these models pertain to climatological averages i.e. on time scales of at least a season or a year. On the other hand the middle atmosphere models, which formerly consisted of one or two dimensional parameterisations, are now started to be formulated fully 3-dimensional, being able to contain as many details on transport as computer space allows.

Both modelling activities describe transport "off-line", which implies that separate models are used (limited area models, weather forecast models or global circulation models extending up to the stratopause) to provide the necessary advective and diffusion transport data. This will be discussed in section 4. Finally a review will be given of some important developments in long range transport models (section 5).

2. MATHEMATICAL CONCEPTS

Here, a brief review will be given of the mathematical framework of transport models. More detailed descriptions can be found in the literature (Pasquill and Smith, 1983; Nieuwstadt and Van Dop, 1982 and Hutzinger, 1985).

2.1. Conservation Equations

The dry atmosphere can be considered as a mixture of ideal gases. The dynamics of atmospheric flow is given by the Navier-Stokes equations, which together with the continuity equation and the energy equation completes the set of atmospheric equations. When c denotes the concentration of a contaminant (in units of mass per unit volume of fluid) the equation of conservation of mass is simply

$$\frac{\partial c}{\partial t} + \frac{\partial uc}{\partial x} + \frac{\partial vc}{\partial y} + \frac{\partial wc}{\partial z} = s \, , \qquad (1)$$

where s denotes the sources and sinks of contaminant c and u,v,w are the components of the wind velocity in a fixed (Eulerian) frame of reference. By decomposing c and the wind velocity components in an average value, and a fluctuating part denoted by an overbar and a prime respectively: $c = \bar{c} + c'$, $u = U + u'$ etc. and averaging noting that $\overline{c'}$, $\overline{u'}$ etc equal zero, we obtain

$$\frac{\partial \bar{c}}{\partial t} + U\frac{\partial \bar{c}}{\partial x} + V\frac{\partial \bar{c}}{\partial y} + W\frac{\partial \bar{c}}{\partial z} = - \frac{\partial}{\partial x}\overline{u'c'} - \frac{\partial}{\partial y}\overline{v'c'} - \frac{\partial}{\partial z}\overline{w'c'} + S \, , \qquad (2)$$

Eqn. (2) is the familiar starting point for numerous approximations. In order to be able to solve eqn. (2), assumptions have to be made for the eddy correlation terms at the right of eqn. (2). The usual approach in ABL modelling is to put

$$- \overline{u'c'} = K_x \frac{\partial \bar{c}}{\partial x} , \quad - \overline{v'c'} = K_y \frac{\partial \bar{c}}{\partial y}$$

$$\text{and} \quad - \overline{w'c'} = K_z \frac{\partial \bar{c}}{\partial z} , \tag{3}$$

a "receipt" which is usually referred to as first order closure or the gradient transfer assumption. This approach is a purely empirical one and based on the analogy of the (3-dimensional) turbulent diffusion process in the ABL with molecular diffusion. This implies that the scales of the diffusive processes are much smaller than the scales of the mean flow. There is some justification for this approach when in eqns. (2, 3) averages over the order of one hour are considered, since the spectrum of ABL turbulence has a "gap" at frequencies of the order of 1/one hour. Substituting eqn. (3) in eqn. (2) yields

$$\frac{\partial \bar{c}}{\partial t} + U \frac{\partial \bar{c}}{\partial x} + V \frac{\partial \bar{c}}{\partial y} + W \frac{\partial \bar{c}}{\partial z}$$

$$= \frac{\partial}{\partial x} (K_x \frac{\partial \bar{c}}{\partial x}) + \frac{\partial}{\partial y} (K_y \frac{\partial \bar{c}}{\partial y}) + \frac{\partial}{\partial z} (K_z \frac{\partial \bar{c}}{\partial z}) + S . \tag{4}$$

The empirical coefficients K_x, K_y and K_z determine the strength of the turbulent exchange and may be functions of the co-ordinates in inhomogeneous flows. This approach is clearly more unsatisfactory in modelling on global scales, where averages over much longer times and corresponding spatial scales are taken. A generalisation of eqn. (3) is

$$- \overline{u_i'c'} = K_{ij} \frac{\partial \bar{c}}{\partial x_j} , \tag{5}$$

where the diffusion coefficients are now replaced by the diffusion tensor K_{ij} (u_i', i=1,2,3 equals u', v' and w' respectively). With this formulation one may describe also countergradient fluxes, whereas the scalar diffusion formulation only describes downgradient fluxes. Nevertheless the first order closure approach remains doubtful when averaging times are such that large atmospheric eddies are contained in the "eddy flux" terms.

2.2. Lagrangian Model Formulation

In the Lagrangian model formulation the mean concentration of a moving polluted parcel of air is considered. The volume of the parcel is assumed large enough, so that concentration changes due to turbulent exchange through the boundaries can be neglected. It is also assumed that during its travel the distortion of the parcel due to the turbulent motion is small, so that the parcel remains an entity. The mean concentration in the air parcel is then simply given by

$$\frac{d\bar{c}}{dt} = S , \tag{6a}$$

where S, as in eqn. (4), contains the source and sink terms. This term should be evaluated along the trajectory of the air parcel, thus $S = S(\underset{\sim}{X}(t),t)$, where $\underset{\sim}{X}$ denotes the parcel location. The problem remains to determine the parcel trajectory. This can be done by using the Eulerian-Lagrangian relationship. In integrated form it reads

$$\underset{\sim}{X}(t) = \underset{\sim}{X}_o + \int_o^t \underset{\sim}{u}_E \, (\underset{\sim}{x} = \underset{\sim}{X}(t'), \, t') \, dt' \; . \tag{6b}$$

The initial position is denoted by $\underset{\sim}{X}_o$ and $\underset{\sim}{u}_E$ is the Eulerian wind velocity field.

In ABL models usually an air parcel is considered which is adjacent to the earth's surface and which has a fixed height, h. Since the wind velocity changes with height the mean transport wind should be determined from a vertical average of the wind velocity over the parcel height h. In practice, the wind velocity is obtained from radiosonde data or from dynamic models of the wind field. The 925 mbar level wind data (~ 800 m altitude) are considered representative for the transport wind in trajectory models.

The solution of the equation (6) for a sufficient number of initial positions provides a fast and relatively easy picture of atmospheric dispersion. Trajectory models can be modified such that they may account for a variable parcel height, or include horizontal diffusion. The Lagrangian approach has been successfully applied in the EMEP study (see section 5) on long range transport and deposition of sulphur dioxide and sulphate on the European continent (Eliassen, 1980).

2.3. Eulerian Model Formulation

The master equation for a number of applications in atmospheric dispersion calculations is eqn. (4). In a few cases the equation has analytical solutions for which we refer to Pasquill and Smith (1983) or Sutton (1943). In mesoscale and long range transport applications non-homogeneity and non-stationarity of wind and turbulence make that only numerical solutions will provide solutions to eqn. (4). Nevertheless we shall briefly discuss some current mathematical simplifications of the transport equation. First, over (more or less) flat terrain vertical transport of diffusion will be dominated by diffusion and so the fourth term at the left of eqn. (4) may be omitted.

Also in moderate wind conditions it is assumed that the horizontal advection dominates the horizontal turbulent diffusion. With these assumptions eqn. (4) reduces to

$$\frac{\partial \bar{c}}{\partial t} + U \frac{\partial \bar{c}}{\partial x} + V \frac{\partial \bar{c}}{\partial y} = \frac{\partial}{\partial z} (K_z \frac{\partial \bar{c}}{\partial z}) + S \; . \tag{7}$$

Besides, Van Dop et al. (1982) show that in the necessary discretization of eqn. (4) the horizontal diffusion will in most cases have sub-grid scale dimensions (when the natural variability of the wind-field is well-represented at the selected grid-size), and therefore, can be neglected.

2.4. Volume Averaged Models

The normal numerical procedure to solve eqn. (7) is to select a vertical (and horizontal) grid size which is small enough to resolve a sufficient amount of details of the atmospheric structure. This usually leads to the choice of an equidistant grid, or a grid of which the gridsize is gradually increased at higher altitudes. The number of gridpoints required in the vertical direction is of the order of 10-20. This can be rather demanding with respect to computer capacity and processing and possibly inefficient, the more so since it may be doubted whether this detail in vertical structure is really needed in long range transport modelling.

An approach which might be worthwhile pursuing instead, is the volume averaging of eqn. (1). It consists of applying the operator

$$\int_{\Delta} \ldots \, dxdydz$$

on eqn. (1), where Δ is a (small) volume in space. Usually the operation is carried out in one (the vertical) direction. The method has some distinct advantages: the atmosphere has a layered structure and it is often sufficient to evaluate layer averaged characteristics. So by selecting a limited number of vertical layers (surface layer, mixed layer, cloud layer, free troposphere layer etc.) a significant reduction of the total amount of vertical layers may be achieved. Of course the dynamics of these layers should be prescribed. The mixed layer for instance may vary in height from 100 to 2000 m. This approach is adopted e.g. in the SAI, KNMI and RIVM model (Meinl and Builtjes, 1984; Van Dop and De Haan, 1983; Van Egmond and Kesseboom, 1983) and is rigorously treated by Lamb (1982).

Also dynamical volume averaged models have been developed. These so-called large eddy simulation models are restricted to the convective boundary layer (Schumann and Friedrich, 1986), where they are quite satisfactory in describing the features of the mixed layer dynamics. However, the large computational requirements make application in GCM's in the near future unlikely.

2.5. Numerical Methods

Solutions of the transport equation can be obtained by numerical methods only. Often practical limitations (computer facility, run-time) determine which method should be followed. However, the primary requirement for a numerical method should be that errors introduced by the method should be smaller than the other uncertainties in the transport problem.

The basic equation is usually a parabolic partial differential equation (eqn. (4)). In case more components are involved, $c_1 \ldots c_n$, we have a set of n of these equations which are coupled by chemical reactions. The numerical implications of the chemistry will not be discussed here so that this section is devoted to the numerical analysis of eqn. (4). There is, however, one important feature of numerical chemical schemes which should be borne in mind: they cannot cope with negative concentrations. This has an important feed-back on numerical schemes for advection, since they usually create (small) negative concentration values.

The ideal advection scheme should have the following properties.

- It should be mass conserving.
- The accuracy must be sufficient, i.e. the contaminant must be transported with the right speed (and in the right direction).
- The scheme should yield positive concentrations everywhere.
- The numerical diffusion should be much smaller than the real atmospheric diffusion.
- Process time and memory requirements should be low.

It is obvious from the large number of publications still appearing on this subject that the ideal scheme has not yet been formulated, though computer capacity increases by an order of magnitude every ten years. A few well-known numerical procedures are the pseudospectral method (Christensen and Prahm, 1976), the method of moments (Egan and Mahoney, 1971), the Chapeau-function method (Chock, 1985) and Smolarkiewicz' scheme (1984). An excellent review with a detailed test of some of the more current schemes is given by Van Stijn et al. (1987).

3. SUB-GRID SCALE PHENOMENA

The presence of a large range of scales in atmospheric transport inhibits a comprehensive description of all processes. The discretisation in the numerical processes cut off a number of phenomena which have a non-negligible impact on the larger scale processes. This impact should be properly parameterised. In this section we give a summary of some of the more important parameterisations.

3.1. The Treatment of Air Pollution Sources

The transport equation should be completed with an initial field, $\bar{c}(x,y,z,t=0) = \bar{c}_o$ and with boundary conditions, which define the in- and outflow over the model boundaries.

The sources are contained in the term S at the right of eqn. (4). They consist of surface sources, which are estimated total emissions over relatively large areas, and point sources. In emission inventories surface sources are usually presented on a grid with a size of comparable dimensions with the numerical grid size. These sources can be easily implemented in transport models. The lateral dimensions of releases from point sources, however, may be sub-grid scale for quite a long time. Implementation of these sources can only be carried out accurately when the "plume" dimension has become at least of the same order as the numerical grid-size. This problem may be tackled by approximating the concentration fields of point sources by analytical Gaussian fields close to the source (cf. Karamchandani and Peters, 1983) in the initial stage of the dispersion.

3.2. Atmospheric Turbulence

All movements which are not contained in the average 3-dimensional windfield are caused by atmospheric turbulence. Though in the ABL the transport in the horizontal direction, parallel to the surface, is dominated by the mean windfield, the transport in vertical direction is mainly the result of turbulent atmospheric motions. The intensity of these motions is closely related to the mean wind shear and the atmospheric stability. These matters are amply discussed in Pasquill and Smith (1983). We shall first review those topics which are relevant for long range transport modelling in the ABL.

In the lower troposphere we may distinguish a few layers: the surface layer, the atmospheric boundary layer (in which the surface layer is included), and the free troposphere.

The turbulence in the ABL is generated by friction of the atmospheric flow with the surface and buoyancy effects due to radiative cooling and heating of the surface.

During day-time a mixed-layer develops which is characterised by a near neutral potential temperature gradient. In horizontally homogeneous and almost cloud free conditions the dynamics of this layer can be formulated (Tennekes, 1973). The development of the mixed-layer height is then roughly known. We may recall that this is an important parameter in vertically integrated transport models. A typical value of the daily maximum mixed-layer height at moderate latitude is 1600 m.

During night-time a radiation inversion develops with a quasi equilibrium height with a typical value of ~ 200 m. Within this layer vertical motions are strongly damped so that the vertical dispersion will be small. It should, however, be noted that buoyant plumes of (large) point sources may penetrate the nocturnal inversion layer, so that pollution is directly injected in the (slightly stable) free atmosphere.

3.3. Clouds and Precipitation

An important element in long range transport modelling is the precipitation: crudely half of the atmospheric pollutants are deposited on the surface by precipitation. The process of cloud formation and precipitation cannot yet be fully described by numerical meteorological models. Intertwained in this process is the air pollution, which is absorbed by (or evaporates from) cloud or rain droplets, or which serves as condensation nuclei for the formation of water droplets when the pollution consists of particulate matter. Representative amounts of precipitation can be estimated from observations or model data. First should be noted that precipitation, be it from single convective clouds or originating from large frontal systems, is highly intermittent and spatially strongly varying, when it is compared with the density in space and time of the synoptic rain collector network. The spatial resolution amounts to 50-100 km over Europe (n.b. there are no observations over sea!), which largely exceeds the size of convective storms or the width of rain bands and cells in frontal systems. The information gathered by the synoptic network is therefore not much more than a random sample. A second problem is that amounts of precipitations are reported as accumulated sums over 6 or 12 hours. This is clearly unsufficient for say a reliable three hourly rain analysis on a grid of 80x80 km.

One may hope to avail of high resolution rain data from (interlinked) radar networks in the future. On the other hand regional meteorological models are able to provide precipitation data with the desired resolution in time and space though their results should be considered with care. An advantage of models is that they also predict precipitation over sea. In the operational ECMWF circulation model the formation of clouds is based on the Kuo convection scheme which relates precipitation amounts to convective cloud cover, C_c, according to

$$C_c = a + b \log P ,$$

where P is the daily average precipitation rate and a and b are constants (Slingo, 1985). Three other types of cloud cover (high, middle and low level) are related to C_c, relative humidity and vertical velocity according to simple empirical relationships.

3.4. Orography

The presence of orography invokes vertical motions which can be felt high up in the troposphere. Because vertical motion is closely related to condensation and precipitation processes, the modelling of orography influences these processes. Apart from some small areas on earth which can be considered flat, orography should be considered in transport models on any scale. And also here some aspects are "sub grid": in the smaller scales models valley winds and local recirculating flows are not well described. In the larger scale models the mountainous areas are not well represented. The Alps for instance with many peaks and ridges over 4000 m have a maximum height of ~1300 m on the 100 x 200 km grid of ECMWF model, and are thus transformed to a large hill with gentle slopes extending far into Germany, Italy and France. This creates a precipitation and cloud pattern which is quite different from reality.

3.5. Chemistry

Large sub-grid scale variations in concentrations may occur. In chemical reactions usually grid averaged concentrations are used, so that reaction rates may be both under and overestimated. Though this subject is not in the scope of this chapter we mention it for completeness' sake.

4. METEOROLOGICAL DATA

In the introduction the paramount importance of meteorological data of the atmospheric boundary layer was already indicated. The major features are the transport by the mean atmospheric flow, the turbulence, and regarding deposition, clouds and precipitation.

4.1. Mean flow

Mean wind data can be obtained either from routine synoptic observations, or from a dynamical model. The former data can be made available easily but have the disadvantage that most data are ground-based and only over land sufficiently dense, while upper air data are spatially and temporally too sparse to give a reasonably detailed analysis of the wind field for mesoscale and long-range transport models. Moreover, it requires considerable effort to derive reasonable wind fields from observations, which are often affected by the local situation, and therefore not very representative for a (grid square) average wind velocity field.

An alternative is offered by generating wind data from regional meteorological models (Anthes and Warner, 1978) or limited area models. These models are primarily designed to be used as regional weather forecast models. Since they calculate meteorological fields with a spatial and temporal resolution which fit with those of mesoscale and long range transport models, they are very suitable as input for dispersion modelling on this scale. For example the limited area model (LAM) which is currently used at KNMI is derived from the ECMWF (European Centre for Medium Range Weather Forecasting). There are 4 layers below 3000 m altitude in the model. Moreover, these models contain simple parametrizations of precipitation (see section 3.3).

In general the ECMWF model seems to provide a sufficient amount of meteorological data to feed transport models on the continental and global scale.

5. CURRENT DEVELOPMENTS IN LONG RANGE TRANSPORT MODELLING

5.1. The EMEP Model

The EMEP (European Monitoring and Evaluation Programme) model is a (Lagrangian) trajectory model which describes the long term average air concentrations and total deposition of SO_2 and SO_4^{2-} over Europe (Eliassen, 1980). It was one of the first operational models and was initially used for the evaluation of the acidification of the Scandinavian lakes. The meteorological data are based on the 850 mbar radiosonde observations. The spatial resolution is 127x127 km^2. Chemical transformation rate and dry and wet deposition are assumed to be linearly proportional to the concentrations. A constant ABL height of 1000 m is assumed in the routine model version. Notwithstanding its simplicity and shortcomings the model proved to be extremely useful after extensive tests and validations. A research version which includes (many) other compounds and a complex chemical scheme is under development (Eliassen et al., 1982).

5.2. The EPA Regional Oxidant Model

This Eulerian grid model, which is developed at the Environmental Protection Agency (EPA), (Lamb, 1982) has the objective to guide the formulation of regional emission control strategies in the United States. The model contains three "dynamic" layers. The horizontal domain is ~1000x1000 km^2 and the grid size approximately 18 km (2500x3 grid points). The model aims at including all relevant chemical and physical processes related to transport, transformation and deposition of air pollution, such as full photochemistry (including slow reactions), night-time chemistry of the products and precursors of photochemical reactions, cumulus cloud physics and chemistry, mesoscale vertical motion, sub-grid scale chemistry and emissions from natural sources of volatile organic compounds (VOC's), NO_x and stratospheric ozone intrusions.

5.3. The NCAR Acid Deposition Modelling Project

This project (NCAR, 1985; Chang et al., 1987) is funded by the US EPA and the National Science Foundation (NSF). Its principal task is to develop a Eulerian regional acid deposition model which is suitable for assessing source-receptor relationships. The prototype model contains 15 layers (at fixed heights) and has a horizontal resolution of 80 km (1700x15 grid points). It covers an area of approximately 3000x3000 km^2. The project is designed for the development of "credible" models, based on recent progress and state of the art knowledge of mesoscale meteorology, tropospheric chemistry and advanced computing.

5.4. The PHOXA Project

The two previous projects pertain primarily to application at the American continent. The PHOXA (Photochemical Oxidant and Acid Deposition Model Application) project was initiated by the German and Dutch Government. The primary goal is to be able to develop different control strategy options in Europe. It attempts to apply a number of existing transport and deposition models. Among these are
- a two-layered trajectory model including chemistry, the MPA model (De Leeuw en Van den Hout, 1985)
- a regional transport model developed by SAI based on their episodic photo oxidant model (Meinl and Builtjes, 1984);
- the transport and deposition of acidifying pollutants-model (TADAP) developed by ERT, a very sophisticated transport model including a detailed analysis of dry and wet deposition of acidifying pollutants.

5.5. RIVM/KNMI Co-operation on Dispersion Modelling

Two Dutch institutes (the Institute for public health and environmental hygiene (RIVM) and KNMI) are combining their research and experience (Van Dop et al., 1983 and Van Egmond and Kesseboom, 1983) on mesoscale and long range transport modelling.

The objectives are to develop and make operational
- a two-layered trajectory model including full chemistry;
- a Eulerian (vertically integrated) transport and deposition model including simple linear chemistry and deposition, to be applied in Europe;

- an episodic photo oxidant model and an acid deposition model, including full chemistry. The transport model is based on the above mentioned model;
- a (partially prognostic) model to be used for the dispersion of accidental releases of toxic or hazardous material (calamities at nuclear or chemical plants).

At KNMI the emphasis is on the meteorological preprocessing, i.e. the preparation of meteorological fields and the prognostic aspects. The more operational activities are carried out by RIVM.

5.6. Global Transport Models

Global transport formulations are usually one or two-dimensional. The 2-D models explain the behaviour of zonally averaged quantities over time scales of a month. There is some justification for this approach in the stratosphere since there is a strong zonal flow which ensures that average concentrations of trace gases are considered over a latitude circle. The formulation of the problem is similar to that in the ABL and consists of a conservation equation for a species $\bar{c}$,

$$\frac{\partial \bar{c}}{\partial t} + \bar{u}_i \frac{\partial \bar{c}}{\partial x_i} = - \frac{1}{\rho_o} \frac{\partial}{\partial x_i} \rho_o \overline{u'_i c'} + s \ , \qquad (8)$$

(cf. eqn. (4)), where an overbar denotes zonally averaged quantities and the primes deviations from it. The larger height scale enforces the inclusion of the mean atmospheric density $\rho_o(z)$.

Unless in ABL transport models, the separation into "mean" and "eddy" transport is arbitrary and though eqn. (8) may be still correct, the successive modelling of the "eddy" term by a gradient formulation (see eqn. (5)) introduces large uncertainties, in view of the long averaging time. The question arises whether the partitioning in advective transport and diffusive transport in zonally averaged models is correct and some alternative formulations have been developed (McIntyre, 1980; Plumb, 1979).

The complex structure of the flow in the troposphere makes it highly improbable that the above approach will work. Here we have the mean Hadley circulation on which are superimposed the horizontal baroclinic eddies and the vertical convective motions, which all contribute to the transport of pollutants (see Fig. 1). Therefore a 3-D formulation seems inevitable. The meteorological data for the 3-D description of transport may be obtained from tropospheric GCM's such as the one from ECMWF. The ECMWF T21 model is capable of describing transport at 16 levels up to 26 hPa with a horizontal resolution of ~6° and a timestep of 45 minutes. The current operational version contains 19 layers with a horizontal resolution of 1½°.

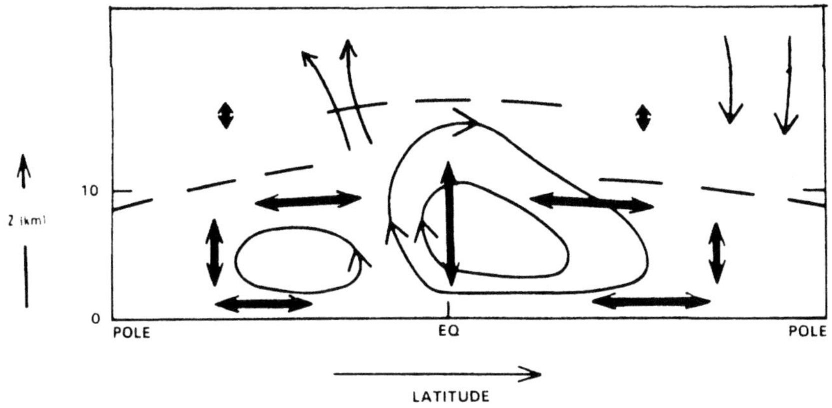

Figure 1. Schematic illustration of zonally-averaged transport in the troposphere. Single arrows: mean circulation; double arrows: diffusive transport. Dashed line: tropopause (source WMO, 1985).

6. CONCLUSIONS

Regional (~ 1000 x 1000km^2) ABL transport models are developed at various places. They work with grids of ~ 100 km and use timesteps of one hour or less. Meteorological data for these models are obtained both from observations and regional dynamical models. They often contain complex multiphase chemistry and detailed description of boundary layer processes. The more complex models can only be run on today's largest computers and simulate the dispersion and chemistry over a few days only. Climatological periods of one year or longer could be tackled by these models but require excessive amounts of computertime, though random sampling of typical weather episodes could reduce computer time significantly and still yield reliable climatological data.

Model output consists of spatial and temporal distributions of a number of species in the region up to a height of a few km.

ABL transport models can be extended so that they cover the whole global troposphere in a trivial way. However, it is not likely that in the near future computer capacity is sufficient to handle these models. Extending the averaging time to one month or so and the gridsize to a few hundred km like in 3-D stratosphere modelling studies may make the modelling more managable, but requires substantial small scale parameterisations both in space and time.

All models discussed here operate "off-line", i.e. meteorological data are provided by (a combination of) a separate circulation model and observations. This seems a sensible approach, also for future global tropospheric modelling. The ECMWF is capable to produce most of the desired data with the proper time and space resolution. One of the advantages of the "off-line" approach is that one is able to make improved parameterisations of (i) convective processes, including cloud cover and precipitation, (ii) the liquid phase (not covered by ECMWF) and (iii) ABL-free troposphere exchange processes.

7. REFERENCES

Anthes, R.A. & T.T. Warner, 'Development of hydrodynamic models suitable for air pollution and other mesometeorological studies', Mon. Wea. Rev., **106** (1978) 1045-1078.

Chang, J.S., R.A. Brost, I.S.A. Isaksen, S. Madronich, P. Middleton, W.R. Stockwell and C.J. Walcek, 'A three dimensional Eulerian Acid Deposition Model: Physical Concepts and Formulation", (1987). Submitted to the Journal of Geophysical Research.

Chock, D.P., 'A comparison of numerical schemes for solving the advection equation, Part II', Atmospheric Environment, **19** (1985) 571-586.

Christensen, O & L.P. Prahm, 'A pseudospectral method for dispersion of atmospheric pollutants', J. Appl. Meteor., **15** (1976) 1284-1294.

De Leeuw, F.A.A.M. & K.D. van den Hout, Modelling of photochemical episodes, Procs. Cost 611 workshop, Bilthoven, 23-25 September 1985, CEC, Bruxelles.

Egan, B.A. & J.R. Mahoney, 'Numerical modelling of advection and diffusion of urban area source pollutants', J. Appl. Meteor., **11** (1971) 312-322.

Eliassen, A., 'The OECD study of long range transport of air pollutants: long range transport modeling', Atmos. Environment, **12** (1978) 479-488.

Eliassen, A., 'A review of long-range transport modeling', J. Appl. Meteor., **19** (1980) 231-240.

Eliassen, A., O. Hov, I.S.A. Isaksen, J. Saltbones & F. Stordal, 'A Lagrangian long-range transport model with atmospheric boundary layer chemistry', Journal of Applied Meteorology, **21** (1982) 1645-1661.
Global tropospheric chemistry: a plan for action. Report of the US National Academy of Science (1985).

Karamchandani, P. & L.K. Peters, 'Analysis of the error associated with grid representation of point sources', Atmospheric Environment, **17** (1983) 927-933.

Lamb, R.G., A regional Scale (1000 km) Model of Photochemical Air Pollution, Part I: Theoretical Formulation. EPA-600/3-83-035, Environmental Protection Agency, Research Triangle Park (NC), 1982.

McIntyre, M.E., 'An introduction to the generalized Lagrangian mean description of wave, mean-flow interaction, Pure Appl. Geophys., **118**, (1980) 152-176.

Meinl, H. & P.J.H. Builtjes, Photochemical Oxidant and Acid Deposition Model Applications (PHOXA), Dornier, Friedrichshafen, 1984.

Monin, A.S. & A.M. Yaglom, Statistical FLuid Mechanics, Vol I, MIT, Cambridge (Mass.), 1971.

NCAR Regional Acid Deposition Model, NCAR Technical Note TN-256-STR, Boulder, Colorado, 1985.

Nieuwstadt, F.T.M. & H. van Dop (eds.), Atmospheric Turbulence Modelling, Reidel, 1982.

Pasquill, F. & F.B. Smith, Atmospheric Diffusion, Ellis Horwood Ltd., 1983.

Plumb, R.A., 'Eddy fluxes of conserved quantities by small-amplitude waves', J. Atmos. Sci., **36**, (1979) 1699-1704.

Schumann, U. and Friedrich (eds.), Direct and large eddy simulation of turbulence, Procs. of the Euromech Colloquium 199, München (1986).

Slingo, J.M., Parameterization of Cloud Cover, procs. of the ECMWF seminar, Shinfield Park, UK, 9-13 September, 1985, 17-46.

Smolarkiewicz, P.K., 'A fully multidimensional positive definite advection transport algorithm with small implicit diffusion', J. Comp. Phys., **54** (1984) 325-362.

Sutton, W.A.L., On the equation of diffusion in a turbulent medium. Proc. Roy. Soc., A 182 (1943) 48.

Tennekes, H., 'A model for the dynamics of the inversion above a convective boundary layer', J. Atmos. Sci., **32** (1973) 992-995.

Van Dop, H., B.J. de Haan & C. Engeldal, The KNMI mesoscale air pollution model, Royal Netherlands Meteorological Institute, Scientific Report W.R. 82-6, 1982.

Van Dop, H. & B.J. de Haan, 'Mesoscale air pollution dispersion modelling', Atm. Environment, **17** (1983) 1449-1456.

Van Dop, H., Atmospheric Distribution of Pollutants and Modelling of Air Pollution Dispersion in O. Hutzinger (Ed.), The Handbook of Environmental Chemistry, 4A, Springer, Berlin, 1985, pp. 107-147.

Van Egmond, N.D. & H. Kesseboom, 'Mesoscale Air Pollution Dispersion Models - I Eulerian Grid Model', Atmos. Environment, **17** (1983) 257-265.

Van Stijn, T.L., J. van Eijkeren and N. Praagman, A comparison of numerical methods for air quality models. KNMI scientific report (1987). Available on request.

WMO 1985 Atmospheric Ozone 1985, Assessment of our understanding of the processes controlling its present distribution and change, WMO Global ozone research and monitoring project Rep. No. 16 (1985).

OZONE ON AN URBAN AND REGIONAL SCALE
- With special reference to the situation in the Netherlands -

Robert Guicherit (Ph.D.)
TNO Division of Technology for Society
P.O. Box 217
2600 AE Delft
The Netherlands

ABSTRACT

Formation of ozone in the boundary layer during photochemical pollution episodes gives rise to an ozone concentration ranging from 100 to more than 400 ppb, with monthly averages showing a maximum in summer. The frequency of such episodes is controlled by meteorology and varies widely in place and time. There is, however, a general tendency for ozone concentrations to be lower in urban areas than in downwind rural areas, because in urban areas some ozone is removed by reaction with other pollutants, mostly nitric oxide. In polluted atmospheres these reactions constitute a major sink for ozone. Conversion rates of NO_x range from about 1% per hour in winter to over 15% per hour during photochemical pollution episodes. Some NO_2 is removed from the boundary layer by dry deposition, up to more than 50% of NO_x in Europe may be converted into peroxyacetyl nitrate (PAN), and the remainder into nitrates and some other peroxy nitrates. In the Netherlands PAN levels have been increasing by more than 10% annually over the last decade.
Dry and wet deposition remove HNO_3 and PAN. Some PAN decomposes into NO_2 and peroxy radicals. This would suggest that the effect of anthropogenic NO_x emissions are felt only on a regional scale, i.e. several hundreds of kilometres.
However, the decomposition rate of PAN depends on temperature. At lower temperatures it is slow enough for PAN to be transported over very large distances away from its source.
We argue that transport of PAN, and therefore of NO_x plus anthropogenic carbon monoxide, methane and less reactive hydrocarbons are responsible for the observed ozone increase in the troposphere in the northern hemisphere, and for the observed PAN and ozone maximum in spring (May) in the northern background (free) troposphere and at remote sites away from man-made pollution sources.

I. S. A. Isaksen (ed.), Tropospheric Ozone, 49–62.

1. INTRODUCTION

Ozone is an important secondary pollutant in the boundary layer, and comes from three sources:

- transport from the stratosphere across the troposphere into the boundary layer during meteorological events which sporadically occur around the jetstream;
- influx from the free troposphere;
- production in the boundary layer itself, especially during periods of photochemical air pollution.

Although it is a complex matter to apportion the relative contributions to ground level ozone concentrations from the stratosphere/troposphere on the one hand and man-induced processes on the other hand, one can state that ozone concentrations in polluted areas ranging from 100 up to over 400 ppb during pollution episodes require a dominant contribution from ozone forming processes within the boundary layer itself.

The major removal processes for ozone and its precursors in the boundary layer are photochemical decomposition, reaction with trace gases, and deposition. Ozone and its precursors have such long life-times that polluted air masses from urban and industrialized areas can affect suburban and rural areas for considerable distances downwind: over 1000 km for ozone, and several thousands of kilometres for some of the precursors.
In fact, elevated ozone concentrations have been measured in many downwind rural areas, where local ozone precursor sources are lacking. In fact, there is little evidence for significant geographical variation in the maximum hourly ozone concentrations recorded at widespread locations over Europe during major continental-scale photochemical episodes. Although elevated ozone concentrations appear to be widely distributed, this does not imply that photochemical ozone is evely distributed across Europe. As an example, ozone concentrations measured and calculated on July 26 th, 1980 (15.00 GMT) are depicted in Fig. 1. Calculations were carried out by members of the project entitled PHOXA (photochemical oxidant and acid deposition model application), which is sponsored by the German Umweltbundesamt and the Dutch Ministry of Housing, Physical Planning and the Environment (Meinel and Builtjes, 1984). The 26th of July was the end of an expisode that lasted from July 22-26. The weather was characterized by an anticyclone moving slowly over northwest Europe towards the nord-central part, and disappearing in northerly direction during the last day of the episode. The maximum recorded hourly ozone concentration was around 160 ppb on July 26th and was observed in the Netherlands. Fig. 1 clearly illustrates the large scale of the episode.
The frequency of photochemical episodes is meteorological controlled. This is the reason why meteorological variability from year to year and over Europe mainly account for the variability of frequency and intensity of episodes from year to year and over different parts of Europe. For the Netherlands this is demonstrated in Table 1. Moreover,

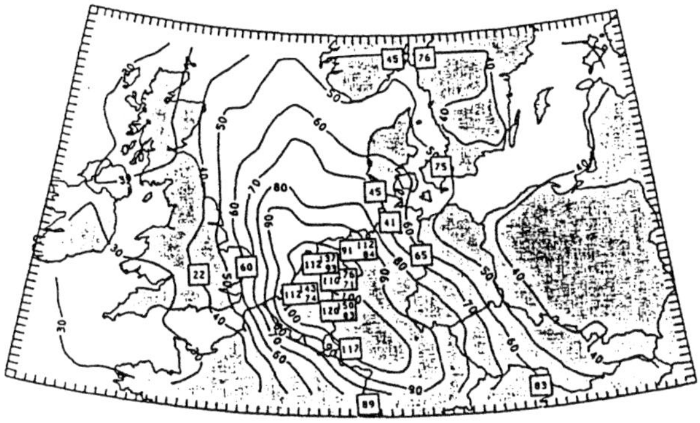

Fig. 1. Calculated and measured O_3 concentrations on July 26 (1980) at 15.00 GMT. The figures in boxes are hourly measured O_3 concentrations. Some boxes also display a range of O_3 values (i.e. minimum and maximum) for a cluster of sites within a block of model grid cells.

Table 1 Ozone data (1976 and 1977) for several locations in the Netherlands.

Year	Station	No. of hours with $O_3 \geq$ 0.06	0.08	0.10 ppm	No. of days with $O_3 \geq$ 0.06	0.08	0.10 ppm	Max 1 hour conc. (ppm)
1976	Delft (urban)	353	145	71	61	29	26	0.20
	Vlaardingen (industrial)	691	298	149	95	48	29	0.27
	Vlissingen (remote/coastal area)	742	327	181	100	50	29	0.22
1977	Delft	102	21	4	26	6	1	0.12
	Vlaardingen	106	31	2	19	5	2	0.10
	Vlissingen	34	13	4	10	4	2	0.11

1976 no. of days with $T_{max} \geq 25°C$(46)
1977 no. of days with $T_{max} \geq 25°C$(7)

percentile values are generally lower in urban areas than in downwind rural areas due to chemical quenching of ozone by other pollutants, of which NO seems to be the most important one. This is demonstrated for the Netherlands in Fig. 2 (van Aalst, 1984). In this figure the daily trend of ozone for the summer months is given for a rural site (Terschelling) and two urban sites (Delft and Schiedam). Although ozone levels at Terschelling are higher, the oxidant levels ($O_x = O_3 + NO_2$) over the country as a whole seem to be more or less the same. However, downwind of areas where very reactive hydrocarbons are emitted, high ozone levels may be observed locally, as for example in the Rijnmond area in the Netherlands, where peak hourly concentrations of up to 270 ppb were measured. Finally, aircraft measurements reveal ozone plumes and troughs where the ozone has been destroyed by large plumes of NO_x emissions (Fig. 3).

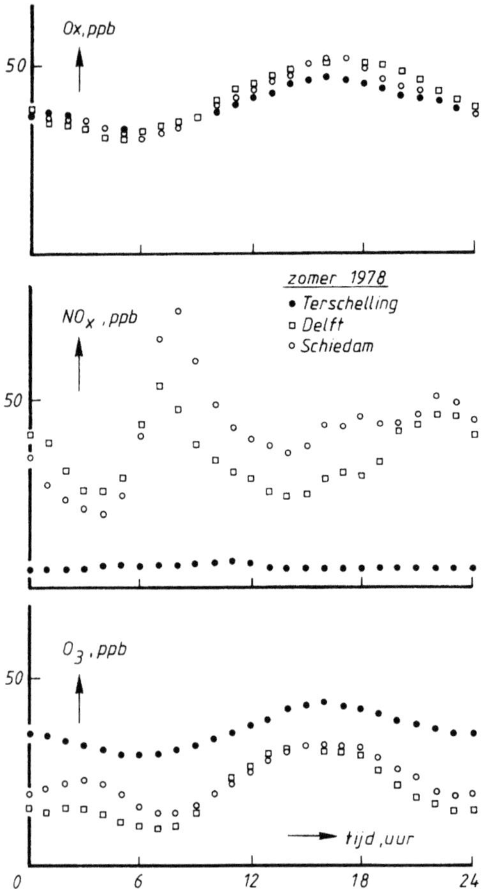

Fig. 2. Daily trend for O_x; NO_x and O_3 (summer 1978).

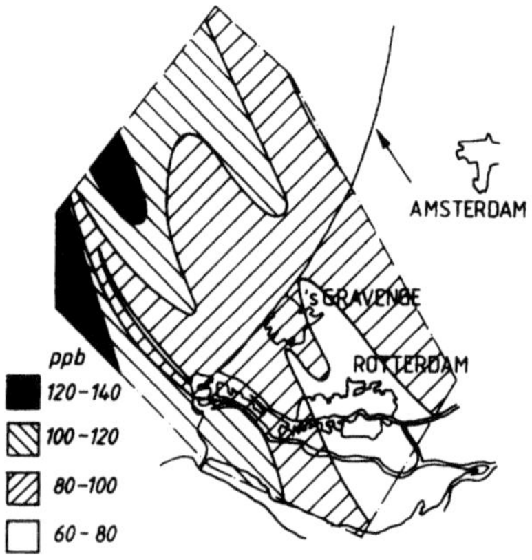

Fig. 3. O_3 concentration at 650 m altitude

2. LONG TERM TRENDS OF URBAN OZONE CONCENTRATIONS AND ITS PRECURSORS

Time series of urban ozone observations in Delft are shown in Fig. 4. Because ozone exhibits a distinct annual pattern, time series have been split up for different quarters of the year. The figure hardly reveals any trend, or at most a slight downward one.
This conflicts with the observations of, e.g., Hartmanngruber, Attmanspacher and Claude (1985); Fuster and Warmbt (1985); Reiter et al. (1983). According to these authors an increase of ozone concentrations has clearly taken place at middle and high latitudes during the last two or three decades. They report changes during the summer months of 20-50%; the current annual rate of increase seems to be 1-2%.

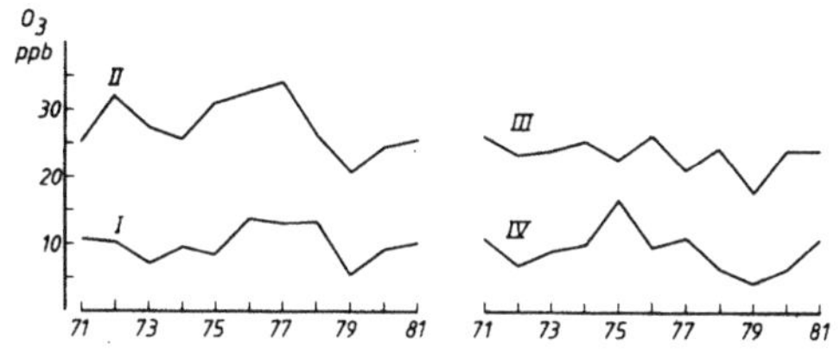

Fig. 4. Quarterly O_3 concentrations measured at Delft.

In order to explain a slight downward trend, we will look at the precursor concentration trends at the ozone observation site, and particularly at the increase of NO concentrations that may have occurred over the same period. Fig. 5 depicts the NO_x trend for Delft. It is more or less representative for the country as a whole in showing an upward trend of NO of about 5% annually. The situation for NO_2 differs in that its concentration rose until the late seventies, levelling off or even slightly declining after 1977. This picture is also reflected by oxidant (O_x = O_3 + NO_3) measurements; an upward trend of O_x till the mid-seventies followed by a more or less constant level from that time on.

Ozone quenching by NO can also be seen in Fig. 6, which illustrates the daily trend of NO_x and ozone for week-days (Mondays through Fridays) and for weekends (Sundays). Ozone concentrations over weekends are on average 10% higher, whereas NO_x and NMHC concentrations are both about 22-24% lower. This picture also holds for the less polluted parts of the country. Maximum ozone concentrations are even up to 15% higher during the weekend.

The behaviour of hydrocarbon precursors is more complex. Many hundreds of different volatile compounds are emitted into the atmosphere both by man-made and by natural sources. Not all these organic compounds are of equal importance in photochemical ozone formation. The importance of each depends on its chemical structure and reactivity. In the Netherlands, major sources of atmospheric hydrocarbons are motor ve-

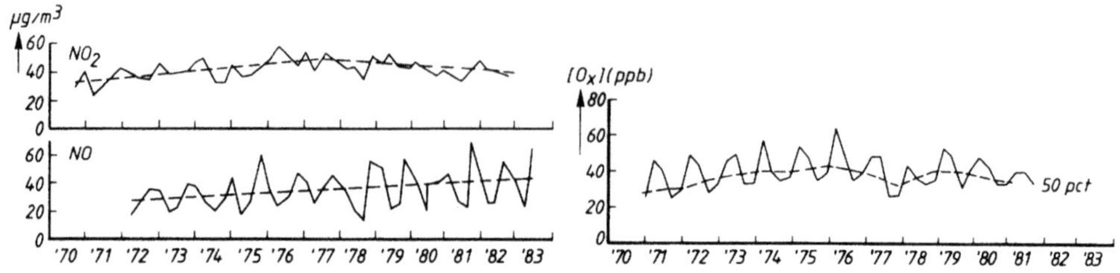

Fig. 5. Yearly trend of the NO_x and O_x concentrations measured in the west of the Netherlands.

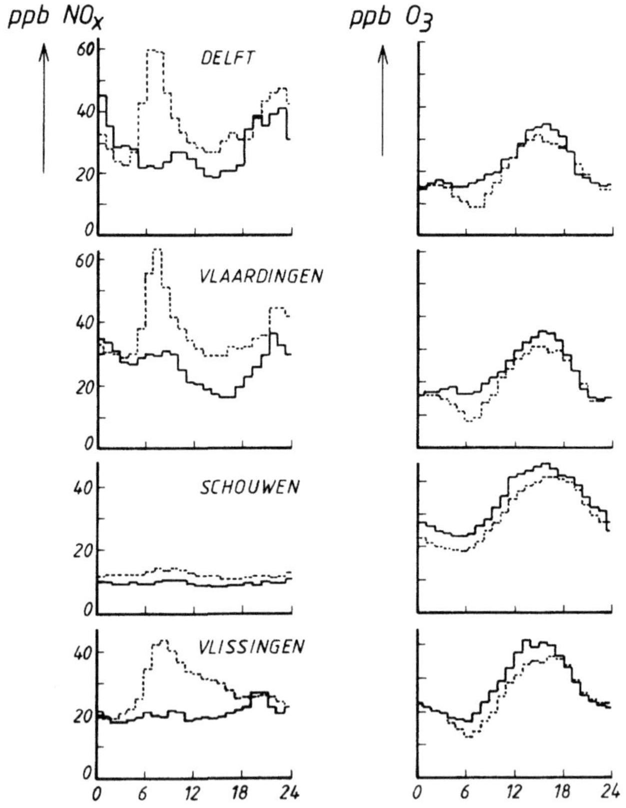

Fig. 6. Daily trends for NO_x and O_3 at various locations in the Netherlands.
Dashed line: Mondays through Fridays;
Drawn line : Sundays.

hicle exhaust gases and evaporation of liquid hydrocarbons. Together these sources are responsible for an annual emission of 220 out of a total of about 480 ktons (base year 1980), or 45%.

Looking at the NO_x and hydrocarbons emission trends (Fig. 7), we find that emissions of volatile organic compounds have somewhat declined since 1970, whereas emissions of NO_x are still increasing, as is the amount of hydrocarbons emitted by traffic. Trends for acetylene are depicted in Fig. 8. Since most acetylene is emitted by mobile sources, an upward trend of 4-5% annually over the last ten years for mobile source emissions of hydrocarbons can be concluded. In view of the relatively long atmospheric residence time of acetylene, this trend does not necessarily reflect an emission trend in the Netherlands alone.

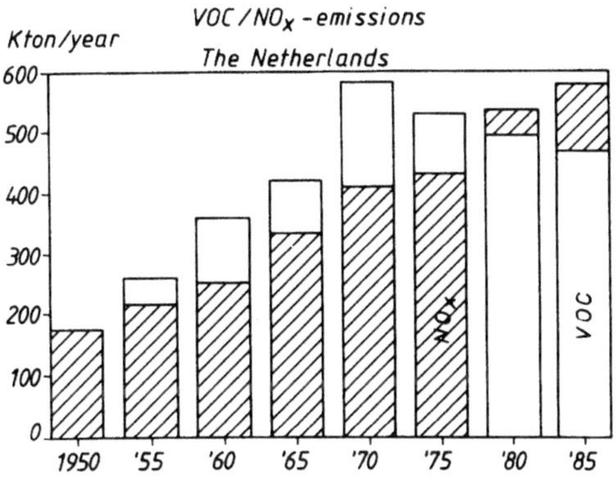

Fig. 7. Historical VOC and NO_x emissions in the Netherlands.
Sources: Dutch Acification Research Programme and the National Emission Registration.

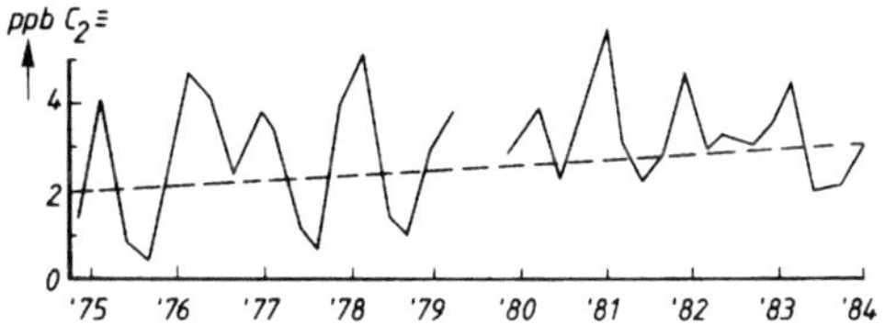

Fig. 8. Yearly trend of the acetylene concentration in the west of the Netherlands.

3. YEARLY TREND OF OZONE IN THE TROSPOSPHERE ABOVE THE BOUNDARY LAYER

The input of ozone into the atmospheric boundary layer by downward transport depends on the concentration of ozone in background (free) tropospheric air. Clearly, changes in the ozone content of the troposphere, e.g. through man-made precursor emissions, will influence ground level ozone concentrations. An indication of the ozone content of the lower free troposphere at the latitude of the Netherlands could be obtained in the following ways.

For high wind velocities accompanied by low pollution levels, when convective mixing is most effective, ozone from the troposphere above the boundary layer is most likely to influence ozone distribution in the boundary layer itself. This is demonstrated in Fig. 9, which is a plot of summer 98% ozone values and ozone values for high wind velocities. High winds tend to distribute ozone evenly, and the distinction between levels in the more and in the less polluted parts of the country is lost.

For 8 years of ozone measurements at Delft we have plotted for each month of the year the ozone concentration as a function of windspeed. The monthly graphs in Fig. 10 show that for windspeed of more than 12 $m.s^{-1}$ the ozone concentration does not vary anymore. If we now plot the monthly maxima, which may be regarded as ozone concentrations representative for the troposphere above the boundary layer, a maximum in May is observed (Fig. 11). This is in contrast to a mid-summer maximum in the boundary layer.

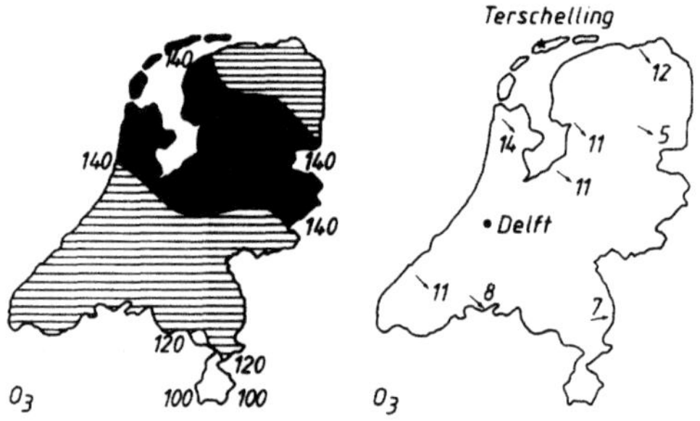

Fig. 9. 98% large-scale O_3 pollution levels for the Netherlands (summer 1982) and O_3 levels around 70 $\mu g.m^{-3}$ on April 28; 1985 (0100 GMT) for nw-wind. O_3 ($\mu g.m^{-3}$) wind velocity ($m.s^{-1}$). Source: Country wide measuring network operated by the National Insitute for Public Health and the Environment (RIVM).

The same trend emerges when we look at ozone concentrations measured in the island of Terschelling, in the north-west of the country, for winds from the sea, where no sources of pollution are to be expected (Fig. 11). The same figure also plots measurements in remote areas of Canade at 53-59°N (Logan, 1985).

From Fig. 11 we have deduced the zonal average ozone concentration at a latitude of 50 to 60°N. Over the year this concentration is given in Fig. 12, which also depicts oxidant levels measured in western Europe. In areas with low levels of NO_x, oxidant levels will be essentially equal to ozone levels. In areas with higher NO_x levels, ozone levels will be lower. The figure shows that yearly average oxidant levels are 26 to 55% higher than so-called background ozone levels. When these data are compared with the annual averages of 10 to 15 ppb recorded at the turn of the century at Montsouris Observatory near Paris, it follows that there has been an increase of some 200%.

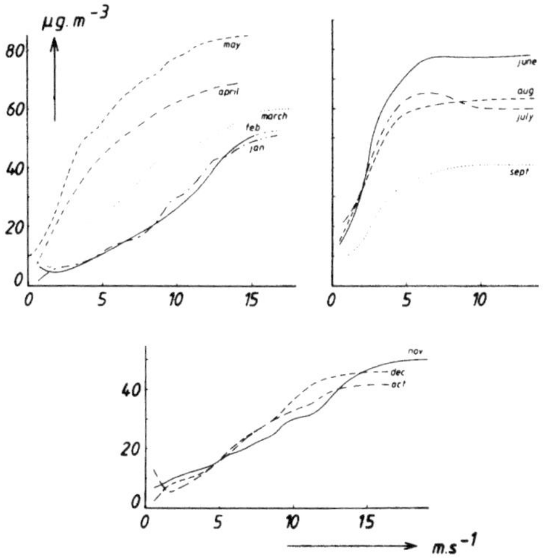

Fig. 10. Monthly average O_3 concentrations for Delft as a function of wind speed.

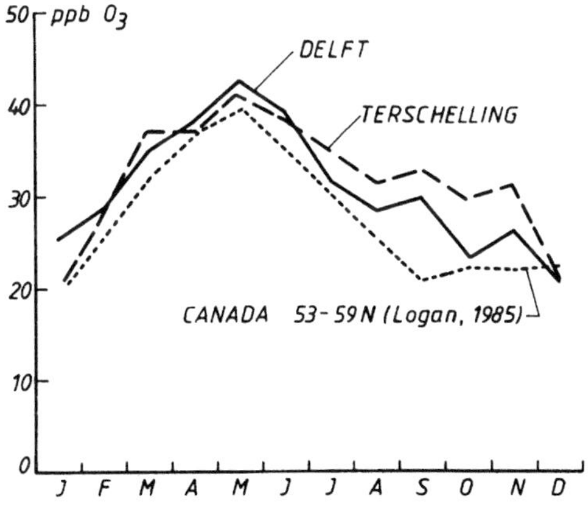

Fig. 11. Background ozone levels.

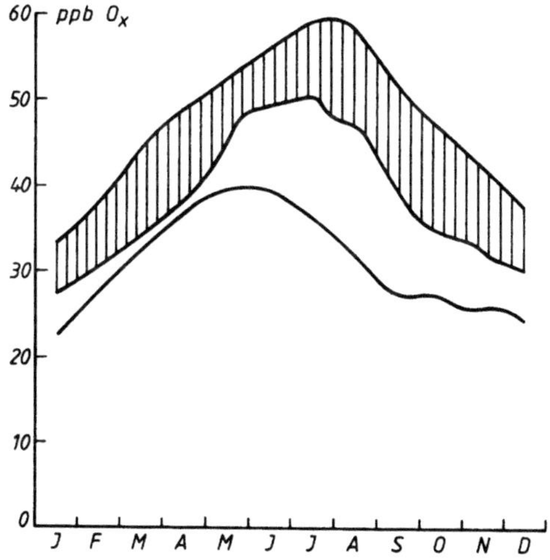

Fig. 12. Background O_3 levels (derived from Fig. 11) and oxidant levels as measured in west Europe (hatched area).

4. THE ROLE OF NO_x CHEMISTRY

In polluted atmospheres, the most readily observed sink for ozone is its reaction with NO. In the atmosphere, nitrogen oxides undergo the following reactions:

$$NO \xrightarrow{O_3} NO_2$$
$$NO_2 \xrightarrow{R.O_3} PAcN$$
$$NO_2 \xrightarrow{O.H} HNO_3$$
$$NO_2 \xrightarrow{O_3} NO_3 \xrightarrow{NO_2} N_2O_5 \xrightarrow{H_2O;AER} HNO_3$$
$$NO_3 \xrightarrow{RCHO} HNO_3$$
$$NO_3 \xrightarrow{OLE} NPA\ (D)N$$

$$NHO_3 \xrightarrow{NH_3} NH_3NO_3$$

PAcN = peroxyacyl nitrates
RCHO = aldehydes
OLE = olefins
NPA(D)N)= nitroperoxyalkyl nitrates and dinitrates (Bandow et al., 1980)

To the nitric acid formed in the daytime by reaction of NO_2 should be added a significant amount of HNO_3 produced at night by N_2O_5 hydrolysis, by reaction of NO_3 with aldehydes, or by heterogeneous chemical processes involving N_2O_5/aerosol reactions.
Conversion rates of NO_x are higher during the daytime, ranging from 1% h^{-1} in winter to over 15% h^{-1} during photochemical episodes in Western Europe (Guicherit et al., 1980).

Figure 13 shows that in the winter months sedondary products of NO_x add up to 10% of all nitrogen compounds in the atmosphere; this figure increases to 35% in summer, and during smog episodes to over 65%.

The average yearly PAN/HNO_3 (v/v) ratio in Delft is 0.3 (0.1-0.2 in the winter vs. 0.4-0.5 in summer months) with maxima of $\geqq 1$ during smog episodes. In polluted atmospheres in the U.S., Spicer (1977) found three times as much PAN as HNO_3. This means that, apart from some NO_2 that is removed from the boundary layer by dry deposition, a considerable amount, up to over 50%, of NO_x is converted into PAN. The remainder is converted into NO_3, either directly by reaction with OH, or indirectly by reactions initiated by O_3. There has been a dramatic increase in PAN concentrations measured at Delft, viz. over 10% annually (Fig. 14). For NHO_3 no data are available.

Dry and wet deposition remove HNO_3 and PAN. Some PAN will decompose into NO_2 and the peroxy radical. Accordingly, the effects of man-made NO_x emissions should only be seen on a regional scale namely several hundreds of kilometres. The decay rate of PAN, however, is temperature-controlled. Especially in cooler air masses, with a low NO/NO_2 ratio, its decay rate is long enough for it to be transported over very large distances. Due to turbulent exchange cooler air with considerable PAN concentrations might be raised in temperature, and NO_x may be formed again by decomposition of PAN. In its turn NO_x may enter the photochemical cycle with CO and hydrocarbons (methane and the more stable hydrocarbons like C_2-C_4; benzene and even toluene) leading to the formation of ozone.

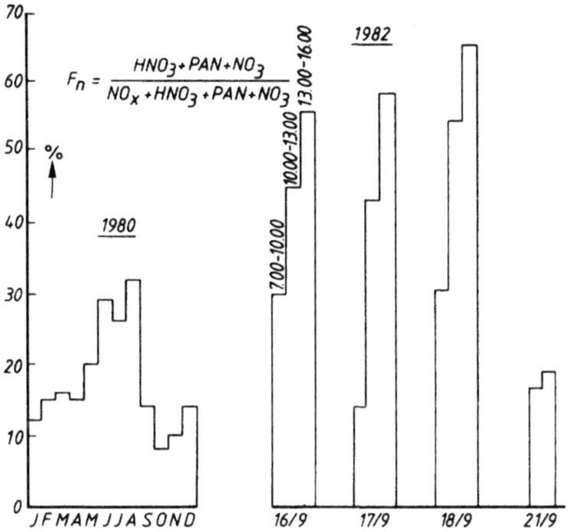

Fig. 13. Ratio of secondary NO_x products and all nitrogen containing species over the year and during smog episodes.

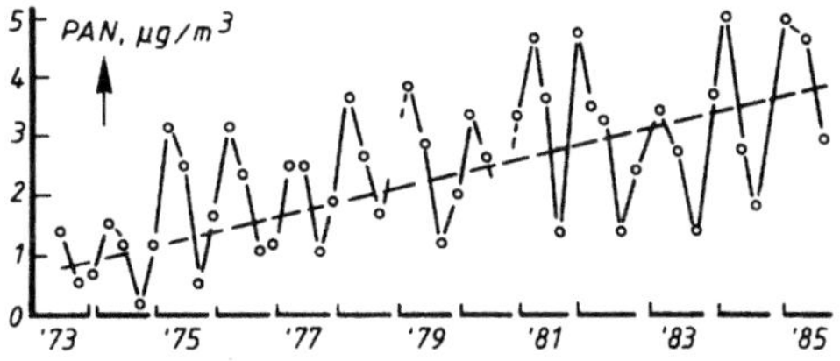

Fig. 14. Yearly trend for PAN in the west of the Netherlands.

Plots of monthly average PAN concentrations for wind directions from the sea show a maximum, as is the case for ozone (Fig. 15). The conventional view regarding the ozone spring maximum was that stratosphere-troposphere exchange is most effective during the late winter and early spring (Danielson 1986, Mahlman and Moxim, 1978). Since the lifetime of ozone in the troposphere is only about one month, one may expect concentrations of tropospheric ozone to be largest in spring. However, tropospheric PAN concentrations also show a peak in spring. Because PAN is only formed by photochemical reactions in the troposphere itself, one might speculate that, at our latitudes, the ozone maximum is also due to photochemical reactions in the free troposphere. There are, in fact, indications that, in winter, PAN and ozone precursors accumulate in air masses at higher latitudes. As soon as there is enough UV light available, which is the case from early spring on, they will begin to react fast, and so lead to the observed ozone and PAN maxima.

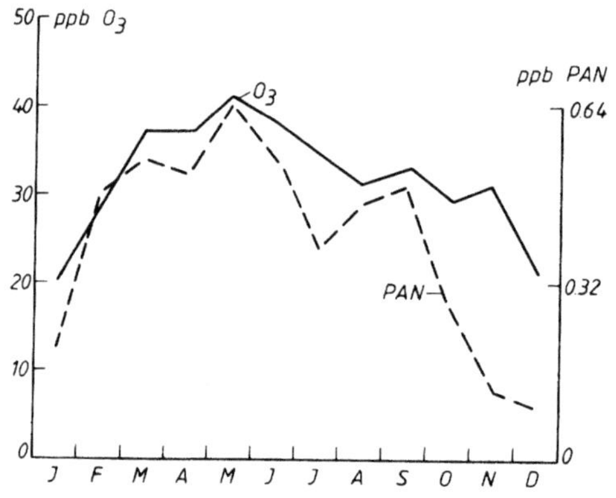

Fig. 15. Background ozone and PAN concentrations in west Europe.

5. REFERENCES

Aalst, R.M. van
Fotochemische luchtverontreiniging in Nederland. Oorzaken, concentraties en effectn op vegetatie.
TNO report R 84/212, 1984.

Attmanspachter, W., R. Hartmannsgruber and P. Lang
'Long period tendences of atmospheric ozone based on ozone measurements started in 1967 at the Hohenpeissenberg Meteorological Observatory.'
Meteorol. Rundsch. 37, 193, 1984.

Bandow, H., M. Okuda and H. Akimoto
'Mechanism of the gas-phase reactions of C_3H_6 and NO_3 radicals.'
J. Phys. Chem. 84, 3604, 1980.

Danielsen, E.F.
'Stratosphere-troposphere exchange based on radioactivity, ozone and potential vorticity.'
J. Atmos. Sci. 25, 502, 1968.

Dütsch, H.U. (1985)
'Total ozone in the light of ozone soundings, the impact of El Chichon.'
In: Zerefos, C.S. and A. Ghazi (Eds.).
'Atmospheric Ozone' Proc. of the Quadrennial Ozone Symposium in Halkidiki, Greece, Sept. 1984.

Guicherit, R., K.D. van den Hout, C. Huygen, H. van Duuren, F.G. Römer and J.W. Viljeer
'Conversion rate of nitrogen oxides in a polluted atmosphere.'
Proc. 11th NATO-CCMS Int. Techn. Meeting on Air Pollution Modelling and its Applications, Amsterdam, Nov. 1980.

Hartmannsgruber, R., W. Attmanspacher and H. Claude (1985)
'Opposite behaviour of the ozone amount in the troposphere and lower stratosphere during the last years, based on the ozone measurements at the Hohenpeissenberg observatory.'
In: Zerefos C.S. and A. Ghazi (Eds.).
'Atmospheric Ozone' Proc. of the Quadrennial Ozone Symposium in Halkidiki, Greece, Sept. 1984.

Logan, J.A.
'Troposphere ozone: seasonal behaviour, trends and anthropogenic influences.'
J. Geophys. Res. 90, 10463, 1985.

Mahlmann, J.D. and W.J. Moxim
'Tracer simulation using a global general circulation model; Results from a mid-latitude instantaneous source experiment.'
J. Atmos. Sci. 35, 1340, 1978.

Meinel, H. and P.J.H. Builtjes
Photochemical oxidant and acid deposition model application (PHOXA) withing the framework of control strategy development.
UBA Berlin and VROM Leidschendam, 1984.

Reiter, R. and H.J. Kanter
'Time behaviour of CO_2 and O_3 in the lower troposphere based on recordings from neighbouring mountain stations between 0.7 and 3.0 km ASL including the effects of meteorological parameters.'
WMO Techn. Conf. on Observation and Measurement of Atmospheric Contaminants (TECOMAC), Vienna, Oct. 1983.

Reiter, R., K. Munzert and H.J. Kanter
'Parameterization of the variation of CO_2 and O_3 in the lower troposphere based on 5-years recordings at 0.7, 1.8 and 3.0 km ASL with consideration of the most important magnitude of meteorology, biomass, and anthropogenic effects.'
WMO Techn. Conf. on Observation and Measurement of Atmospheric Contaminants (TECOMAC), Vienna, Oct. 1983.

Spicer, C.W.
'Photochemical atmospheric pollutants derived from nitrogen oxides.'
Atm. Environ., 11, 1089, 1977.

Warmbt, W.
'Results of long term measurements of near surface ozone in GDR.'
Z. Meteorol. 29, 24, 1979.

OZONE MEASUREMENTS IN HISTORIC PERSPECTIVE

D. Kley, A. Volz, and F. Mülheims
Institut für Chemie 2: Chemie der Belasteten Atmosphäre
Kernforschungsanlage Jülich, GmbH, Postfach 1913, FRG

ABSTRACT. Of the pre-1900 data on ozone only the Montsouris series provides accurate (± 20 %) information on historic levels of tropospheric ozone in middle Europe. The large amount of measurements obtained with the Schönbein technique is only of a qualitative nature. A consideration of ozone mixing ratios from the Montsouris series and other measurements made in 1940 in combination with a back extrapolation of two long-term series of ozone measurements made in Germany points to a time around 1940 as the beginning of the ozone increase that is being observed today in many parts of the northern hemisphere.

1. INTRODUCTION

The concentrations of many tropospheric trace gases are increasing today. Ozone, as measured at Arkona since 1956 [Feister and Warmbt, 1987], shows an increase of 2.6 % per annum. Increases of a similar magnitude have been found for ozone in the free troposphere [Attmannspacher et al., 1984].

Ozone is the primary, chemically active, trace gas in the troposphere. It is the precursor to hydroxyl and perhydroxy radicals which govern the oxidation of reduced trace gases such as hydrocarbons and acid precursors. Ozone is also active as a greenhouse gas. Finally, ozone is toxic at ppm levels and, furthermore, causes chronical effects to plants at concentrations that are commonly observed in the northern hemisphere.

In view of this, it is important to carefully review historic literature on ozone and to verify the techniques that have been used so that the accuracy of the older measurements can be assessed.

Ozone was discovered in 1839 by C.F. Schönbein and soon after its discovery measured at many locations around the world. A great deal of information was gathered and published. In this work we review some of the older ozone measurements that were obtained by using iodometric, i.e. chemical, techniques. Future work needs to be done with regard to the early measurements by optical techniques.

Many of the pre-1900 ozone data have been recently reviewed by Bojkov (1986). They indicate mixing ratios that are considerably lower than those commonly observed in rural areas today.

I. S. A. Isaksen (ed.), Tropospheric Ozone, 63–72.

2. THE MONTSOURIS SERIES

A quantitative method for the determination of ozone in air was used from 1876 to 1910 at the Meteorological Observatory of Montsouris at the south perimeter of Paris. This method utilizes the iodide/iodine catalyzed oxidation of arsenite (AsO_3^{---}) to arsenate (AsO_4^{---}) by ozone in solution which was already known to Soret in 1854.

$$
\begin{array}{lll}
(O_3)_{aq} + I^- & \longrightarrow & IO^- + O_2 \\
AsO_3^{---} + IO^- & \longrightarrow & AsO_4^{---} + I^- \\
\hline
\text{net: } O_3 + AsO_3^{---} & \longrightarrow & O_2 + AsO_4^{---}
\end{array}
$$

An apparatus consisting of a wash bottle through which ozone was sucked for 24 hours was built [Marié-Davy, 1876]. It was filled every day with 20 cm^3 of 5×10^{-4} molar K_3AsO_3 and 2 cm^3 of a 3 % KI solution. After an amount of approximately 2-3 m^3 of ambient air had been sucked

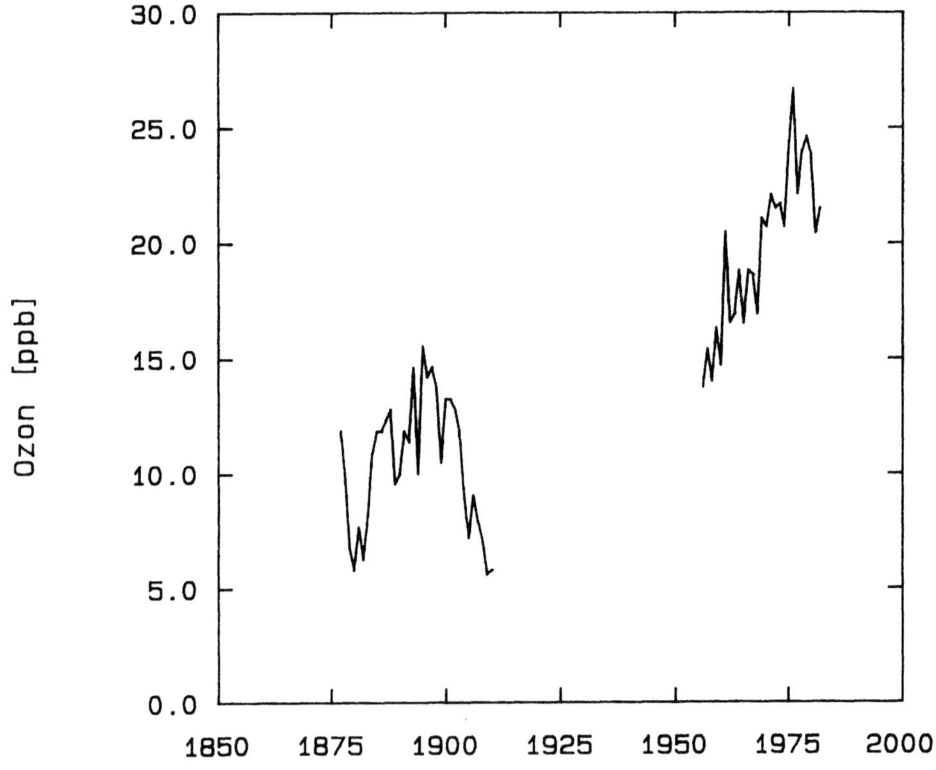

Figure 1. Annual averages of ozone mixing ratio at Montsouris adjusted for SO_2 interference, (Volz and Kley, 1987) and for Arkona (Feister and Warmbt, 1987).

through the solution the remaining arsenite was titrated with iodine. Results of ozone determinations for 34 continuous years beginning in 1876 have been published [Albert-Lévy, 1878, 1906, 1907; Miguel, 1908, 1909, 1910]. Some of the results from this diligent work have been recently reviewed [Volz et al. (1985a, b); Bojkov (1986)].

Volz and Kley (1987) have rebuilt the Montsouris apparatus and compared its performance with a modern UV-photometer. They found that the Montsouris method is accurate to within 3 % in measuring ozone in air. They also analyzed the first 10 years (1876-1886) of the original daily measurements made at Montsouris that, together with meteorological information, were discovered in the literature.

From an analysis with respect to wind direction they concluded that SO_2 interferences due to the vicinity of Paris rendered the O_3 data from Montsouris low by 2-3 ppb on the average. A similar correction of 3 ppb was derived for the years 1905-1910 from interference measurements [Albert Lévy, 1906, 1907; Miguel, 1908, 1909, 1910]. It should be noted that the correction is an additive constant rather than a multiplicative factor as applied by Bojkov (1986). The corrected annual mean ozone concentrations are reproduced in figure 1. As Volz and Kley argued, the Montsouris series should be accurate to within ± 2 ppb.

3. THE SCHÖNBEIN METHOD

One of the most widely used methods of analysis for atmospheric ozone was provided by Schönbein himself. His technique involved filter or blotting paper that was soaked in a solution of potassium iodide and starch in water. After drying, strips of this "Schönbein" paper were exposed to air for 12 to 24 hours, well shielded from sun and rain. After exposure, the strips were moistened and the bluish color that developed, if ozone had been present in the air, was measured on a chromatic scale and used as a relative measure for the ozone concentration.

Schönbein's test for ozone involves the following reactions

$$O_3 + 2\,I^- + 2\,H^+ \longrightarrow I_2 + O_2 + H_2O$$
$$I^- + I_2 \longrightarrow I_3^-$$

$$\text{net: } O_3 + 3\,I^- + 2\,H^+ \longrightarrow I_3^- + O_2 + H_2O$$

O_3 oxidizes iodide to the tri-iodide ion which, in the presence of starch, forms a complex with the amylose helix. The complex has a strong absorption in the yellow part of the spectrum and, hence, is colored blue.

Evidently, the Schönbein test involves ionic processes and should be dependent on humidity. This was already well known more than 100 years ago. Fox (1873) has reviewed the large body of information on atmospheric ozone and its measurement. Most of the techniques that were used up to then employed the Schönbein paper. There is no need to repeat all points of the discussion on the performance of the Schönbein method. The central question is the following: Are the Schönbein data accurate enough to serve as another benchmark for comparison with today's ozone values? This is doubtful from the very beginning because of the questions that had been raised already by 1873. However, going through individual records from that time it seems as if some of the features such as seasonal dependence are quite similar to what is observed today [Linvill et al., 1980; Logan, 1985]. In order to be able to assess the Schönbein data it was necessary to compare the method against today's more accurate techniques. We have, therefore, set up a test facility at which the necessary comparison was performed.

The set up is shown in figure 2. Humidified and thermostated air was mixed with ozone and passed through a flow tube at atmospheric pressure. Several Schönbein paper-strips could be hung in series from small hooks. Ozone mixing ratios were measured upstream and downstream from the sites for the papers. The Schönbein papers were purchased from Riedel de Haen.

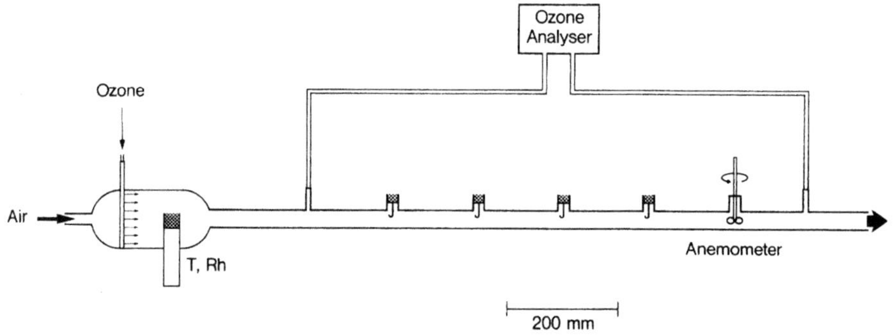

Figure 2. Test set-up for the evaluation of Schönbein papers. Sensors for temperature and relative humidity indicated by T and Rh, respectively.

Experiments were performed to investigate the response of the papers to wind speed, humidity, ozone concentration and duration of exposure. A typical result is displayed in Fig. 3. The ordinate is the "Schönbein-number", i.e. the degree of coloration that was obtained after exposure of the strips for the time indicated in the figure. The coloration was measured on a chromatic scale where zero corresponds to blank and # 10 to the deepest tint which was attained by dipping the strips into dilute H_2O_2 solution. As figure 3 shows, Schönbein numbers different from zero were obtained at exposure times of already a few minutes. The coloration went to a broad maximum at exposure times ranging from a few 100 to about 1000 minutes. Higher O_3 mixing ratios produced maxima at shorter times than lower mixing ratios. At exposure times of 1000 minutes and beyond, depending on the O_3 mixing ratio, the strips became progressively discolored and finally reached zero again on the Schönbein scale.

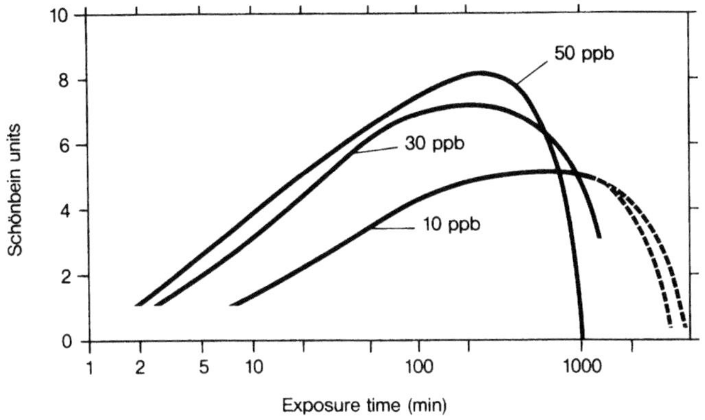

Figure 3. Results of test runs for Schönbein papers at 80 % relative humidity. Ozone mixing ratios are indicated in the figure.

The initial part of the curves in figure 3 could be fitted by a relation $dS/dt=E[O_3]$, where S stands for Schönbein number. The sensitivity factor E was strongly dependent (positively) on humidity but showed little dependence on wind speed.

Two effects conspire to make historic ozone measurements by the Schönbein technique of little use. The first of these is the strong, more than linear, dependence of the sensitivity constant on relative humidity. Much worse seems to be the situation with regard to exposure time. As fig. 3 shows exposure times of 10 minutes up to one or two hours should have been used because for these times, according to the relation $dS/dt=E[O_3]$, the resulting Schönbein numbers would have been approximately proportional to the total dose ($\int[O_3]dt$). A literature search failed to identify such experimental conditions. Rather, exposure times of 8 to 24 hours were commonly used. This is amazing since air chemists at that time knew about the effects of long exposure time: "The want of any definite rate of progression in the deepness of the tints assumed by iodized starch tests, in proportion to the number of hours during which they have been exposed, has tended to shake the confidence of observers in this test considered as a measure of quantity" [Fox, 1873, page 290].

A likely explanation for the blanking effect is easily found (and was also known in 1873): The iodine liberated from the iodide oxidation by ozone has a high vapor pressure ($\approx$ 0.3 mb at 25^oC) and can thus be lost from the papers by evaporation. This effect can a) cause a plateau to exist such as observed, simply by the establishement of a steady state situation between liberation of iodine by iodide oxidation and loss of iodine by evaporation; b) cause the iodine content of the paper to approach zero after the iodide that was initially present had been oxidized by ozone.

We have made the following test to confirm this hypothesis. The ozone mixing ratio in the air flowing over the test strips was set to zero at a time when the maximum blue color should have occurred.

Further exposure for some hours under ozone free conditions caused the iodine to disappear as evidenced by the lack of coloration at time of analysis. However, the time constant for iodine loss from the paper strips was always longer than that when ozone was present. Therefore, it is concluded that an unknown reaction of iodine occurs in the presence of ozone in addition to the evaporative loss. No attempts were made to gain further insight into this situation.

The effect eluded to under b) above makes it difficult if not impossible to compare our laboratory results to actual field measurements from 100 years ago and also renders intercomparison of Schönbein - derived ozone data nearly impossible. The reason is that the total amount of iodide initially present on the paper will determine the length of time for the plateau to exist and the time constant for the decay as well.

Knowing about the deficiencies of the Schönbein method, while using the accurate Montsouris AsO_3^{---}/AsO_4^{---} technique, Albert-Lévy and his coadjutors did parallel sampling with the arsenometric and the Schönbein method over the course of one year. Their calibration curve is replotted in fig. 4. Since the sampling time was 24 hours this calibration applies to the plateau region of fig. 3. But it must be emphasized that the field calibration comprises the range of humidities encountered in Paris during one year. In fig. 4 we have also plotted our laboratory test results for plateau conditions, i.e. at variable exposure time, versus ozone mixing ratio. While the initial slope of the curve from the laboratory test is consistent with the Montsouris calibration it becomes strongly non linear at ozone mixing ratios above 10 to 15 ppb. Two effects may have contributed to the

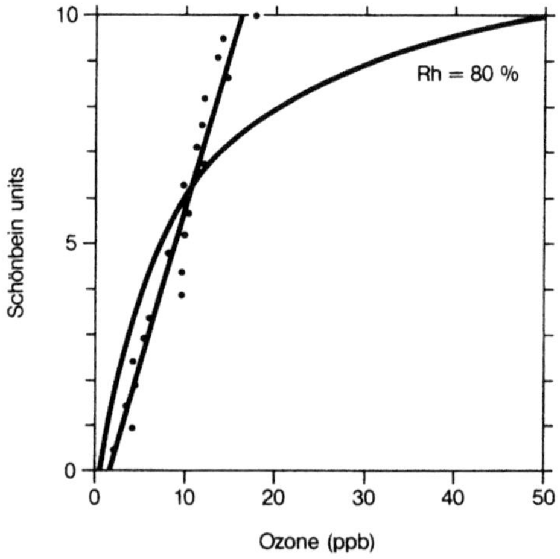

Figure 4. Montsouris calibration of Schönbein method against arsenometric method (straight line and filled dots). Curved line represents laboratory evaluation of Schönbein method against ozone as measured by UV photometry (this work).

discrepancy. First, our laboratory curve was obtained for a relative humidity of 80 %, the largest that could be achieved in the test set-up. Since the dependence of the Schönbein no. on humidity is very strong, higher than 80 % humidity at the upper range of ozone mixing ratios would have straightened our curve. Secondly and more important, the test papers used by Albert-Lévy most likely had a higher iodide content than those used in our laboratory study.

In a simple mechanistic model for the formation of iodine and loss by evaporation a relation of the type

$$\frac{d[I_2]}{dt} = [O_3]\,[I^-]k - [I_2]\,L$$

can be written where L is first order evaporative loss rate.
Under plateau conditions ($d[I_2]/dt = 0$):

$$[I_2] \approx \frac{k}{L}[I^-]_o\,[O_3]\,(1 - [O_3]t)$$

where $[I^-]_o$ is the initial iodide concentration and the second term in the parenthesis accounts for the changing iodide concentration due to oxidation by ozone. Normally $[O_3]t \ll 1$ but under conditions of insufficient capacity of the paper this term can no longer be neglected. The above relation then leads to a non linear calibration curve as observed.

In summing up it is clear that, although at Montsouris a linear regression between ozone collected for 24 hours and resulting Schönbein number existed and absolute ozone concentrations from the Schönbein technique were obtained, none of these results can be applied for deriving absolute values at the other sites where Schönbein's method was used. We did not find information on the paper weight in the literature, i.e. iodide amount, and we are thus unable to decide if a "Montsouris-type" calibration, i.e. linear regression or non-linear relationship as our laboratory calibration indicates, was applicable. Chances are that the latter conditions were prevalent [see Fox (1873)].

4. OTHER MEASUREMENTS

A careful study of ozone by using iodometry was performed on the Pfänder mountain (altitude 1064 m) near lake Konstanz (Bodensee) by Ehmert and Ehmert (1941 a,b). They used a thiosulfate variant of the technique. Sodium thiosulfate was added to the iodide solution and the oxidation by ozone proceeded at pH=5 (boric acid buffer). This reaction sequence can be written as

$$\begin{array}{lll} & O_3 + 2\,I^- + 2\,H^+ \longrightarrow & O_2 + I_2 + H_2O \\ & I_2 + 2\,S_2O_3^{--} \longrightarrow & 2\,I^- + S_4O_6^{--} \\ \hline \text{net:} & O_3 + 2\,S_2O_3^{--} + 2H^+ \longrightarrow & O_2 + S_4O_6^{--} + H_2O \end{array}$$

The thiosulfate is oxidized to tetrathionate by the liberated iodine. The analysis was done by titration with standard iodine solution. In many ways, this is similar to the Montsouris method. For example, it overcomes the I_2 volatility. However, as the net reaction indicates, the thiosulfate technique is dependent on pH while the Montsouris method is not.

Ehmert and Ehmert did parallel measurements. They sampled ozone through the same inlet line into a pH=5 buffered solution (boric acid) and a neutral solution, respectively. Assuming an I_2/O_3 stoichiometry of 1.0 for the neutral solution, they found a stoichiometry of 1.5 for the boric acid buffered solution. The field measurements were made with buffered solutions at pH=5.

The effect of pH on the stoichiometry of the iodometric determination of ozone was also investigated by Dietz et al. (1973) between pH 4 and 14. They found an I_2/O_3 stoichiometry of 1.06 at pH=4 changing to 0.8 at pH=14. For ozone sampling with a solution that was initially neutral, we have calculated a change of less than 0.2 pH units due to ozone under the experimental conditions of Ehmert. This is too small to change the stoichiometry significantly. Therefore, the change from 1 to 1.5 as observed by Ehmert and Ehmert must be related to other reasons.

One possible explanation can be found in the formation of the iodate ion (IO_3^-) apart from I_3^- under neutral or alkaline conditions [Grunwell et al. (1983)], which in acidic solution is rapidly converted to triodide (I_3^-). It is, therefore, conceivable that the experiments of Ehmert rather had a stoichiometry of 1 for the acid buffered solution and a stoichiometry of < 1 for the neutral solution. If this was the case, then the correction applied by Ehmert and Ehmert needs to be removed.

The measurements were performed at a mountain site (altitude 1064 m) near Bregenz in southern Germany during 2 weeks in September of 1940. The mean value was 15 ppb based on a stoichiometry of 1.5. In accordance with the above discussion, this value should be conservatively multiplied by 1.5 to give 22 ppb as an upper limit while the 15 ppb given by Ehmert represent the lower limit.

These 2 weeks of data are valuable because of the location and altitude of the site. It is close to and at the same altitude as the Hohenpeissenberg observatory where continous measurements are performed since 1971. The annual mean values at Hohenpeissenberg, if represented by a linear regression between 1971 and 1984, are

$$\chi = 27.8 + 0.413 \cdot T \text{ (ppb)}$$

where T is the time in number of years and T = 0 for 1970. For the time series as a whole the ratio of mean summer maximum values to annual mean is 1.45.

A back extrapolation of the Hohenpeissenberg data to 1940 gives a value around 22 ppb for the summer ozone mixing ratios. The exact equivalence of the numerical value with the upper limit from the Ehmert measurements is certainly just fortuitous.

5. DISCUSSION

The Schönbein data are too qualitative in nature to serve as historic ozone reference data. However, there is enough information from other, more quantitative, measurements. The Montsouris series is the oldest accurate data set. For the time 1876-1910 the mean ozone level was just slightly above 10 ppb. Annual averages ranged from 5 to 15 ppb.

Recent long-term observations in middle Europe were not started before 1956. The Arkona data set [Feister and Warmbt (1987)] and the Hohenpeissenberg series [Attmannspacher et al. (1984)] all indicate significant increases of surface ozone. If analyzed by linear regression, ozone at Arkona increases by 0.37 ppb • a^{-1} and at Hohenpeissenberg by 0.41 ppb • a^{-1}, respectively. These two figures are very close and suggest a common cause. It is difficult to know when these increases began. The Montsouris series shows that it started after 1910. The values for 1940 measured at the Pfänder mountain site [Ehmert and Ehmert (1941)] are consistent with a linear back extrapolation of the Hohenpeissenberg summer data. Similarly, a back extrapolation of Arkona values to 1940 is consistent with the Montsouris 30 year mean. Taken together, the collected information points to 1940 as the onset of the secular ozone increase

REFERENCES

Albert-Lévy, 1878, *Annuaire de l'Obsavatoire de Montsouris*, p. 495-505

Albert-Lévy, 1900, *Annales de l'Observatoire Municipal* **1**

Albert-Lévy, 1906, *Annales de l'Observatoire Municipal* **7**, p. 368

Albert-Lévy, 1907, *Annales de l'Observatoire Municipal* **8**, pp. 15, 119, 124

Attmannspacher, W., Hartmannsgruber, R., und Lang, P., 1984, 'Langzeittendenzen des Ozons der Atmosphäre aufgrund der 1967 begonnenen Ozonmeßreihen am Meteorologischen Observatorium Hohenpeißenberg', *Meteorol. Rdsch.* **37**, 193-199

Bojkov, R.D., 1986, 'Surface ozone during the second half of the nineteenth century', *J. Climate and Appl. Meteor.* **25**, 343-352

Ehmert, A und Ehmert H., 1941a, 'Über die chemische Bestimmung des Ozongehaltes der Luft', Forsch. u. Erf. Ber. RWD A Nr. 13 (1941) Reprinted as *Ber. Dt. Wetterdienst US-Zone*, **Nr. 11**, 67-71

Ehmert A. und Ehmert H., 1941b, 'Über den Tagesgang des bodennahen Ozons', *Ber. Dt. Wetterdienst US-Zone*, **Nr. 11**, 58-62

Feister, U. and Warmbt, W., 1987, 'Long-term measurements of surface ozone in the German Democratic Republic', *J. Atmos. Chem.* 5, 1-21

Fox, C.B., 'Ozone and Autozone', *Churchill Publ.*, *London*

Linnvill, D.E., Hooker, W.J., and Olson, B., 1980, 'Ozone in Michigan's environment 1876-1880', *Monthly Weath. Rev.* **108**, 1883-1891

Logan, J.A.,1985, 'Tropospheric ozone: seasonal behavior, trends and anthropogenic influence, *J. Geophys. Res.* **90**, 10463-10482

Marié-Davy, M.H., 1876, 'Dosage De L'Ozone De L'Air, *Bulletin Mensuel de l'Observatoire De Montsouris* **52**, 57-61

Miguel, P., 1908, *Annales de l'Observatoire Municipal* **9**, 8, 242

Miguel, P., 1909, *Annales de l'Observatoire Municipal* **10**, 6, 228

Miguel, P., 1910, *Annales de l'Observatoire Municipal* **11**, 6-7

Volz, A., Smit, H.G., and Kley, D., 1985a, 'Klimatologie und Chemie des troposphärischen Ozons: Natürliche Bilanz und anthropogener Einfluß; in: *Wege und Wirkungen von Umweltchemikalien*, Herausgeber: *Arbeitsgemeinschaft der Grossforschungsanlagen* (AGF), Bonn-Bad Godesberg

Volz, A., Smit, H.G.J., and Kley, D., 1985b, 'Ozone Measurements in the 19th Century: The Montsouris Series, *EOS* **66**, 815

Volz, A. and Kley, D., 'Ozone in the 19th Century: An Evaluation of the Montsouris Series', *Nature* (submitted 1987)

LARGE SCALE DISTRIBUTIONS OF O3, CO AND CH4 IN THE TROPOSPHERE FROM SCIENTIFIC AIRCRAFT MEASUREMENTS (STRATOZ III)

A. Marenco
Laboratoire d'Aérologie (CNRS - LA 354)
Université Paul Sabatier
31062 Toulouse Cedex
France

ABSTRACT. During the STRATOZ III experiment (June 1984) aimed to study some minor air constituents, continuous O3, CO and CH4 measurements were made during a series of flights aboard a "Caravelle 116" scientific aircraft, along meridional tracks between Greenland and Patagonia. The tropospheric contents are found significantly higher in the northern hemisphere than in the southern one, and reflects : (1) for O3, an important photochemical formation from anthropogenic pollution of industrialized areas (North America; Europe) at this period of the year; (2) for CO and CH4, the meridional distribution of their sources and the influence of tropospheric photochemistry. Except for West Africa, the tropical area constitutes an important sink for tropospheric ozone (photochemical losses and air ascents). A prominent CO difference between the two hemispheres is observed in the continental air, while the oceanic air is much more homogeneous. In the case of CH4, the decrease from the north to the south is regular, but the results indicate the existence of important natural sources in the southern hemisphere (Amazonia; South Pacific).

1. INTRODUCTION

The development of photochemical theories has shown that the cycle of tropospheric ozone is closely related to those of nitrogen oxydes (NO, NO2) and carbon derivatives (CO, CH4, hydrocarbons). The effects of anthropogenic pollution on atmospheric ozone are presently raising dramatic problems : (a) in the stratosphere where a destruction of the O3 layer may be anticipated; (b in the troposphere where an eventual long term increase of tropospheric ozone may occur as a consequence of the pollution growth (carbon and nitrogen compounds).

While the stratosphere has been rather well studied, there is a lack of large scale investigations in the troposphere. 2D-models of the troposphere are of first generation and presently need to be updated.

Studies of O3, CO and CH4 distributions in the background troposphere are sparse, even rarely based on concurrent measurements and their scope is often limited. The most comprehensive results were obtained by Seiler [10, 2], Ehhalt [9, 1], Routhier [8] and Heidt [3],

I. S. A. Isaksen (ed.), Tropospheric Ozone, 73–81.

from specific flights along meridian tracks.

The evaluation of an eventual long term increase of tropospheric ozone requires an up-to-date knowledge of its distribution, as well as that of its precursors, and a precise location of its sources. There is also a need for a verification of photochemical theories on a larger scale. The existence of a gap following the publication of the latter data prompted us to perform new ozone measurements and to extend the observations to the principal compounds which are closely related to the O3 cycle in the troposphere (CO, CH4, NOx, PAN, Hydrocarbons...). This was the matter of the STRATOZ program in its extended tropospheric version, developed since 1983.

2. EXPERIMENTAL

2.1. STRATOZ Program

The Stratoz program is currently developed by four european laboratories (EERM of Toulouse; Laboratoire d'Aérologie of Toulouse; KFA of Jülich; IMGF of Frankfurt). It consists of vertical and meridional measurements in the troposphere, from an instrumented scientific aircraft "Caravelle 116" of the Centre d'Essais en Vol of Brétigny (France). During the STRATOZ III campaign (June 4-26, 1984), 21 flights ranged between 0 and 12 km altitude have been made on a meridional track (from 70°N to 60°S) over Europe, North America (east coast), Atlantic Ocean, South America (east and west coasts), Africa (east coast) and Europe (cf Figure 1).

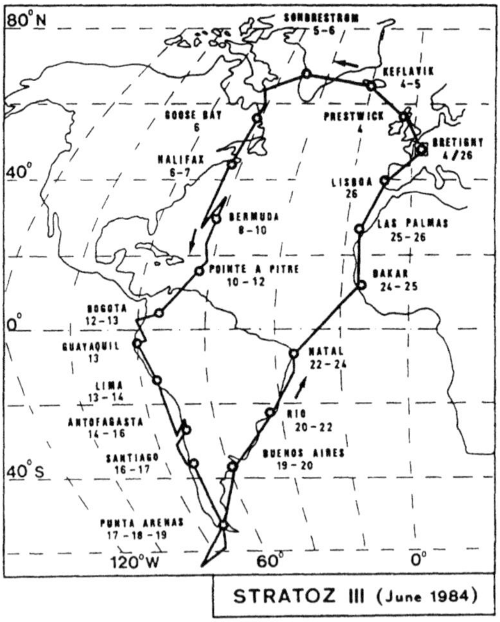

Figure 1. Flight track of the STRATOZ III mission (June 4-26, 1984)

2.2. Experiments

A variety of compounds, amongst the more important ones involved in the tropospheric chemistry have been measured by the different laboratories, including principally :

- O_3, CO and CH_4 continuously measured by the Laboratoire d'Aérologie of Toulouse;
- NO, NO_2, NOy (continuous measurements); PAN, NMHC, CO, CH_4 (discontinuous measurements) by the KFA of Jülich;
- SO_2 (discontinuous measurements) by the IMGF of Frankfurt.

Ozone was measured using an UV absorption analyser (sensitivity 2 ppb, precision 1 ppb, accuracy 3%), specially maintened and calibrated prior the mission. CO and CH_4 were directly measured during flight operation, using an automatic gas chromatograph, manufactured in the laboratory. This apparatus is sensitive (sensitivity 4 ppb, precision 2 ppb), rapidly operational (40 minutes) and well adapted to the simultaneous measurement of CO and CH_4 (2 minutes cycle). The analyzer calibration was checked every 6 analyses by using standards (intercalibration series with KFA indicating deviations of 5% for CO and 0.2 % for CH_4).

3. RESULTS : VERTICAL PROFILES

More than 10000 O_3 measurements, and 2000 concurrent measurements of CO and CH_4, were made in the troposphere and in the lower stratosphere during the 21 flights of the campaign, yielding to 42 vertical profiles (ascents and descents) along with horizontal distributions mainly recorded at cruising altitudes [6, 7]. The profiles may be classified according to their main characteristics.

[Brétigny to Bermuda] - Relatively high mixing ratio are observed in the troposphere (O_3 : 40-70 ppb; CO : 150-250 ppb; CH_4 : 1.80 ppm), with important peaks in altitude. The O_3 peaks (often higher than 100 ppb) are sometimes related to residues of stratospheric air and more often the result of photochemical O_3 formation in polluted air masses (correlation with CO and CH_4).

[Bermuda to Lima] - These profiles correspond to the tropical belt of America. The data are caracterized by a net regulation of the vertical profiles and an important decrease of the tropospheric contents, particularly for O_3 (10-30 ppb). The CH_4 is no longer related with CO. Near Guayaquil (2°S), a characteristic increase of CH_4 is observed in altitude (1.65 to 1.75 ppm), probably related to the occurence of continental air masses coming from the east above the amazonia forest.

[Lima to Rio] - The O_3 profiles show little variability (20-30 ppb) and a decrease of CO and CH_4 is observed southwards. The CO content is lower in the oceanic air masses (60-70 ppb) than in the continental ones (90 ppb), and remains constant. The CH_4 content, more and more depressed (1.60 ppm), exhibits sometimes marked variations in the troposphere for oceanic air coming from the southern Pacific Ocean. Noticeably, the CH_4 peaks are often positively correlated with those of O_3, which may be indicative of important natural exhalations of CH_4 and light hydrocarbons in some oceanic areas, resulting in a photochemical formation of ozone.

[Rio to Natal] - Here we are back to the tropical belt of the northern regions of South America, with the same O3 decrease as previously observed between Bermuda and Lima.

[Dakar to Brétigny] - The crossing of ITCZ is characterized by a net increase of O3 and CO contents. The vertical O3 profiles obtained in the tropical belt along Africa are significantly higher (40-50 ppb) than those obtained at the same level near the american continent. An explanation could be a photochemical O3 generation from the combustion of vegetation (bush and forest fires) still active at this period of the year, via NMHC and NOx. Northway, towards Europe, the tropospheric contents are increasing up to their average level of the mean latitudes of the northern hemisphere, with important peaks in altitude from diverse origins.

4. RESULTS : MERIDIONAL CROSSECTIONS

From the whole results of STRATOZ III, we have established the meridional and vertical distributions (2D representation) which are displayed in six crossections corresponding to southway and northway paths (cf Figure 2). The first ones (southbound flights) are perfectly meridional and have been obtained within a time period of 12 days. The others (northbound flights) are also slightly transversal and juxtapose two sets of data separated by a lag of three weeks (beginning and end of the mission).

4.1. Ozone

[Northbound flights]

Due to the limitation of maximum flight level at 12 km, the two stratospheres (NH, SH) are only observed at mean and higher latitudes where they are at lower altitude. They are characterized by high O3 contents which are significantly more important in the northern hemisphere (310 ppb) than in the southern one (200 ppb). Moreover, they present a very characteristic horizontal stratification of the isolines (constant mixing ratios). Though the season of the southern hemisphere is winter, the altitude of the chemical tropopause goes up for high latitudes (50-60°S).

The two tropospheres (NH and SH) differ significantly in the content and structure of O3 distributions. O3 contents are much higher in the northern hemisphere, reaching values which are twice those of the southern hemisphere for mean latitudes (NH : 50-60 ppb; SH : 25-30 ppb).

In the lower troposphere over tropics, we observe the important O3 sink (Caribbean sea, north of South America) which goes up to the tropopause level.

The isolines are approximatly vertical in the northern hemisphere, while O3 contents are maximum at all altitudes of the mean latitudes. This area corresponds to the highly industrialized regions of North America (east coast) for which maximum photochemical O3 generation is observed. By contrast, in the southern hemisphere, where the isolines

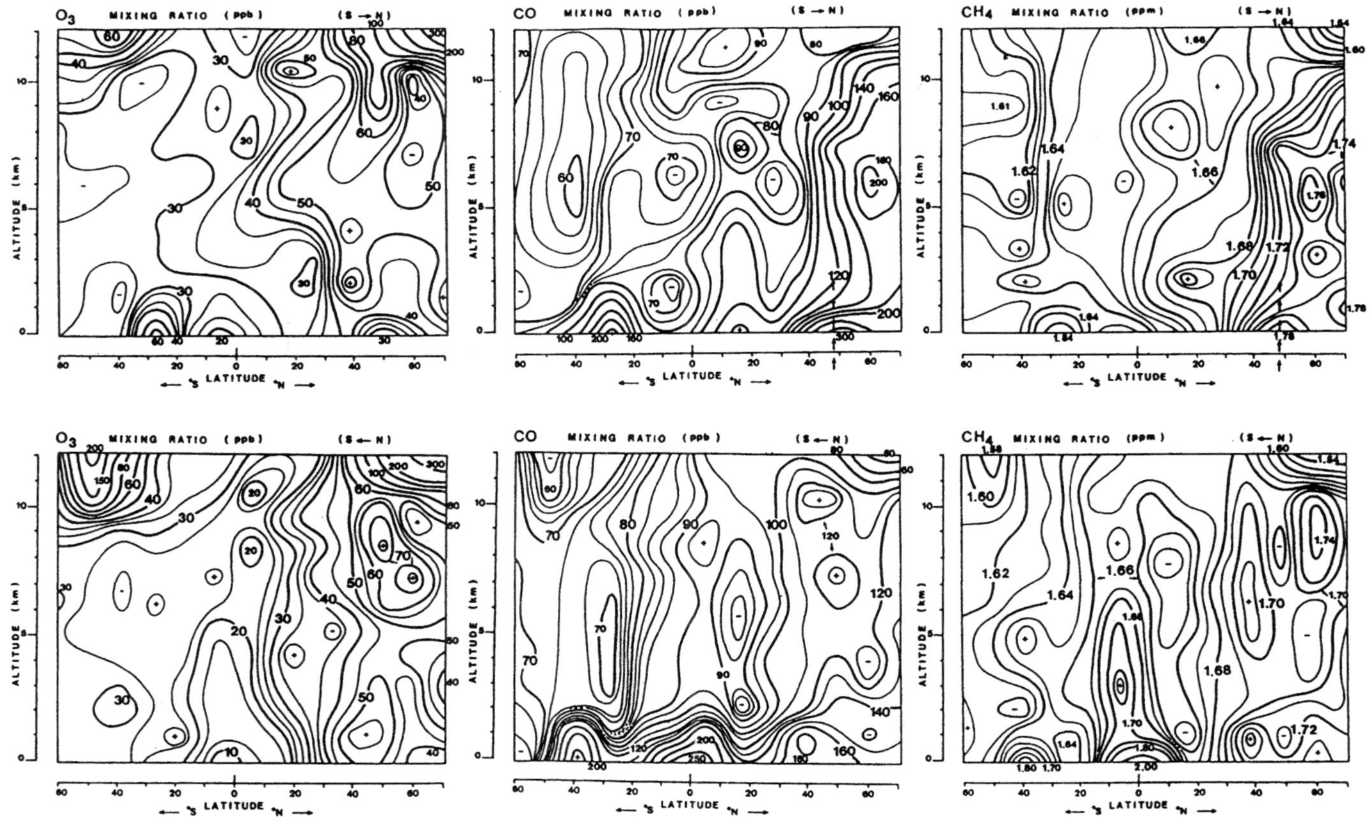

Figure 2. Latitude altitude crossection of O3, CO and CH4 mixing ratios for northbound (upside) and southbound flights (downside), derived from STRATOZ III data (June 4-26, 1984)

are horizontal or at 45°, the O3 distributions are much more homogeneous and constant. The transition from the high stratospheric values (higher latitudes) to the low tropical values (lower troposphere) involves a large tropospheric domain characterized by a low gradient.

This distribution, as a whole, indicates the existence of three large tropospheric domains : (1) Northern hemisphere (high O3 content) with maximum photochemical O3 sources related to polluted areas in the summer season [5] and to eventual stratospheric contributions; (2) Southern hemisphere (low and homogeneous O3 content) with minimum photochemical sources, due to non polluted air in the winter season and to probable stratospheric contributions; (3) Tropical zone showing an important O3 sink (photochemical losses) mainly localized in lower layers, but also perceptible up to the stratosphere, seemingly corresponding to a region of important vertical air ascents.

[Southbound flights]
This second meridional and vertical section displays, again, the large domains described above : stratospheric domain (though the stratosphere of SH is hardly visible here) and tropospheric domains with the NH troposphere showing higher O3 content than the SH one.

By contrast, the tropical sink is very localized and corresponds to the extreme east coast of South America (5°S).

The characteristic high values observable at the mean latitudes of the NH (photochemistry in polluted air of western Europe) are seemingly extended to a part of the SH at the altitude of the middle troposphere. This is likely related to the characteristics of the areas located along the flight path in the two sections (meridional and vertical circulation of air; photochemical O3 formation of natural and anthropogenic origin). The influence of the continental origin of air on tropospheric O3 contents is seen : for South America air in the SH and principally for Africa air in the tropical regions (photochemical formation?). The O3 sink, located in the area of Caribbean sea and northern regions of South America, does no longer exists here and the difference is perceptible as far as the mean latitudes of the northern hemisphere (isolines of O3 tilled or horizontal).

4.2. Carbon Monoxide
The two distributions are characterized by : (1) a horizontal stratification of isolines (constant mixing ratios) in lower layers (0-2 km), with a decreasing gradient from the surface level; (2) a vertical stratification of isolines in the mean and higher troposphere, with a decreasing gradient between the high latitudes of the northern hemisphere and those of the southern one.

This indicates a surface location of natural or anthropogenic CO sources in the two hemispheres. Their predominance in the northern hemisphere is relevant to the presence of large continental areas which are also amongst the most industrialized in the world.

Nevertheless, the asymmetry in the CO mixing ratios (NH : 90-160 ppb; SH : 60-80 ppb) is less prominant that one would expect with regards to the difference of sources. This is related to the occurence of meridional air circulation and to the photochemistry of CO.

Accordingly, considering the seasonal variation of CO which is maximum in winter and minimum in summer [5, 10] the period investigated here corresponds to the minimum contrast between the two hemispheres.

The two distributions differ in the CO contents of the northbound crossection; as compared with the southbound section, they are : (1) lower for the mean and higher troposphere southwards the north tropic; (2) higher for the whole troposphere at the high latitudes of the NH.

The first difference is related to the characteristics of the air along the flight path (continental or oceanic origin). The second difference is related to the presence of more polluted air masses in the northern branch of the northbound flights. This must be interpreted upon consideration of the time separation between the two parts of the section; it also reflects the occurence of a longitudinal rather than latitudinal distribution.

At the stratospheric level, mainly visible in the northern hemisphere due to marked differences with the troposphere, the isolines recover a horizontal stratification with lower CO mixing ratios (45 to 50 ppb).

4.3. Methane

The two CH4 distributions are mainly characterized by a vertical stratification, except for : (1) the NH stratosphere (horizontal stratification for the 2 crossections, and low mixing ratios: 1.52 ppm); (2) some areas of the lower troposphere, in relation with the existence of local sources.

This is typical for the tropical belt (southbound flights) where possible CH4 releases from petroleum fields (Mexico, southern North America) and from the amazonian forest are detected up to the higher troposphere, likely resulting from an "organized" vertical air ascent, as also indicated by the O3 data. The influence of the polluted area around Rio (northbound flights) is also significant, but limited at the lower troposphere.

With reservations, also mentioned for the analysis of CO data, it can be observed that the tropospheric CH4 contents in the northbound crossection are lower in the whole tropical belt and higher for the high latitudes of the NH, compared to the southbound one.

The CH4 mixing ratios are higher in the northern hemisphere, with a rather regular gradient between the higher values found in lower layers for the high latitudes of the NH (1.75-1.80 ppm), and the lower values found in higher layers for the high latitudes of the SH (1.60 ppm).

This reflects the superiority of the CH4 sources in the NH, with locations being often different from those of the CO sources. The relatively low, but highly significant difference between the two hemispheres is certainly due to the long tropospheric residence time of CH4 (8-10 years; [3, 4]) but might be also related to a meridional distribution of the sources less pronounced than for CO.

5. CONCLUSION

The results obtained during the STRATOZ III campaign provide a new and coherent representation of the large scale distributions of O3, CO and CH4. Though the crossections reported here depict a temporary seasonal situation (summer in the NH) along well defined meridional paths, they provide valuable informations on the main feature of O3 and precursors distributions in the troposphere and lower stratosphere.

A more thorough investigation of the origins and distributions of O3, CO and CH4 in the troposphere, will be soon presented in light of the results obtained for other trace gases (submitted for publication), and with the help of more detailed meteorological studies.

Nevertheless, the distributions reported here already constitute a precise and coherent set of valuable data which can be efficiently used in the two dimensional models of the troposphere.

ACKNOWLEDGEMENTS

The authors wish to thank F. Karcher from EERM , the crew of the airplane of CEV, and all participants for their kind assistance. This work was sponsored by the french Ministry of Environment.

REFERENCES

(1) Ehhalt D. H., Rudolph J., Meixner F. and U. Schmidt (1985) 'Measurements of selected C2-C5 hydrocarbons in the background troposphere : vertical and latitudinal variations' J. Atmos. Chem., 3, 29-52.

(2) Fishman J. and Seiler W. (1983) 'Correlative nature of ozone and carbon monoxide in the troposphere; implications for the tropospheric ozone budget' J. Geophys. Res., 88, 3662-3670.

(3) Heidt L. E., Krasnec J. P., Lueb R.A., Pollock W. H., Henry B. E., and Crutzen P. J. (1980) 'Latitudinal distributions of CO and CH4 over the Pacific' J. Geophys. Res. 85, 7329-7336.

(4) Khalil M.A. and Rasmussen R. A. (1983) 'Sources, sinks, and seasonal cycles of atmospheric methane' J. Geophys. Res., 88, 5131-5144.

(5) Marenco A. (1986) 'Variations of CO and O3 in the troposphere : evidence of O3 photochemistry' Atmos. Envir., 20, 911-918.

(6) Marenco A. and Said F. (1987) 'Meridional and vertical ozone distribution in the background troposphere (70°N-60°S; 0 - 12 km altitude) from scientific aircraft measurements during stratoz III experiment (June 1984)' submitted to Atmosph. Envir..

(7) Marenco A., Macaigne M. and Prieur S. (1987) 'Meridional and vertical CO and CH4 distributions in the background troposphere (70°N-60°S; 0-12 km altitude) from scientific aircraft measurements during STRATOZ III experiment (June 1984)' Submitted to Atmosp. Envir..

(8) Routhier F. and Davis D. D. (1980) 'Free tropospheric/boundary layer airborne measurements of H2O over the latitude range of 58°S to 70°N : comparison with simultaneous ozone and carbon monoxide measurements' J. Geophys. Res., 85, 7293-7306.

(9) Schmidt U., Khedim A., Johnen F. J., Rudolph J. and Ehhalt D. H.

(1982) 'Two dimensional meridional distributions of CO, CH4, N2O, CFCl3 and CF2Cl2 in the remote troposphere over Atlantic ocean' Proc. 2nd Symp. on composition of the nonurban troposphere, May 25-28, Williamsburg, VA., USA, 52-55.

(10) Seiler W. and Fishman J. (1981) 'The distribution of carbon monoxide and ozone in the free troposphere' J. Geophys. Res., 86, 7255-7265.

OZONE CHANGES AT THE SURFACE AND IN THE FREE TROPOSPHERE

Rumen D. Bojkov
Atmospheric Environment Service
Downsview, Ontario
Canada

ABSTRACT. Analysis of the available long-term ozone records of surface and ozonesonde observations from rural stations confirm the continuous increase of ozone concentration, by about 1% per year, during the last two decades. Although the data are very sparse and do not represent a global average, they reveal some interesting seasonal variations. It appears that the surface ozone during November-January is increasing at a greater rate than during May-October. This could be a confirmation to recent modelling studies highlighting the important combined effect of anthropogenic emissions and natural level of concentrations of NO_x, non-methane hydrocarbons (NMHC), as well as CO and CH_4, on the ozone production in the troposphere. The ozone in the upper part (500-300 hPa) of the troposphere usually shows a lower rate of increase than the ozone in the lower part (850-500 hPa). The monthly deviations from the mean of the lower stratospheric ozone compared with those of the surface ozone show some general concurrence of the shape of the long-term fluctuations, but also show some differences, which together with the opposite (by sign) trends of both series may indicate domination of processes affecting the surface ozone regime, other than solely the cross-tropopause transfer.

1. INTRODUCTION

Triggered by concern of possible ozone destruction caused by human activity, considerable attention has been given to stratospheric ozone studies since the early 1970's. With improvement in the understanding of ozone photochemistry and the role of OH, NO_x, CO, CH_4 and non-methane hydrocarbons (NMHC), it has been suggested that gas-phase generation and destruction of ozone, even in clean tropospheric air, may be more important in contributing to the level of the tropospheric ozone concentration than the transport from the stratosphere (Crutzen, 1973; Chameides and Steadman, 1977; Fishman et al., 1979; Liu et al., 1980, 1987). It was realized that improved knowledge of the regime of tropospheric ozone would be beneficial for: projecting its effect on the increase of the global surface temperature (WMO, 1982; Bojkov 1983; Wang, 1984); clarifying its role in the alarming decline of central Europe's coniferous forests (Grennfelt and Schjoldager, 1984) and, last but not least, for better understanding of the complex tropospheric

I. S. A. Isaksen (ed.), Tropospheric Ozone, 83–96.

chemistry interactions (NRC, 1984).

A global increase of tropospheric ozone, based on analysis performed in WMO by Angell (1979) was first documented in the WMO report to the UNEP Coordinating Committee on the Ozone Layer (Bojkov, 1979), and again reported at the Ozone Symposium in Boulder (Angell and Korshover, 1980). The first statistical analysis establishing that the ozone increase in the free troposphere (850-400 hPa) over the northern hemisphere since late 1960's is (1.12 $\pm$ 0.54)% per year was presented by Bojkov and Reinsel (1984). The surface ozone increase in GDR (Warmbt, 1979; Feister and Warmbt, 1987), in the troposphere above Hohenpeissenberg (Attmannspacher, 1982), and at four NOAA stations (Oltmans and Komhyr, 1986) are some of the more specific observational studies on this issue. Tiao et al. (1986) has so far offered the most detailed statistical trend analysis of the available ozonesounding data.

Furthermore, it was recently established that during the last 30 years of the 19th century in North America, the average daily maximum of the surface ozone partial pressure was approximately 1.9 mPa, and that the European measurements between the 1850's and 1900 were mostly in the range 1.7 to 2.3 mPa (Bojkov, 1986). All of these values are only about half of the mean of the daily maximum of the observations taken during the last 10-15 years, which may indicate that surface ozone has been on the increase for longer than only the last two decades.

Although increases in the surface and tropospheric ozone seems to be well established, its magnitude and possible causes are still a controversial issue due mainly to a lack of enough reliable data. The purpose of this paper is to present a brief update through 1986 of observational results from a few surface and ozonesounding stations located in non-urban surroundings (and therefore not under the direct influence of industrial pollution). The results reveal interesting seasonal characteristics and confirm the existence of a substantial positive and statistically significant trend.

2. DATA AND METHODOLOGY

The surface ozone data from continuous recordings at Hohenpeissenberg - Bavaria (47°48'N, 11°E; 975 m), Point Barrow - Alaska (71°N, 157°W; 11 m), Mauna Loa - Hawaii (20°N, 155°W; 3400 m) and South Pole (90°S; 2800 m) were extensively used. Discussion of their instrumentation and quality can be found in Attmannspacher and Hartmannsgruber (1979) and Oltmans and Komhyr (1986). Summary characteristics were available for Zugspitze - Bavaria (47°38'N, 11°E; 2964 m) courtesy of Dr. R. Reiter, and for Arkona (55°N, 13°E; 42 m) and Fichtelberg (50°N, 13°E; 1213 m) from the study of Feister and Warmbt (1987) together with private communications with Dr. W. Warmbt. It will be noted that all stations are located on mountain top areas except Barrow and Arkona which are coastal. From Mauna Loa the data set collected during downslope wind between 9 p.m. and 4 a.m. are used since according to Oltmans and Komhyr (1986) they better represent the ozone in the lower troposphere.

The ozonesonde data were considered only if the ascent reached at least 30 hPa, and the total ozone normalization factor was between 0.9

and 1.3 as an assurance for a good quality profile. Particular attention was given to the frequency of the soundings since no meaningful monthly mean could be established with less than three soundings per month. In actual fact some of the autumn months at Edmonton and Goose Bay did not meet the latter criteria and the trend estimates for this season at these locations is under question.

Ozone soundings at Hohenpeissenberg, Thalwil-Payerne (47°N, 7°E; 491 m), Resolute (75°N, 95°W; 64 m), Goose Bay (53°N, 60°W; 44 m) and Edmonton (53°N, 114°W; 668 m) were considered. Hohenpeissenberg, with more than 1500 MAST (Brewer) ozonesoundings in the 1967-86 period and all released at 08 hour local time after strict preflight preparations, are the best source of information. Similar number of releases of the same type of sonde at Thalwil (near Zurich) and since 1969 at rural Payerne are considered ideal for stratospheric studies, but changes in the local time of release of the sondes required a correction for the concentrations in the tropospheric boundary layer. This was derived from general knowledge of the diurnal course of the ozone there. It seems that further work on the application of monthly corrections is desirable. With the Canadian stations one is facing a consistency problem due to a switch from using MAST (Brewer) to ECC (Komhyr) sondes during 1979 (1980 at Goose Bay). A height dependent correction reflecting the surplus of ozone usually reported by ECC (Komhyr) sonde of about 4 to 12% in the troposphere, 2-3% in the lower stratosphere and a deficiency of greater than 10% above 26 km was derived from about two hundred simultaneous soundings by both types and applied accordingly. Tiao et al. (1986) have introduced a statistically derived "intervention", however, for the same conversion they applied a different correction for each of the four Canadian stations.

It should be mentioned that the total ozone records of Hohenpeissenberg, Resolute, Edmonton and Goose Bay were revised in accordance with all past Dobson instrument calibrations, and their vertical ozone profiles recalculated accordingly. Although, the corrections are usually less than one percent, they may have an important bearing on trend determination and some differences were obvious between our estimates and the trends published by the otherwise precise study of Tiao et al., (1986). A description of the corrections is currently being prepared for publication.

Finally, for the convenience of the reader let us recall that at normal atmospheric sea-level pressure, the ozone partial pressure in mPa (unit used in this paper) is equal to the ozone concentration expressed in ppbv divided by 10.

3. ANNUAL AND DIURNAL OZONE CHANGES

Figure 1 offers an instant overview of the mean annual surface ozone changes at the stations under consideration here. A summary curve of the weighted means of 5 GDR stations which show an overall 2.3% per year increase are also shown. All stations except the South Pole (which has the shortest record) show a steady increase. Their estimated trends are listed in Table I and are all statistically significant at the .99

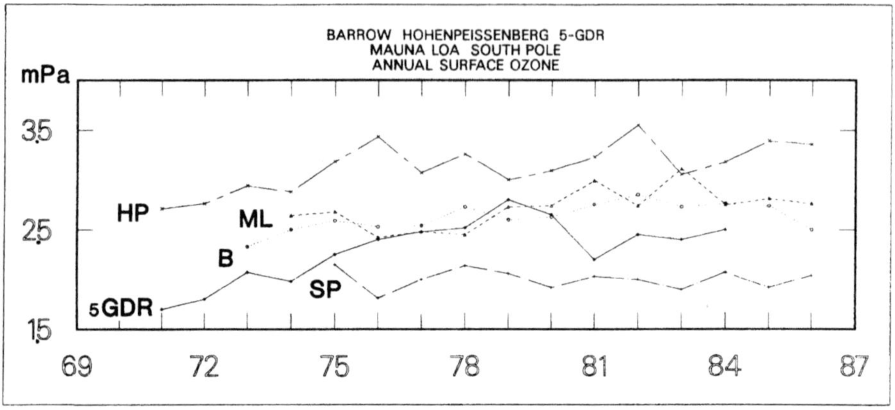

Figure 1. The course of the mean annual ozone partial pressure measured at the surface at South Pole (—.—), 5GDR stations (———), Barrow (....), Mauna Loa (----) and Hohenpeissenberg (— —) since early 1970's. All except South Pole show steady increase; their exact trend is discussed in the text. The standard deviation of each annual mean is less than 0.4 mPa.

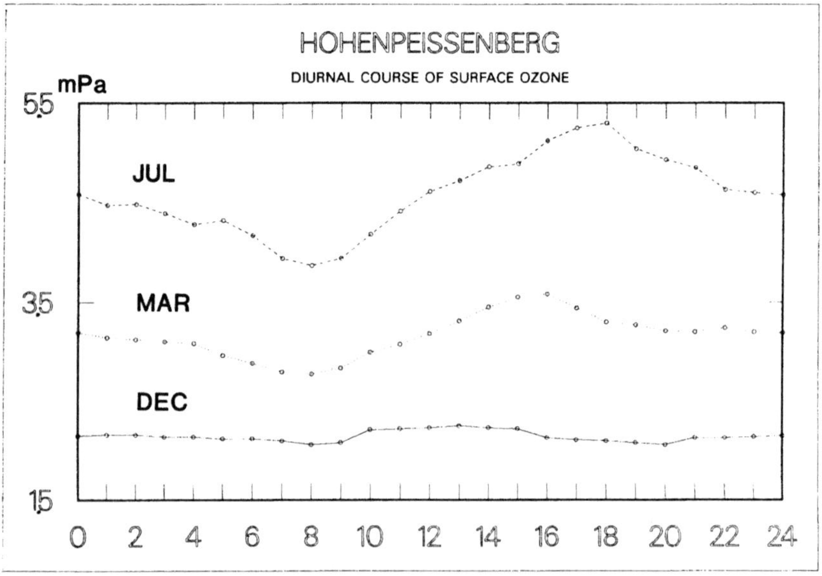

Figure 2. Diurnal course of surface ozone partial pressure for July, March and December at Hohenpeissenberg. Note that the occurrence of the daily maximum shifts from ~ 13:00 in December, to ~ 16:00 in March (and September) and to ~ 18:00 in July.

level. The different behaviour of the South Pole surface ozone may have to do with its location near the top of the Antarctic ice cap. In addition, a reduced strato-troposphere ozone influx may have resulted as a consequence of the spectacular spring total ozone decline observed over Antarctica during the past seven years. In support of the latter speculation comparisons of the monthly deviations of surface and total ozone indeed shows close correspondence. Also listed in Table I are the trends at 850 hPa (700 hPa for Zugspitze) of corresponding ozonesounding stations. They show good agreement, although the trends estimated by the soundings are somewhat higher.

TABLE I
Trends of ozone partial pressure (% per year) and their standard error of estimate at the surface and in a few cases at 850 hPa from nearby ozonesoundings.

Station	mASL	Surface	Period	850 hPa	Period
Barrow	11	0.75 ± 0.20	1973-86	1.60 ± 0.23	1966-86
Mauna Loa	3400	1.04 ± 0.23	1973-86		
Fichtelberg	1200	1.02 ± 0.50	1972-84		
Hohenpeiss.	1000	1.10 ± 0.17	1971-86	1.81 ± 0.26	1967-86
Zugspitze	3000	(2.00)	1974-83	2.18 ± 0.20	1967-86
South Pole	2800	0.15 ± 0.22	1976-86		

Figure 2 illustrates the very different amplitude which the diurnal course of surface ozone exhibits in the northern middle latitudes. The appearance of the minima is at about 08 or 09 a.m., while the daily maxima shifts toward late afternoon, in phase with the extended duration of sunshine hours. The latter could be an indication of the increasing presence of photochemically produced ozone. The differences in the diurnal amplitudes are most obvious; in December it is only 0.2 mPa, in March (and September) it is 0.8 and it is greater than 1.4 mPa in July. Observations from Arosa, Hohenpeissenberg and Arkona reveal an increase, by more than 50% of the diurnal amplitudes during the summer months over the past three decades (Bojkov, 1983; Feister and Warmbt, 1987).

Figure 3 shows the annual course of ozone partial pressure at a few tropospheric levels and one stratospheric level. The annual maximum is reached in June-July everywhere between 1.5 and approximately 8 km; in the lower stratosphere (12 km) it is in April. At approximately 10 km (the annual average position of the tropopause above Hohenpeissenberg) it occurs also in April but with a much reduced magnitude and it shows a secondary maximum in June when 10 km is totally within the troposphere. If one considers the occurrence of the annual maxima at these same levels, on a year-by-year basis a shift from May to June and July could be noted.

The standard deviations (σ) of the montly mean of the lowest four levels (1.5 to 8 km) is 0.7 to 0.4 mPa; in the two top levels σ shows a strong seasonal dependence and is lowest in the fall (approximately 1.4 mPa) and summer (1.8 - 2.2 mPa), but increases up to approximately 3.3 mPa in the winter-spring. A fact obviously related to the more frequently observed cross-tropopause circulation perturbations.

Although the above discussion was drawn from Hohenpeissenberg data we are aware of some differences, for example, in the occurrences of the annual surface ozone extremes; while Hohenpeissenberg minimum is in January and maximum in July, at Barrow they occur in April and November; at Mauna Loa in September and April and at the South Pole in February and July respectively (Oltmans and Komhyr, 1986).

Figure 4 shows monthly mean and monthly maximums of the surface ozone smoothed by b'=(a+2b+c)/4. Remarkable differences in the fluctuations during various years are evident; the 1975 maxima for example has not been surpassed and during 1983, a year with extremely low total ozone, the surface ozone maxima is low but not the lowest on record. If one traces the tops of the annual maxima and the lowest points of both series, one realizes that the course of the maxima shows twice the variability from year-to-year from that of the low points. The long-term positive trend demonstrated by the 12 month running mean is nearly the same; 1.03 vs. 0.88% per year, both statistically significant at the .99 level.

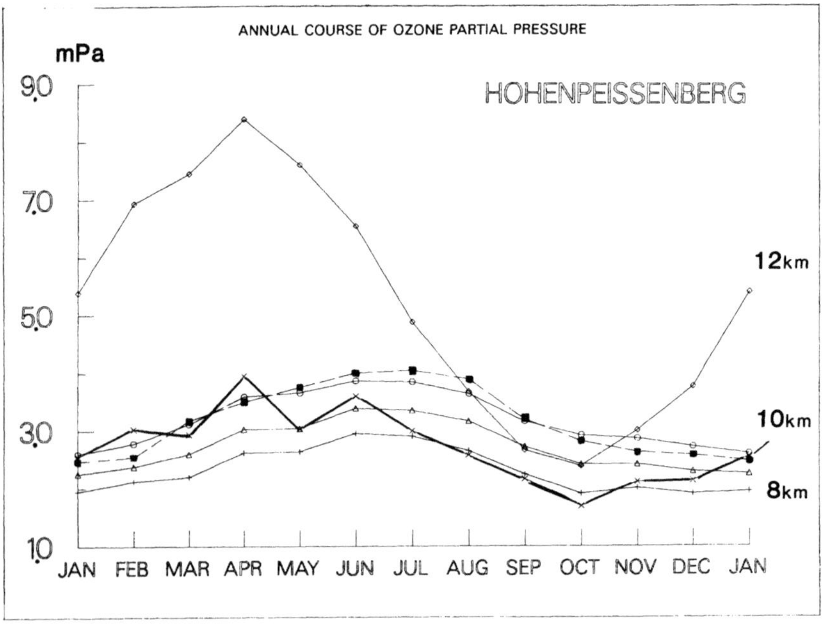

Figure 3. Annual course of the ozone partial pressure at four levels in the free tropospherre (1.5, 3.5, 5.5 and 8 km), one in the lower stratosphere (~ 12 km) and one with transitional position (~ 10 km which in winter is in the stratosphere but in summer is in the troposphere (the heavy line interrupted by an x). The heavy square marks indicate the 1.5 km, the triangles the 5.5 km level.

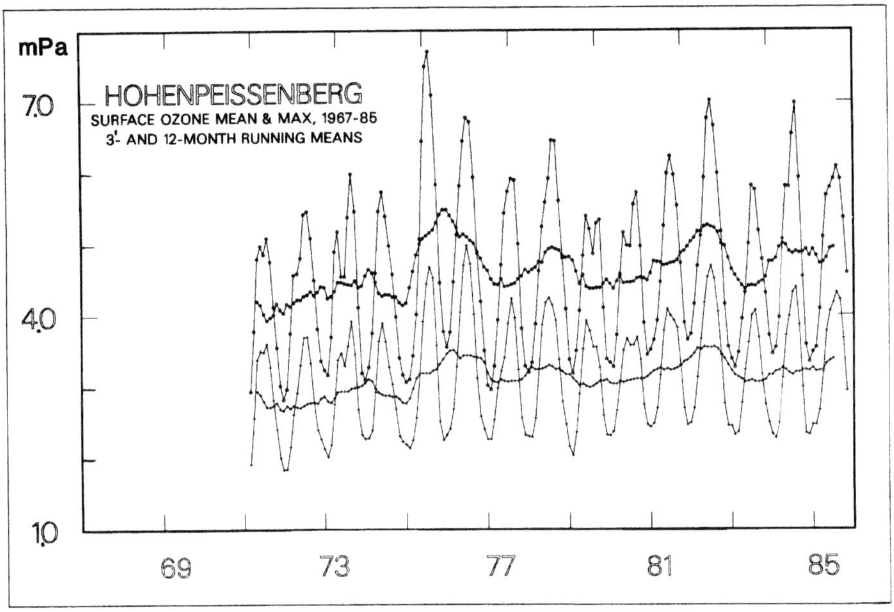

Figure 4. Monthly mean of the surface ozone partial pressure and of its maximums at Hohenpeissenberg, smoothed by the running mean operator b'=(a+2b+c)/4, show details of the year-to-year fluctuations. These are overlayed by 12-month running means indicating a substantial increase since 1971.

Surface ozone monthly means smoothed by b'=(a+2b+c)/4 and overlayed with their 12 month running means are plotted for four stations on Figure 5. The annual amplitude of the surface ozone is strongest (1.9 mPa) at Hohenpeissenberg, smaller but nearly equal (~ 1.4 mPa) at Barrow and Mauna Loa and even smaller still (0.9 mPa) at the South Pole. This is very clearly seen in Figure 5. The annual cycles of all four stations are out of phase with each other due to a shift introduced by different times of occurrence of their annual maxima and minima mentioned before.

4. CONCURRENT SURFACE AND TROPOSPHERIC OZONE CHANGES

The course of annual average ozone partial pressure at the surface (heavy solid line), two lower and one middle tropospheric level at Hohenpeissenberg are shown on the lower panel of Figure 6. The same information for Barrow (surface) with the tropospheric ozone from Resolute appears on the upper panel. The standard deviation of the annual values of the middle tropospheric level is ~0.5 mPa and increases down through the troposphere to about 1.0 mPa at 1.5 km.

The year-to-year changes of ozone at the surface and tropospheric levels are in general agreement with very few exceptions. An increase of both surface and lower tropospheric ozone is obvious, with the latter increasing at a greater rate. Indeed, the trend of the surface ozone is (1.1 $\pm$ 0.2) and (0.8 $\pm$ 0.2)% per year at Hohenpeissenberg and Barrow respectively, while the increase in the entire layer between 1.5 and 5 km is (2.1 $\pm$ 0.2) at Hohenpeissenberg and (1.4 $\pm$ 0.2)% per year at Resolute; the increase in the 1.4 to 1.5 km (850 hPa level) is (1.8 $\pm$ 0.3) and (1.6 $\pm$ 0.2)% per year respectively.

Figure 7 shows the changes in surface and upper tropospheric ozone during December through March - here called winter, and June through August - called summer seasons. From winter to summer the surface ozone change is a striking 65% on average whilst in the upper troposphere it

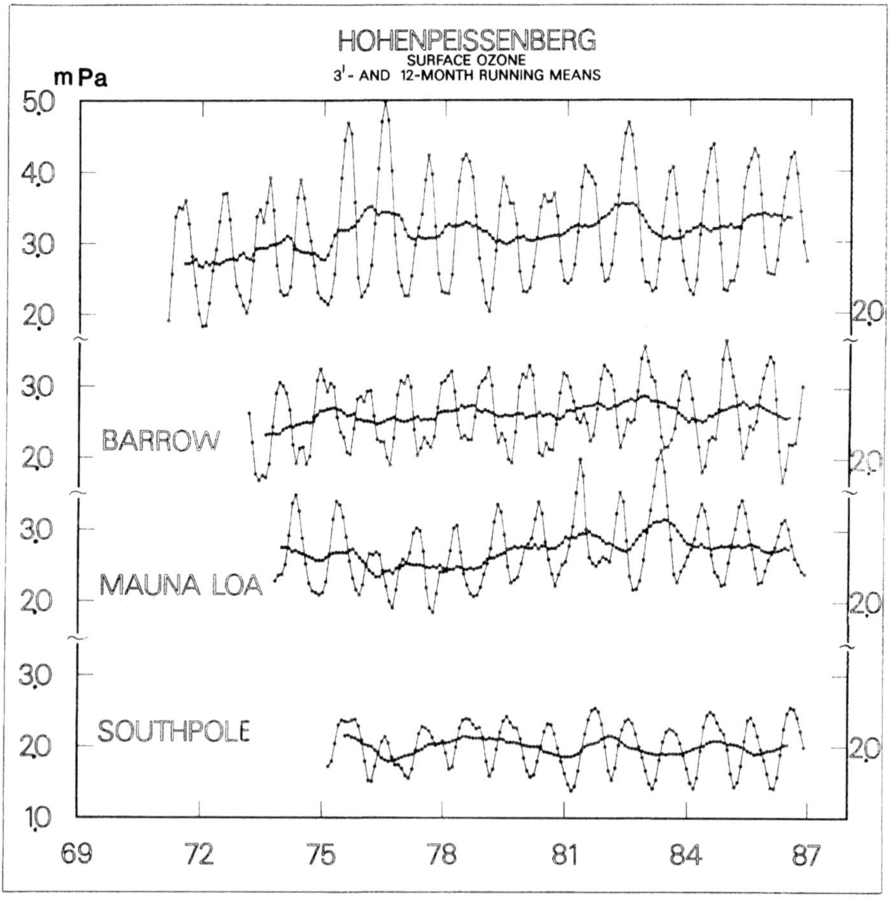

Figure 5. Surface ozone monthly means smoothed by b'=(a+2b+c/4 and 12-month running mean operators for four stations from various latitudes for intercomparison of their amplitudes and long-term changes. Note the 2.0 mPa mark for each station on the right-hand scale.

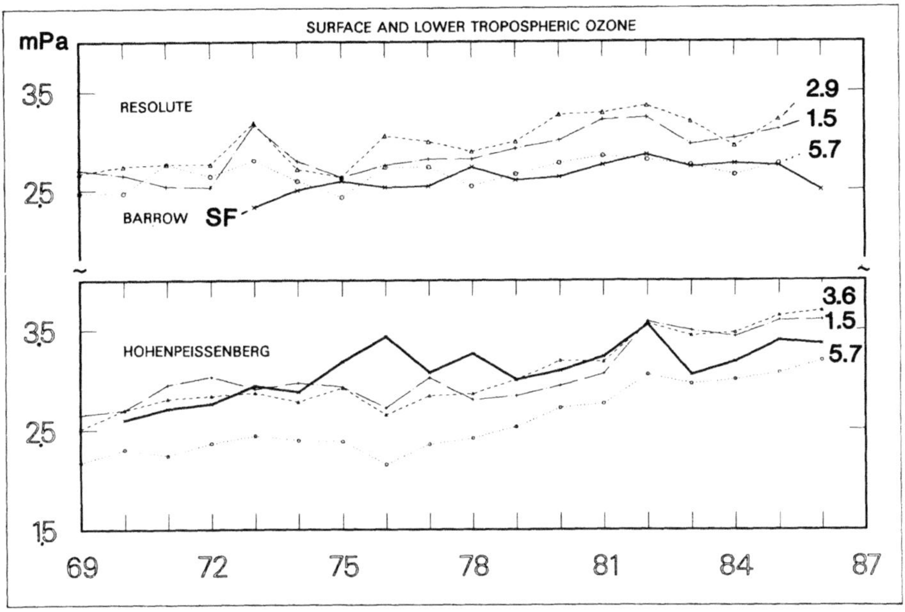

Figure 6. The course of annual average ozone partial pressure at the surface (the heavy continuous line) and three tropospheric levels (their heights in km are indicated by the bold numbers on the right-hand side) at Hohenpeissenberg during the last two decades. The same information is given for Barrow (SF) with the tropospheric ozone values taken from Resolute.

is a more modest 40%. It seems that, in the upper troposphere, in December through March, ozone only started to increase after about 1975 (although rather low values were recorded as early as 1971), while its June-August values have been increasing more regularly from the start of the observations. The trend at the surface is (1.7 ± 0.4) for the four winter months and (1.0 ± 0.6)% per year for the three summer ones. Between 7.5 and 8.0 km the trend is stronger e.g. (2.6 ± 0.4) and (2.3 ± 0.2)% per year respectively. It is pertinent to mention that in the Arkona (on Rugen Island just off the Baltic coast of GDR) the surface ozone trend is also stronger during December-March, i.e. 2.6 vs. 2.1%per year in the summer (significance .99). The record at Mauna Loa exhibits the same characteristics i.e. (1.3 ± 0.7) vs. (1.1 ± 0.8) per year, but the difference is not as great and the level of significance of the trend is below .95.

Figure 8 presents surface ozone trends calculated for each month. Note that the left hand scale is in percent per decade. At Hohenpeissenberg, the trends are positive for all months, although only in the period November through March are they greater than 1.6% per year and significant at the .99 level. The remaining months, i.e. in the

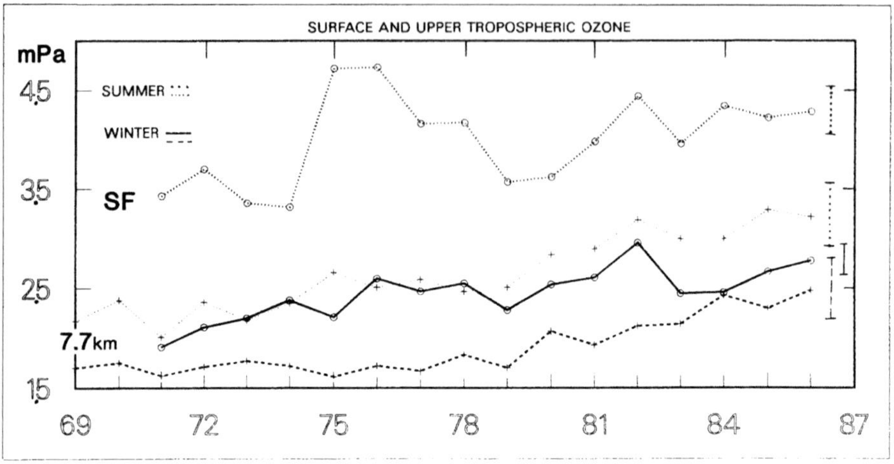

Figure 7. The course of the ozone partial pressure at the surface (o) and in the upper troposphere (+) at ~ 7.7 km above Hohenpeissenberg for December through March, labeled winter (the two lowest heavy lines), and for June through August labeled summer (light dotted lines). The vertical bars on the right-hand side indicate the spread of one standard deviation for each of the four curves.

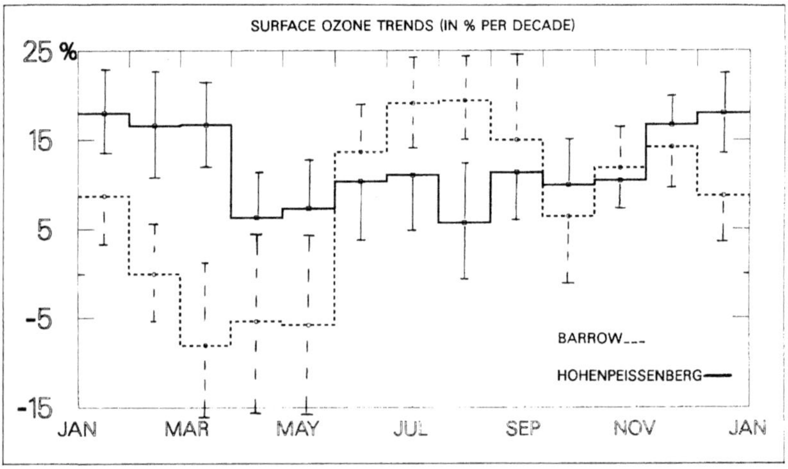

Figure 8. Surface ozone trends (in % per decade) calculated on a month-by month basis at Hohenpeissenberg (——) and Barrow (----). The vertical bars represent 2 standard errors of estimate of each monthly trend; in a few cases, only one side of the error-bar has been plotted to avoid overlapping.

warmer part of the year, the trend is nearly uniform but about one third smaller by magnitude. At Barrow, a much more northern station, the strongest positive trend is exhibited in June through September (greater than 1.6% per year) while March through May are with a negative but statistically insignificant trend. There are a few questionable monthly mean data in March and April at this station and a further careful assessment of the observational records may be needed before firm conclusions can be drawn.

TABLE II
Ozone trends and their standard error of estimate in the upper (500-300 hPa) and lower (850-500 hPa) troposphere at stations with more than 30 sondes per year; from data sets adjusted for total ozone calibrations and for type of ozonesonde.

	all months	Jan-April	May-Sept	Oct-Dec
		Resolute	**(1966-1986)**	
Upper	0.53 $\pm$ 0.58	1.10 $\pm$ 0.58	0.72 $\pm$ 0.42	0.23 $\pm$ 0.67
Lower	1.40 $\pm$ 0.22	1.40 $\pm$ 0.30	1.62 $\pm$ 0.33	1.54 $\pm$ 0.56
		Edmonton	**(1973-1986)**	
Upper	1.08 $\pm$ 0.49	0.99 $\pm$ 0.81	0.51 $\pm$ 0.86	2.47 $\pm$ 0.74 *
Lower	1.30 $\pm$ 0.36	1.46 $\pm$ 0.54	1.36 $\pm$ 0.68	1.33 $\pm$ 0.62
		Goose Bay	**(1970-1986)**	
Upper	-1.53 $\pm$ 0.65	-1.25 $\pm$ 0.86	-0.59 $\pm$ 0.80	-0.38 $\pm$ 1.02 *
Lower	0.55 $\pm$ 0.35	1.05 $\pm$ 0.65	0.32 $\pm$ 0.52	1.19 $\pm$ 0.74
		Hohenpeissenberg	**(1967-1986)**	
Upper	2.10 $\pm$ 0.30	2.35 $\pm$ 0.52	2.14 $\pm$ 0.29	2.50 $\pm$ 0.55
Lower	2.16 $\pm$ 0.22	1.92 $\pm$ 0.36	2.32 $\pm$ 0.19	2.42 $\pm$ 0.37
		Thalwil-Payerne	**(1967-1984)**	
Upper	0.45 $\pm$ 0.30	0.65 $\pm$ 0.35	0.24 $\pm$ 0.35	-0.10 $\pm$ 0.50
Lower	0.40 $\pm$ 0.30	0.55 $\pm$ 0.36	0.44 $\pm$ 0.40	0.30 $\pm$ 0.40

Table II summarizes calculations of ozone trends for the upper (500-300 hPa) and lower (850-500 hPa) troposphere for five ozonesounding stations. There are four columns listing the trends of all months together with three seasonal groupings. In general lower tropospheric trends are more positive than those of the upper troposphere. Payerne is an exception (in all months rate) and the magnitude of its trends is much smaller than in Hohenpeissenberg although from the same synoptic region. We should refer back to the comments made in section 2 where the need to derive a better way to correct the record to compensate for the irregular time of ozonesonde releases over the years was mentioned. The October-December trends for Goose Bay and Edmonton should likewise be considered with great care(*).

Finally, Figure 9 shows the Hohenpeissenberg monthly deviations of the surface ozone and of the ozone partial pressure in the 16-19 km stratospheric layer. The latter is known to best reflect changes in the total ozone. An increasing trend of the surface ozone of (1.13 $\pm$ 0.11)% per year and a decrease of the ozone in the 16-19 km layer by (-0.65 $\pm$ 0.05)% per year is clearly noticeable.

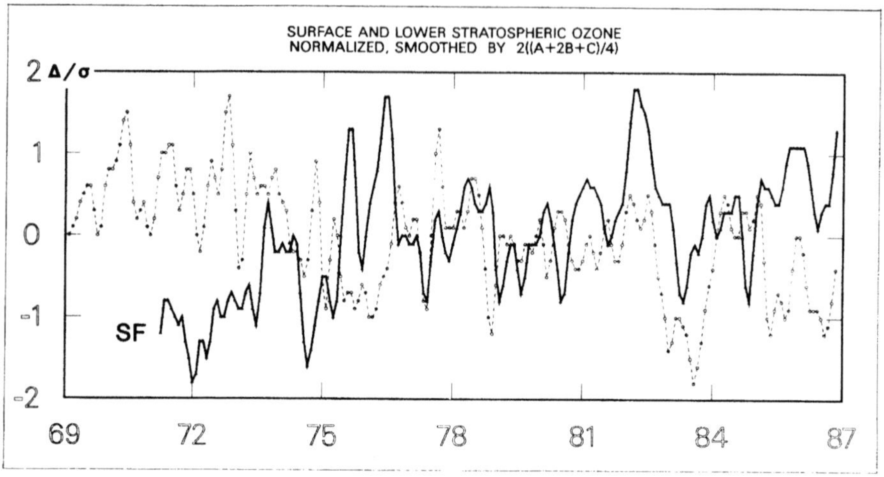

Figure 9. Monthly deviations of the ozone partial pressure at the surface (——) and in the 16-19 km (----) stratospheric layer above Hohenpeissenberg. The deviations are normalized by dividing them by their respective month's standard deviation, and are smoothed twice by the running mean operator b'=(a+2b+c)/4 which reduces the amplitude by about one third. An increasing trend of the surface ozone and a decrease in the 16-19 km is clearly noticeable.

A careful examination of the course of the two curves reveals that the large-scale fluctuations of the surface ozone corresponds to the shape, up to some degree, if not the magnitude, to the fluctuations of the ozone concentration in the lower stratosphere. There are however, a few exceptions (e.g. 1974-76, 1980), which together with the opposite long-term trend of both series, may indicate existence of processes affecting the surface ozone, other than solely the cross-tropopause transfer.

5. CONCLUDING REMARKS

The brief analyses of data from late 1960's through 1986 of some of the most reliable non-urban ozone measuring stations in the Northern Hemisphere confirms a continuous increase of ozone concentrations at the surface by about 1% per year; the increase is even stronger in the lower troposphere and can even be detected in the upper troposphere, but at somewhat lower rates and not always statistically significant. This increase is compensating for part of the decrease in the stratosphere and should be considered in any interpretation of estimates of total ozone trends.

Considering that approximate estimates of the monthly maximum values for the second half of the last century are less than half of the values observed during the past 10-15 years, the presently discussed ozone increase may have started, albeit at a lower rate, earlier than two decades ago.

The diurnal and inter-annual amplitudes of the surface ozone during the summer months in Europe show noticeable increases and a shift of the summer maxima from May-June toward June-July.

Both the surface and lower tropospheric ozone show a greater rate of increase during November-January than during May-October.

Comparisons of the ozone monthly deviations in the lower stratosphere with those at the surface suggest the possibility of processes affecting the surface ozone, other than solely the cross-tropopause transfer. They also clearly indicate that the surface ozone increase was not a result of a few single extremes but a consequence of a steady process.

All the above mentioned observational results concur with some of the main results of the recent modelling study by Liu et al. (1987), namely that the winter ozone may be mostly of anthropogenic origin especially in the lower troposphere in the mid and high latitudes, and that this ozone contributes substantially to the observed spring maxima and to the higher rate of ozone increase. Their conclusion that the photochemical lifetime of tropospheric ozone in winter is of the order of 200 days offers an explanation for the observed increase in non-urban sites.

For a better understanding of the surface and tropospheric ozone budget and changes thereof, the consequences of which will become more important for the human environment in the coming decades, it is most desirable to encourage continuation of existing observational programmes and to put more emphasis on the quality of the collected data.

REFERENCES

Angell, J.K., 1979: Analysis of ozone data performed in WMO (private communication).

____, and J. Korshover, 1980: Update of ozone variations through 1979, Proc. Ozone Sympos., Boulder, 393-396.

Attmannspacher, W., 1982: The behaviour of atmospheric ozone during the last 15 years, based on results of ozone soundings at Hohenpeissenberg, Proc. Workshop on Bilog. Effects of UV-B Radiation, BPT-Berich. 5/82, Munchen, 12-19.

____, and Hartmannsgruber, R., 1982: Intercomparison of instruments measuring ozone near the ground at the Hohenpeissenberg Observatory. 1 October 1978 - 30 April 1979. Ber. Deut. Wetterd., Nr. 161.

Bojkov, Rumen D., 1979: Preliminary results of study of trends in ozone concentration, Appendix to WMO presentation, at UNEP/CCOL-III/3/Add. 3, Paris, Nov. 1979, 7 pp.

____, 1983: Tropospheric ozone, its changes and possible radiative effects. Proc. WMO Conf. Observat. Atmos. Contamin. Vienna, Oct. 1983, WMO Environ. Rep. No. 16, 94-127.

____, 1986: Surface ozone during the second half of the nineteenth century. J. Clim. Appl. Meteorol., 25, 345-352.

____, and G.C. Reinsel, 1984: Trends in tropospheric ozone concentration. Proc. Ozone Sympos., Halkidiki, Reidel Publ. Co., 775-781.

Chameides, W.L., and D.H. Stedman, 1977: Tropospheric ozone. Coupling transport and photochemistry. J. Geophys. Res., 82, 1787-1794.
Crutzen, P.J., 1973: A discussion of the chemistry of some minor constituents in the stratosphere and troposphere, Pure Appl. Geophys., 106, 1385-1399.
Feister, U., and W. Warmlt, 1987: Long-term measurements of surface ozone in the German Democratic Republic. J. Atmosph. Chem., 5, 1-21.
Fishman, J., S. Solomon and P.J. Crutzen, 1979: Observational and theoretical evidence in support of a significant in situ photochemical source of tropospheric ozone. Tellus, 31, 432-446.
Grennfelt, P., and J. Schjoldager, 1984: Photochemical oxidants in the troposphere: A mounting menace. AMBIO, 13, 61-67.
Liu, S.C., D. Kley, M. McFarland, J.D. Mahlman and H. Levy II, 1980: On the origin of tropospheric ozone. J. Geophys. Res., 85, 7546-7552.
_____, M. Trainer, F.C. Fensenfeld, D.D. Parrish, E.J. Williams, D.W. Fahey, G. Hubber and P.C. Murphy: Ozone production in the rural troposphere and the implications for regional and global ozone distributions. J. Geophys. Res., 92, 4191-4207.
NRC, 1984: Global Tropospheric Chemistry - A plan for action. National Research Council, Washington, DC, 194 pp.
Oltmans, S.J., and W.d. Komhyr, 1986: Surface ozone distributions and variations from 1973-1984 measurements at the NOAA-GMCC baseline observatories. J. Geophys. Res., 91, 5229-5236.
Tiao, G.C., G.C. Reinsel, J.H. Pedrick, G.M. Allenby, C.L. Mateer, A.J. Miller and J.J. DeLuisi, 1986: A statistical trend analysis of ozonesonde data. J. Geophys. Res., 91, 13121-13136.
Wang, W.C., 1984: Climatological effects of atmospheric ozone - a review. Proc. Ozone Sympos., Halkidiki, Reidel Publ. Co., 98-102.
Warmbt, W., 1979: Ergebinsse langjahriger Messungen des bodennahen Ozons in der DDR. Z. Meteor., 29, Heft 1, 24-31.
WMO, 1982: Potential climatic effects of ozone and other minor trace gases, WMO Ozone Project Rep., No. 14, Geneva, 35 pp.

TROPOSPHERIC OZONE LIDAR MEASUREMENTS

G. Ancellet, A. Papayannis, G. Mégie, J. Pelon
Service d'Aéronomie du CNRS
BP 3 91371 Verrières le Buisson Cédex

Abstract : Remote measurements of the vertical distribution of ozone up to 15 km are essential to study the ozone budget in the troposphere. Exchanges between the troposphere and the stratosphere were studied using the Differential Absorption Lidar (DIAL) technique and are presented to illustrate the potential of lidar systems to quantify ozone transferts with high spatial and temporal resolutions. A new generation of lidar systems to be operated either in a regional ground based network or in an aircraft is also discussed.

I. Introduction

A study of the ozone budget in the troposphere mainly aims at assessing the respective contributions of the dynamical exchanges through the transition zones and of the photochemical processes. To address theses questions, one has to rely, today, on a multiinstrumental approach (balloon, aircraft, groundbased measurements) and to direct the observation programs as well as the modeling effort towards studies at different temporal and spatial scales. It seems clear that the ozone vertical distribution is one of the key parameter to be measured although not the only one needed. Remote sensing techniques offer two advantages for the determination of this distribution :

- a high vertical resolution to estimate the ozone variations at several altitude levels in the troposphere.
- a measurement continuity which makes possible the observations at different temporal scales ranging from few minutes to one or two days.

An horizontal coverage is also necessary for a description of large scale processes (> 10 km) controlling the ozone distribution in regions such as frontal zones or cut-off lows. To add this new dimension to the remote sensing capabilities, an ozone sounding network has to be implemented and airborne missions should be considered for mesoscale observations.

I. S. A. Isaksen (ed.), Tropospheric Ozone, 97–110.

In this paper, we will focus on the UV DIAL lidar technique which will be described in section II. Up to now, tropospheric ozone measurements using the lidar technique have been conducted to study the exchanges of ozone between the stratosphere and the troposphere. The results of both ground based and airborne instrumentations will be presented in section III. In section IV, we will discuss our prospective studies leading to the conception of future lidar systems in term of automatization and reliability. This new step in the ozone lidar development is essential to the integration of lidars in an observation network or in aircrafts.

II. The DIAL measurement technique

The differential absorption technique has been extensively studied by various authors since it has been proposed for the first time by Schotland (1964). The purpose of this section is thus only to emphasize the major features relevant to the measurements of O_3 and especially the major error sources. In a DIAL measurement the pollutant concentration n_i at a given range R along the line of sight is directly deduced from the temporal - i.e. range resolved - analysis of two backscattered signals S_1 and S_2. They correspond to two laser emitted wavelengths which are respectively absorbed (λ_1) or free of absorption (λ_2) by the constituent under study :

$$n_i\ (R) = \frac{1}{2\ \Delta\ \sigma_{12}} \frac{d}{dR} \ln \frac{S_2\ (R)}{S_1\ (R)} \qquad (1)$$

where $\Delta\sigma_{12}$ is the differential absorption cross section (cm^2) equal to $\sigma_i\ (\lambda_1) - \sigma_i(\lambda_2)$, $\sigma(\lambda)$ being the wavelength dependent absorption cross section of the constituent n_i. The main assumption made in the derivation of (1) is that the two wavelengths λ_1 and λ_2 are chosen close enough so that the spectral variations of the scattering and extinction coefficients of air molecules and aerosol particles over the range (λ_1, λ_2) can be neglected.

The accuracy of the measurements ϵ depends on a large number of parameters and can be expressed as the sum $\epsilon_1 + \epsilon_2$ of two terms :
- ϵ_1 is the statistical error due to the signal and background noise fluctuations and can be evaluated from the system characteristics and ozone absorption coefficients.
- ϵ_2 is a systematic error due to the wavelength dependence of the scattering and absorbing (other than ozone) properties of the atmospheric medium which have been neglected in the derivation of equation (1).

Using the expressions of the various uncertainties, as given for example in Mégie et al. (1985), the error analysis of the DIAL measurement taking into account both experimental parameters and atmospheric characteristics leads to consider two altitude ranges.

a) Boundary layer and lower tropospheric measurements

Below 2 km, ϵ_2 is the most important error term as the molecular and aerosol extinctions are large and as the backscattered signal is high enough so that the statistical error ϵ_1 can be reduced below 10^{-2}. By using average values of the molecular extinction coefficient (Rayleigh scattering), the differential extinction for a wavelength interval $\Delta\lambda = \lambda_1 - \lambda_2 = 5$ nm can be as large as 0.1 inthe wavelength range considered (270 - 310 nm) and can thus not be neglected. A correction has to be made by measuring the ground-level pressure and temperature to derive the atmospheric density in the first kilometers. An uncertainty of a few percents in this determination will reduce the error due to molecular extinction to less than 5×10^{-3}.

The remaining error in ϵ_2 is then related to the aerosol extinction, and a further distinction should be made between ozone measurements performed in rural (non polluted) or urban areas.

Rural areas : Using aerosol particle concentrations as given by Elterman (1970) and typical rural conditions at a 1 km vertical range, the optimization of λ_1 and $\Delta\lambda$ in the wavelength range 265-285 nm leads to values of ϵ_2 less than 2 % for $\Delta\lambda$ as large as 10 nm and $\lambda_1 < 285$ nm (Pelon et al., 1982). This error can be further reduced if needed by using the experimental procedure adopted for urban or polluted areas and described in the following subsection.

Urban or polluted areas : Owing to the large aerosol particles concentrations in the boundary layer, ϵ_2 can reach values as large as 10^{-1}, and a complementary measurement might be needed to eliminate the aerosol differential extinction. This can be done, for example, by using three wavelengths in the 265-285 nm range with the same wavelength interval $\Delta\lambda \simeq 5$ nm. The aerosol differential extinction can then be substracted, if one assumes a linear variation for the aerosol scattering and extinction properties over 10 nm. Inverting the lidar equation for the non absorbing wavelength could also provide an estimation of the aerosol distribution, assuming a knowledge of the aerosol backscatter wavelength dependence and of the aerosol backscatter to extinction ratio. One of these two techniques will avoid the use of a theoretical model that depends on the nature and shape of the particles. Taking into account the experimental errors and the modelling uncertainties, the estimated upper limit of ϵ_2 will then be reduced to 2-3 %.

b) Measurements in the free troposphere

Above 2 km, ϵ_2 decreases rapidly with altitude, for average values of higher tropospheric and stratospheric aerosol content. The choice of the operating wavelengths is then determined by minimizing the value of ϵ_1. Mégie and Menzies (1979) have shown that this corresponds to an optimum value of 1.28 for the optical depth due to the ozone absorption. For measurements up to 10-12 km, the accuracy, given by ϵ_1, stays between its optimum value ϵ_{opt} and 1.2 ϵ_{opt} with $\Delta\lambda = 5$ nm and $\lambda_1 < 290$ nm.

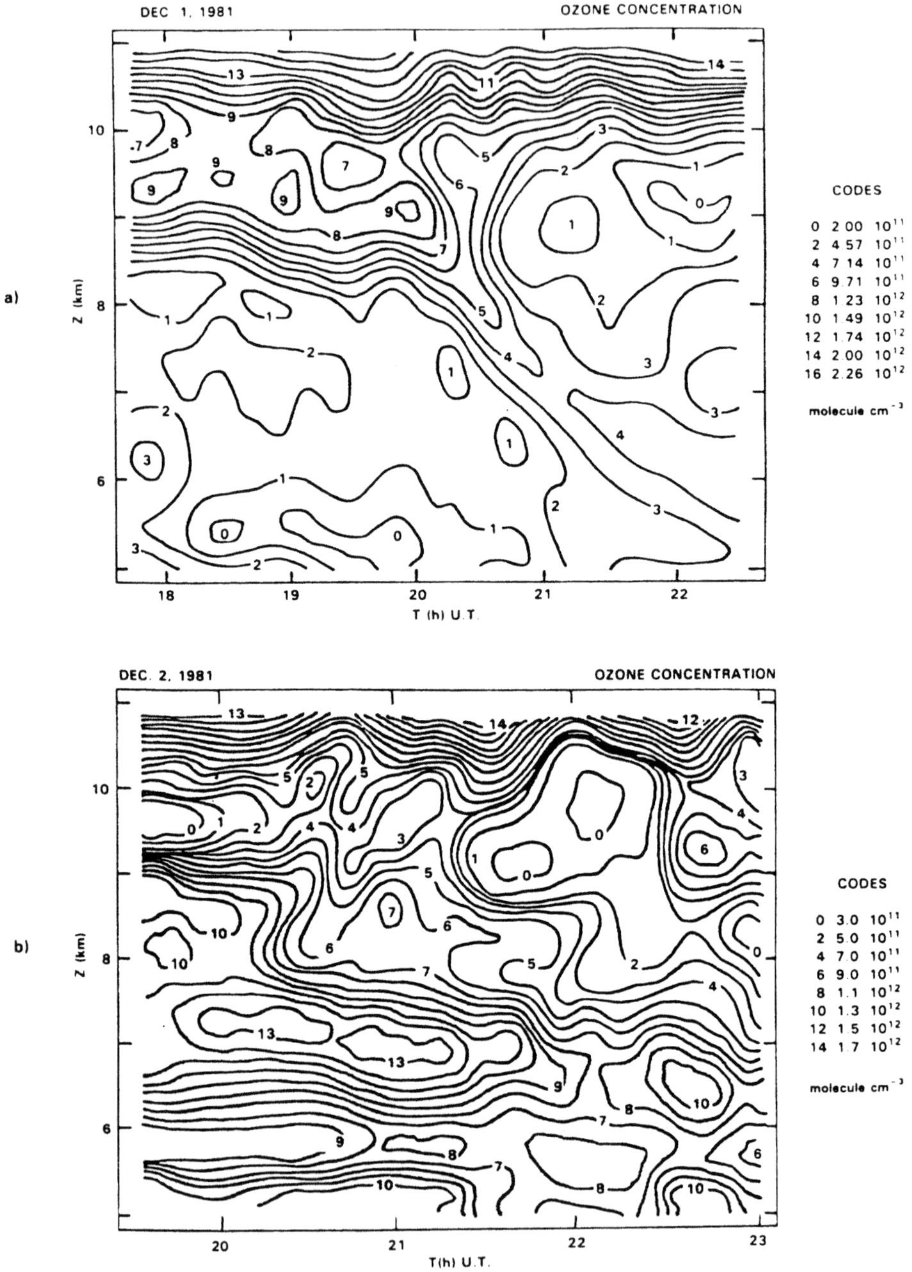

FIGURE 1 : Ozone number density isocontours ans a function of time (horizontal axis) and altitude (vertical axis) for the nights of December 1st and 2nd 1981. The hatched areas correspond to ozone concentrations larger than 10^{12} cm^{-3}.

II. Lidar measurements of tropospheric ozone

Lidar measurements of the ozone vertical distribution in the troposphere and lower stratosphere using the DIAL technique have been demonstrated both for ground based stations (Pelon et al., 1982) and from airborne platforms (Browell et al., 1983). The scientific studies performed have, up to now, been mainly focussed on the exchange of ozone between the stratosphere and the troposphere which occurs in the mid latitude region in association with tropopause folds which develop by a steepening of the tropopause at a jet core, followed by a downward and southward streching of the jet in the northern hemisphere. The result is an irreversible transfer of stratospheric air from the polar reservoir to lower latitudes. The quantitative knowledge of the ozone transfer in such events will obviously rely on a multiinstrumental approach (satellites, aircraft, ground based measurements) and a modelling effort which includes synoptic analysis of potential vorticity and potential temperature cross sections. Due to its ability to perform continuous and high vertical resolution measurements, lidar systems can improve our experimental basis for such studies as illustrated by the following examples.

2.a Ground based studies

The figure 1 shows examples of lidar observations of the ozone vertical distribution taken on December 1 and 2, 1981 (Pelon, 1985). During this time period a meridional circulation was established over France due to a developing wave pattern in the jet stream associated with the descent of a low over southern Europe and the presence of an anticyclonic region over the Atlantic. This resulted in a main front located over Southern France at the 500 mb level during the second night of observation (December 2-3). The measurements reported on figure 1-b are thus representative of a vertical cross-section through the frontal zone in the south west part of the low. The observed altitude decrease in the ozone number density maximum corresponds to the displacement of the front during the measurement period. At the lower edge of the front, simultaneous measurements using the star scintillometry techniques (Scidar), gave evidence for the presence of intense turbulent layers, at 5-6 km altitude just below the peak of the ozone distribution ($1.5 \times 10^{+12}$ mol cm^{-3} i.e., 120 ppbv). If for the previous night (December 1-2) the shapes of the ozone isocontours (figure 1-a) are similar to the fingered structure observed in tropopause folding events, the observed evolution should not be attributed directly to the influence of the main front : in order to explain the origin of the high ozone content observed at tropospheric levels in both cases and to follow its ensuing evolution, air mass trajectories calculated from radiosonde observations, by the Météorologie Nationale at the European Center have been used (figure 2). Such trajectories are necessarily crude, especially in the region of frontal surfaces. The trajectories ending at the Observatoire de Haute Provence (OHP) on December 1 (6 pm and midnight UT) at the 300 and 500 mb levels (figure 2-a) show that the origin of the air masses

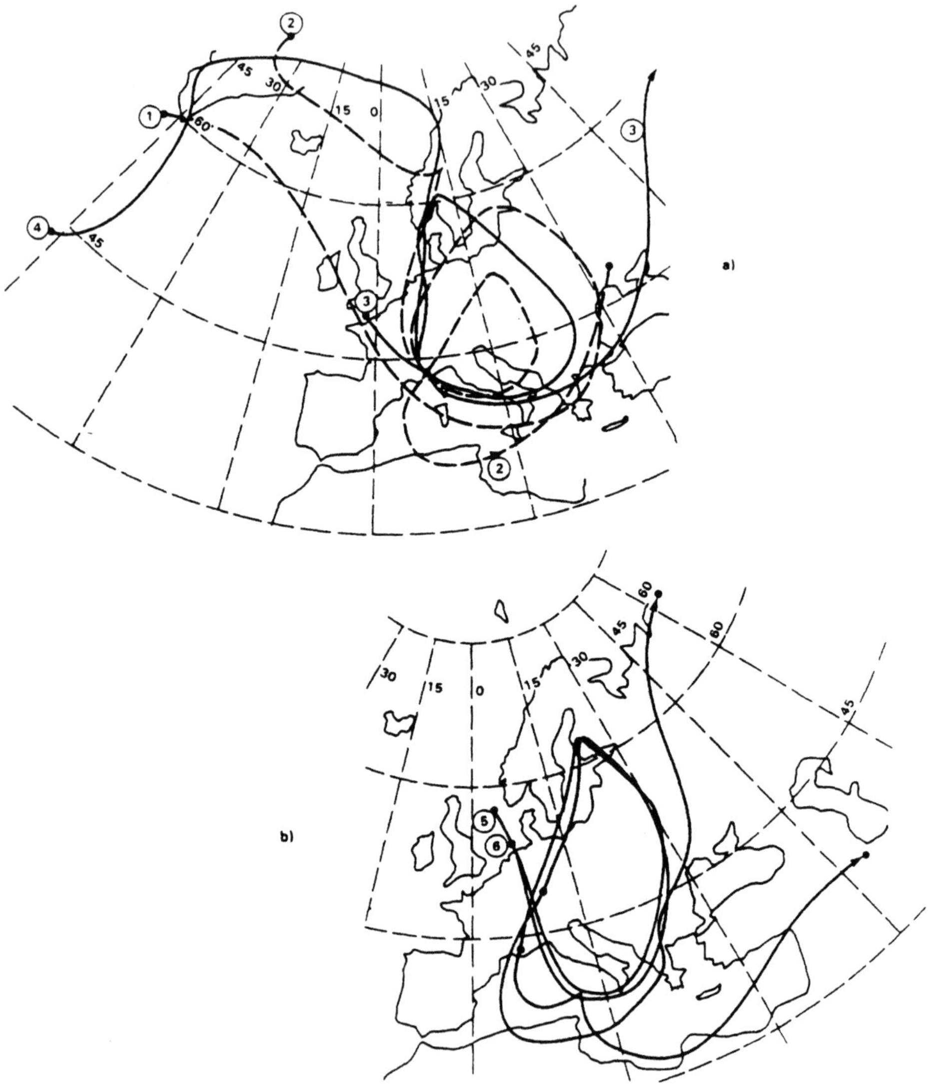

FIGURE 2 : Air mass trajectories ending at the Observatoire de Haute Provence on December 1st, 6 pm (1) 500 mb pressure level ; (2) 300 mb and midnight ; (3) 500 mb ; (4) 300 mb). b) Air mass trajectories in the frontal zone at the 500 mb level originating at points 5 and 6 and ending at 49°N, 10°E and at the OHP (44°N, 5°E) on December 2nd (midnight) and their ensuing evolution.

four days before (November 28) are quite different. The one observed at 300 mb in the early night originated above Greenland at upper levels (225 mb) in the stratosphere (high ozone content), while those observed six hours later come from tropospheric levels at lower latitude. The synoptic situation on November 28 (0 h UT) corresponds to a very well developed frontal zone stretching from north-west of Southern England to the South of Iceland with a low over northern England and Western Scandinavia and a high over the Azores. The time resolved measurements at the OHP corresponds thus to a transverse cross-section through this frontal zone : observed air masses originate at various altitudes with different ozone contents further modified by dispersion and small scale turbulence along the trajectories. Similarly, figure 2-b shows the trajectories ending on December 2 (midnight) at two points located at the 500 mb level in the frontal zone. Their origins are in the cyclonic zone above the North Sea four days before at the 400 mb level (points 5 and 6). The ozone rich air masses coming from this region have thus turned around the low in the associated frontal zone before reaching southern France (figure 1-b).

Considering the air mass trajectories calcultated for the days following the observations, evidence is given that the air coming from points 2 and 3 on figure 2-a (December 1) is going back to higher latitudes (49°N and 57°N) and upper levels for points 2, whereas air originated from point 3 remains at the 500 mb level. On the second night (December 2-3), the trajectories originating at points 5 and 6 on figure 2-b are quite different : the air masses coming from point 5 continue to travel down to lower altitude (37°N) and higher pressure levels (630 mb). On the contrary air masses originating from point 6 reach a latitude of 68°N on December 7 at the 450 mb level. Such a behaviour illustrates the complexity of synoptic stratosphere - troposphere exchange processes : the air masses observed in the frontal zone can either be definitely transferred into the troposphere at mid-latitudes, or return to the stratosphere at higher latitudes in association with the polar jet stream.

This example shows the difficulty in quantifying the ozone transfer related to mid-latitude stratosphere-troposphere exchange processes. It does not give evidence for a large direct ozone transfer at 45° N from the stratosphere to the troposphere but rather indicated that such transfer occurred at higher latitude. The larger part of the transported ozone remains in the frontal zone for several days with a likely possibility for part of it to return to the polar stratosphere. Evaluation of the magnitude of the ozone transfer is also complicated by turbulent exchange processes in the frontal zone which are more intense than in the free troposphere. Such a study is obviously limited by the operation of a single observing station. It shows however that the use of ground based instrumentation (ozone sondes, lidar) in a regional scale network, together with presently available satellite data could allow an evaluation of the ozone transfer within frontal zones or cut-off lows.

2.b. Airborne observation

Airborne lidar measurements of the ozone and aerosol vertical distributions across a tropopause fold have been recently performed by Browell et al. (1987). The DIAL measurements across a fold which occured on April 20, 1984 over southern Nevada and California, show a 2.0 km deep layer with high ozone concentrations and aerosol content that slopes downward from 36° N to the top of the planetary boundary layer at 33° N. Mixing ratios larger than 200 ppbv were measured in the fold, which decrease by 25 % along the central axis of the fold. An analysis of the potential vorticity performed using radiosonde data led to a 0.89 correlation between ozone and potential vorticity, with an average ratio of 50.2 ppbv/10^{-5} cm^2 $Kg^{-1}s^{-1}$ (figure 3). The potential vorticity is conserved in the fold and the observed dilution of ozone and potential vorticity down the fold is then consistent with a spreading by adiabatic irreversible mixing in larger volumes. Observations of the fold at the top of the boundary layer give evidence for a layer oscillation due to forcing by convective plumes and a possible increase of ozone mixing into the boundary layer.

Such a study obviously demonstrates the potential of airborne lidar for high horizontal and vertical resolution studies of specific events in which the ozone transfer could be quantify. Further observations by ground based network and satellite monitoring would then lead to a statistical approach of the time average ozone transfer within frontal zones and cut off lows.

IV. Future lidar systems

The next generation of lidar systems will be operated either in a ground based network or in an aircraft. Therefore the instrumentation will have to be simple, ruggedized and reliable. A new laser source (solid state or gas laser) is highly desirable since the Nd Yag pumped dye lasers, presently used, are complicated and unefficient wavelength conversion chains. The receiving optics will also be modified to separate the simultaneously emitted wavelengths and optimized for daytime and nightime operations. A simple system has been designed to

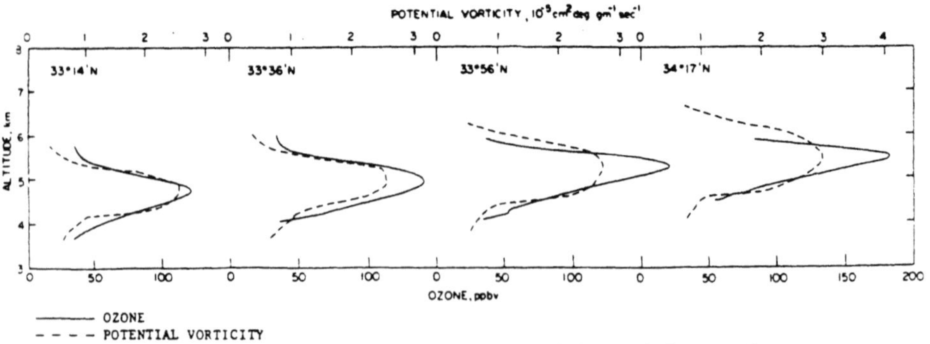

FIGURE 3 : Comparison of O_3 and potential vorticity profiles at various locations across the tropopause fold.(Browell et al,1987)

perform measurements of tropospheric ozone profiles with the following performances :

- altitude range 0.1 - 15 km
- vertical resolution 0.1-1 km
- integration time 1-15 min
- accuracy better than 5 %.

One can then consider two laser systems which use high output energy laser source in the UV (exciplex, or Nd Yag laser fourth harmonic) as the pumping source for the generation of emission lines by non resonant Stimulated Raman Scattering (SRS) in hydrogen and deuterium :

- Nd Yag laser fourth harmonic (266 nm) generating the wavelength pair (289-299 nm) in the first Stokes lines.
- KrF exciplex laser (248 nm) generating the wavelength pair (277-291 nm in the first Stokes lines.

Table 1 : Available wavelengths and output energy using fixed frequency lasers and Stimulated Raman Scattering

KrF LASER SYSTEM

		λ (nm)	Conversion Efficiency (%)	Energy (mJ)	Absorbing molecules
H_2	S_1	277.1	30	< 60	O_3
	S_2	313.2	15	< 30	SO_2.
(4156 cm^{-1})	S_3	360.1	1	2	
D_2	S_1	268.4	25	< 50	O_3
(2986 cm^{-1})	S_2	291.8	10	< 20	O_3 & SO_2
KrF(200 mJ)		248.4	25	< 50	

Nd : Yag (X 4) LASER SYSTEM

		λ (nm)	Conversion Efficiency (%)	Energy (mJ)	Absorbing molecules
H_2	S_1	299.06	30	< 30	O_3
	S_2	341.51	15	< 15	
(4156 cm^{-1})	S_3	397.51	1	1	
D_2	S_1	289.0	25	< 25	O_3 & SO_2
	S_2	316.24	10	< 10	
(2986 cm^{-1})	S_3	349.21	1	1	
Nd : Yag x 4 (100 mJ)		266	25	< 25	O_3

The table I summarizes the available wavelengths and output energies for both laser sources and indicates also the species being possibly detected.

The remainder of the lidar system includes transmission and receiving optics (30 cm diameter telescope and a grating spectrometer to separate the simultaneously emitted wavelenghts), time resolving electronic systems (transient waveforms recorder) and a microcomputer for data acquisition, experiment control and real time data analysis.

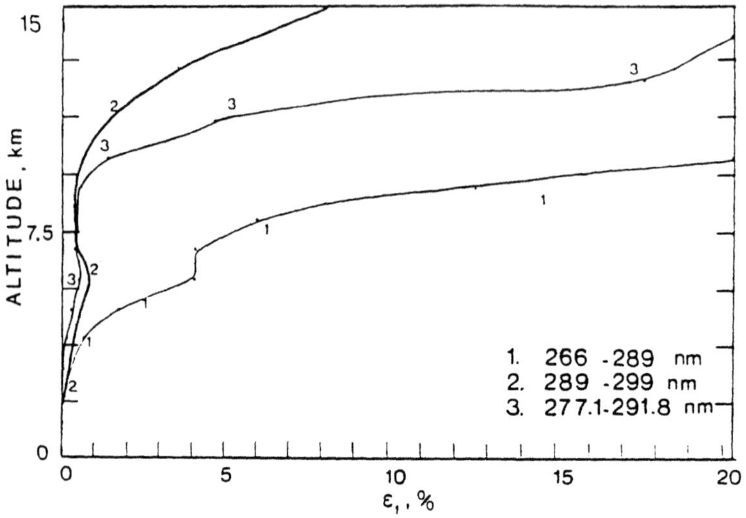

FIGURE 4 : Statistical error ε_1 as a function of altitude for ground based measurements using the indicated wavelength pair.

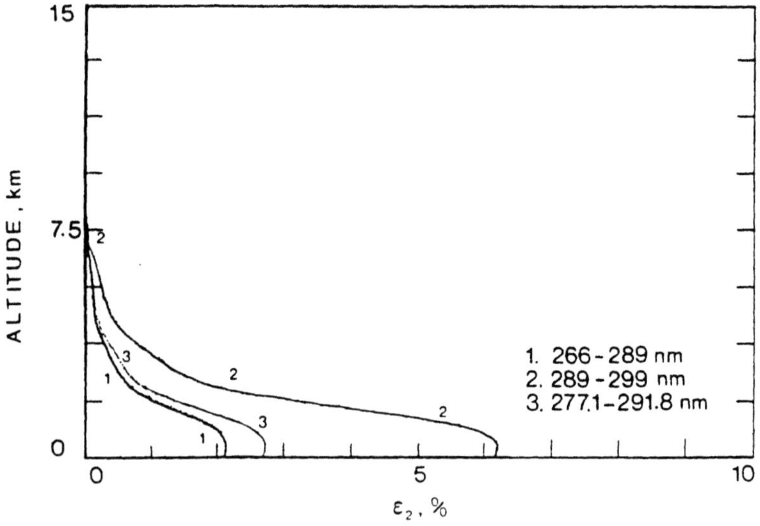

FIGURE 5 : Systematic error ε_2 as a function of altitude for ground based measurements using the indicated wavelength pair.

In the following discussion, the appropriate wavelengths for ground based and airborne measurements will be determined by minimizing simultaneously the statistical (ϵ_1) and systematic (ϵ_2) errors (ϵ_1 is calculated for an integration time of 100 seconds or 10^3 laser shots on each wavelength).

a) Ground based measurements

Considering the results of Table I and following the error analysis presented in section II,, three wavelength pairs seem to be the better adapted for ground based ozone measurements

a) λ_1 = 277.1 nm and λ_2 = 291.8 nm (KrF)
b) λ_1 = 266 nm and λ_2 = 289 nm (Nd : Yag)
c) λ_1 = 289 nm and λ_2 = 299 nm (Nd : Yag)

The statistical error ϵ_1, stays generally below 10 % over the whole altitude range (0-15 km) except for case (b) (figure 4). Case (c) seems to be the best choice for measurements at 15 km altitude. The wavelength pair (a) provides the lowest value of ϵ_1, but is limited to altitudes below 10 km.

The uncertainty is also determined by the systematic error, ϵ_2 and is thus related to the wavelength interval between the on and off wavelengths. The main contribution to ϵ_2, is the Rayleigh extinction which can be as high as 10-30 % of the total absorption in the lowest altitude ranges. However, this effect can be corrected using in situ pressure and temperature measurements which will allow the determination of the density with an accuracy better than 1-2 % leading thus to a residual error of less than 0.6 % in the first kilometers of probing. The error due to Mie extinction is shown on figure 5 for the three cases. The wavelength pair 266-289 nm gives the best results for all the altitude ranges, as ϵ_2 is less than 3 % down to the ground. Errors as high as 6 % are obtained for the case (c) (289-299 nm) in the boundary layer whereas ϵ_2 stays below 2 % above 2 km.

Considering the above results the wavelength pair (289-299 nm) is adapted for measurements at all the ranges since it gives the largest signal-to-noise ratio at 15 km and an acceptable value of ϵ_2 in the rural boundary layer. However for measurement in the PBL only, the pair 266-289 nm which can be obtained with the same laser source, is the best choice because of the large ozone differential absorption at these wavelengths. For measurements in polluted areas, the sulphur dioxide differential absorption at the pair (289-299 nm) is as high as 0.51 the differential absorption due to ozone alone, assuming the same mixing ratios for both species. Consequently SO_2 concentrations will have also to be measured simultaneously using a three wavelength system (see Conclusion).

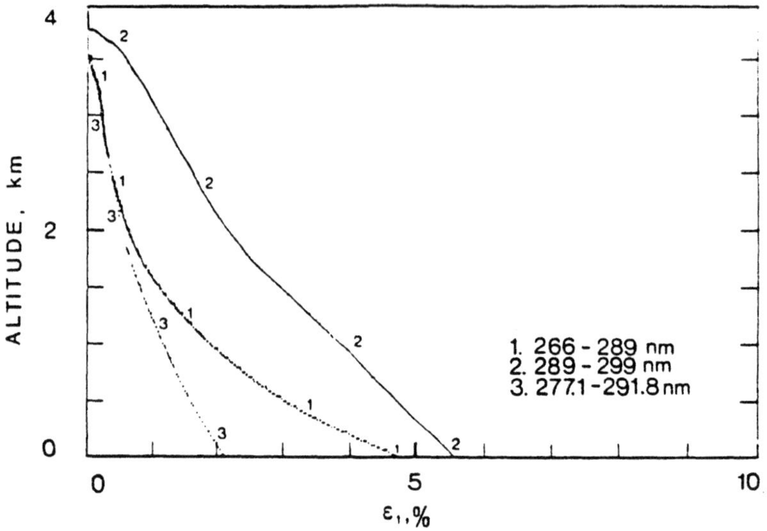

FIGURE 6 : Statistical error ϵ_1 as a function of altitude for airborne measurements using the indicated wavelength pair.

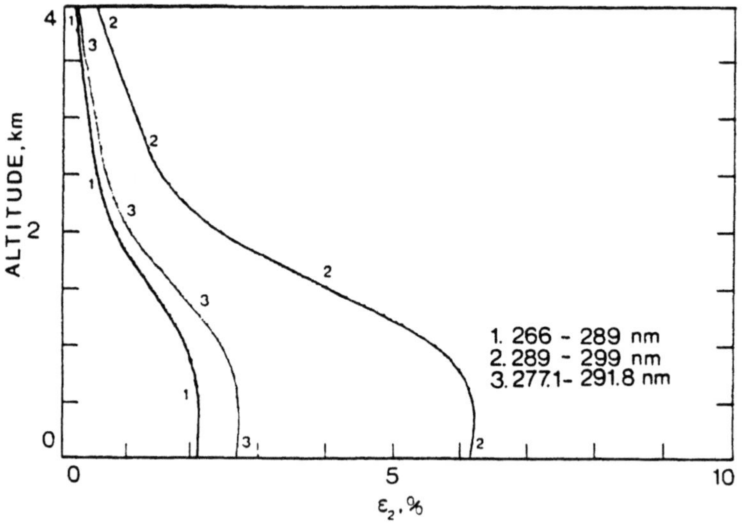

FIGURE 7 : Systematic error ϵ_2 as a function of altitude for airborne measurements using the indicated wavelength pair.

b) Airborne measurements

When considering the same wavelength pairs as before, the statistical error ϵ_1 stays generally below 6 % over the whole altitude range whereas some differences can be observed (figure 6) : case (a) seems to be the best choice as for case (b) the error increases rapidly with decreasing altitude due to the strong absorption at 266 nm, and for case (c) ϵ_1 is not optimized due to the lower absorption coefficient at 289 nm. Nevertheless for a down looking system the overall uncertainty is mainly determined by the systematic error ϵ_2. The error due to the Mie extinction is shown on figure 7 for the three cases. Here again the wavelength pair 277.1 - 291.8 nm which corresponds to the smallest values of $\Delta\lambda$ gives a similar result as the wavelength pair 266-289 nm leaving ϵ_2 smaller than 3 % down to the ground.

Considering the above results, the wavelength pair (277.1-291.8 nm) seems to be the best adapted since it minimizes both ϵ_1 and ϵ_2. Even though, statistical errors as high as 5 % are obtained for the case (b) (266-289 nm) in the boundary layer, this wavelength pair might alternatively be selected since it gives the smallest value of ϵ_2.

V. Conclusion

Studies of ozone exchanges between the stratosphere and the troposphere were conducted using both a ground based lidar and an airborne lidar. They demonstrated the potential of such systems to quantify ozone transfer with a high horizontal and vertical resolution. Since the future systems will be used in a regional scale network and in an aircraft, we have been designing a more simple and ruggedized lidar. Considering two possible laser sources, a Nd Yag laser has been selected as a pumping source (4th harmonic) for generation of emission lines by Stimulated Raman Scattering in hydrogen and deuterium. For a ground based lidar we propose ozone measurements at three wavelengths simultaneously (λ_1 = 266 nm, λ_2 = 289 nm, λ_3 = 299 nm). Wavelength pair 266-299 nm will be used to obtain a first estimation of the ozone profile, the pair 289-299 nm to determine sulfur dioxide vertical profiles in the PBL. Lidar inversion technique (Klett, 1981 ; Browell et al. 1987) using the backscatter aerosol return signal at λ_3 = 299 nm, is applied to determine the aerosol vertical distribution in regions of spatially in homogeneous aerosols. Using calculated aerosol and sulfur dioxide vertical profiles, another calculation is performed with the wavelength pair (266-289 nm) to obtain the final ozone profile. By this iterative process, the ozone vertical distribution can be measured with an accuracy less than 5 % from 0.3 km to 13 km and up to 17 km by increasing the integration time to 15 min.

For an airborne lidar, our analysis leads to the conclusion that the use of a KrF exciplex laser will be best adapted for measurements in the boundary layer. However a subsidiary can be the use of the Nd Yag laser emitting simultaneously three wavelengths to allow a first correction of the sulphur dioxide and aerosol extinction as explained previously. In this case the two systems will be almost equivalent for airborne applications.

REFERENCES

BROWELL E.V., S. ISMAIL, S. SHIPLEY, "Ultraviolet DIAL measurements of O_3 profiles in regions of spatially inhomogeneous aerosols", *Appl. Opt.*, 17, 2827 (1985).

BROWELL E.V., E.F. DANIELSEN, S. ISMAIL, G.L. GREGORY, S.M. BECK, "Tropopause fold structure determined from airborne lidar and in situ measurements", *J. Geophys. Res.* D2, 2112 (1987).

ELTERMAN L., "Relationships between vertical attenuation and surface meteorological range", *Appl. Opt.*, 9, 1804 (1970).

KLETT J., "Stable analytical inversion solution for processing lidar returns", *Appl. Opt.*, 2, 211 (1981).

MEGIE G. and R.T. MENZIES, "Complementarity of UV and IR differential absorption lidar for global measurements of atmospheric species", *Appl. Opt.*, 19, 1173 (1980).

MEGIE G., G. ANCELLET and J. PELON, "Lidar measurements of ozone vertical profiles", *Appl. Opt.*, 21, 3454 (1985).

PELON J., G. MEGIE, "Ozone monitoring in the troposphere and lower statrosphere : evaluation and operation of a ground based lidar station, *J. Geophys. Res.*, C7, 4947 (1982).

PELON J., "Distribution verticale de l'ozone dans la troposphère et la stratosphère : Etude expérimentale par télédétection laser et application aux échanges troposphère-stratosphère", Thèse de doctorat, Université Paris VI (1985)

SCHOTLAND R.M., "Errors in the lidar measurement of atmospheric gases by differential absorption", *J. Appl. Meteo.*, 13, 71 (1974).

TROPOSPHERIC OZONE FROM SATELLITE TOTAL OZONE MEASUREMENTS

Jack Fishman
Atmospheric Sciences Division, Mail Stop 401B
NASA Langley Research Center
Hampton, Virginia 23665-5225
UNITED STATES

ABSTRACT. A method for determining the amount of ozone in the tropical troposphere from concurrent sets of satellite data is presented. This prodedure is applied only in the Tropics and the results indicate that a significant longitudinal gradient is present at low latitudes and that the highest amounts of tropospheric ozone are located west (i.e., downwind) of Africa and South America. Such a distribution suggests that biomass burning, or another stationary source of continental origin, is the largest source of ozone in the Tropics. The integrated amount of ozone in the Tropics that is derived from this analysis is comparable to or only slightly less than the amount of ozone that is present in northern mid latitudes. If such an interpretation is valid, then it suggests that the magnitude of the amount of tropospheric ozone resulting from in situ photochemical production in this region is comparable to the in situ photochemical source from industrialized emissions.

1. INTRODUCTION

The recent study of Fishman et al. (1986) has suggested that photochemical ozone production in the troposphere can be observed in the Tropics from an analysis of satellite-based total ozone measurements. In that study, enhancements of total ozone amounts were found to be coincident with the presence of widespread biomass burning episodes in the Amazon Basin of Brazil. In addition, it was shown that the longitudinal gradient of total ozone at low latitudes was strongly positively correlated with the distribution of tropospheric carbon monoxide measurements which had been obtained from the MAPS (Measurement of Air Pollution from Satellites) experiment aboard a Space Shuttle platform (Reichle et al., 1986). Particularly noteworthy in the Fishman et al. study was the finding that very high levels of total ozone in the Tropics were generally present over the coasts of Africa. Pursuing this observation in more detail, Fishman and Larsen (1987) showed from an analysis of two years of TOMS (Total Ozone Mapping Spectrometer) data that the highest total ozone amounts in the Tropics were located over the west coast of Africa.

I. S. A. Isaksen (ed.), Tropospheric Ozone, 111–123.

In principle, it should be possible to derive a climatology of tropospheric ozone by examining concurrent measurements of total ozone, which should be comprised of the sum of the amount of ozone in the stratosphere and troposphere, and subtracting the amount of ozone in the stratosphere from the total ozone amount. The stratospheric component of the total can be determined from ozone profiles obtained from the SAGE (Stratospheric Aerosol and Gas Experiment) instrument aboard the Atmospheric Explorer Satellite which operated between February 1979 and November 1981 (McCormick et al., 1984). In the Fishman and Larsen (1987) study, we were hesitant to present the results of such an exercise since we knew that the TOMS data that we had been using had to go through a reprocessing procedure (Bhartia et al., 1984) using updated ozone-absorption cross sections. Although we have received some of the reprocessed TOMS data and we intend to utilize them with the concurrent data base available from the SAGE II archived measurements (which is still making measurements from the time it was launched in 1984), corrected TOMS measurements from the 1979-1980 timeframe (i.e., the period of investigation in Fishman and Larsen) still have not been made available through the data archive. On the other hand, P.K. Bhartia (personal communication) notes that the corrections made to the archived TOMS data set used in the Fishman and Larsen study should be relatively constant at low latitudes. Thus, in the study presented here, we have assumed a universal correction to the TOMS data by taking into account the observed 6.6 percent bias (Bhartia et al., 1984), and then calculating a tropospheric residual by subtracting the measured amount of ozone above 100 mb that had been obtained from the SAGE data archive. The results are consistent with the conclusions from the Fishman and Larsen (1987) study and suggest that high levels of tropospheric ozone in the Tropics are found downwind (west) of the coasts of Africa and South America. This finding is consistent with the explanation that biomass burning is the largest source of ozone in the tropical troposphere.

2. SATELLITE DATA

2.1. TOMS Data

Total ozone amounts have been measured globally on a daily basis since 1978 by the TOMS instrument aboard the polar-orbiting Nimbus 7 satellite. It works on the principle of backscattered ultraviolet radiation at several wavelength pairs in the ultraviolet and visible portions of the spectrum (Heath et al., 1975) and can provide a spatial resolution of better than 100 km. Through its choice of wavelengths, TOMS should be able to measure the integrated amount of ozone in both the stratosphere and troposphere although Klenk et al. (1982) have shown with a radiative transfer model that TOMS may underestimate the amount of ozone in the lowest few km of the atmosphere by as much as 40 percent under certain clear-sky conditions and very low surface albedos.

An example of the distribution of total ozone over southern Africa for October 9, 1984, is shown in Figure 1. The units on the contours in this figure are Dobson Units (D.U.), where 1 D.U. = 2.69 x 10^{16} molecules of ozone cm^{-2}. This depiction has utilized the gridded TOMS archived data base (using corrected ozone-absorption cross sections) which has a resolution of 1° latitude by 1.25° longitude in the Tropics. Of particular interest are the two regions of high total ozone amounts on both the east coast and the west coast of the continent. The eastern maximum of 286 D.U. is centered over Kenya near 2°S and 38°E. The western maximum (282 D.U.) is close to the Congo- Gabon border near 3°S and 13°E.

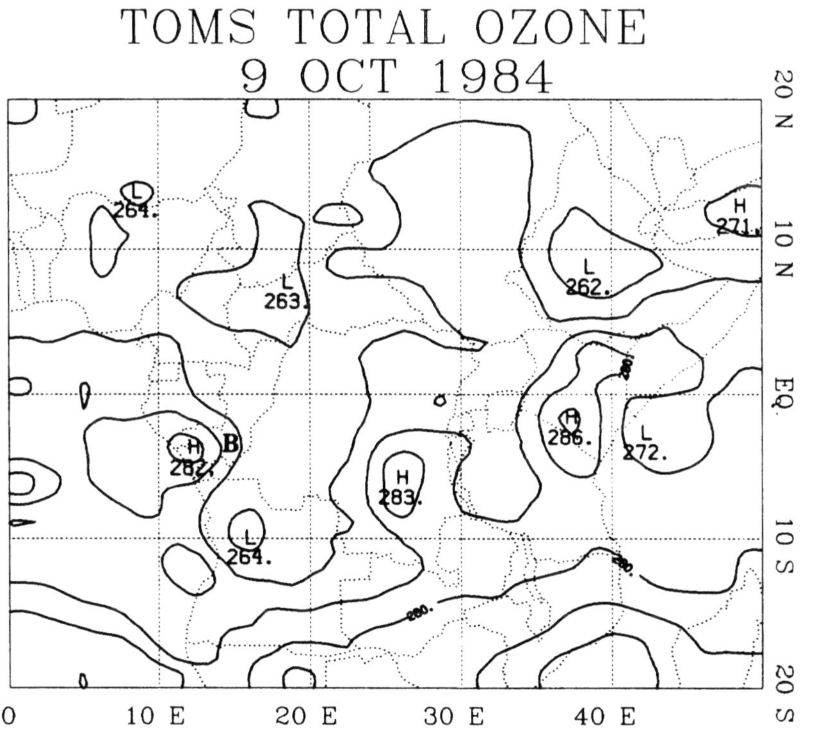

Figure 1. The distribution of total ozone obtained from TOMS data over Africa is shown for October 9, 1984. Contour intervals are 5 D.U.

On this particular date, MAPS was again measuring carbon monoxide (CO) in the troposphere during its second flight aboard the Space Shuttle. A preliminary analysis of the data obtained during this flight indicates that elevated concentrations of CO were present at the same locations near the African coasts where the high total ozone is shown in Figure 1 (Reichle, 1986). Additional analysis of 35-mm photography acquired from an on-board camera during this Space Shuttle mission indicates that widespread biomass burning was also present over the east coast of Africa on October 9 (Connors et al., 1986). Thus, the visual observation of large quantities of smoke and the presence of high CO concentrations near the east coast of Africa support the premise that the high total ozone found over Kenya is a result of in situ photochemical generation of tropospheric ozone.

During the time of this MAPS flight, Cros et al. (1987) report the existence of an air pollution episode in Brazzaville, Congo (4°S, 15°E), which lasted from October 8 to October 10, 1984. This location (marked by "B" in Figure 1) is slightly east of the total ozone maximum observed near the west coast of Africa on October 9. In addition, it should be noted that a total ozone maximum was present in the TOMS depiction for October 8 and was located at 5°S and 20°E. Thus, it appears that this center of high total ozone passed over Brazzaville between the 8th and the 9th, consistent with the onset of high ozone concentrations at the surface that Cros et al. observed. Furthermore, they note that these elevated ozone concentrations (which exceeded 70 ppbv) were accompanied by a "dry haze from the east," further supportive of the existence of a source of ozone which may have originated from widespread biomass burning to the east (i.e., upwind) of Brazzaville. The MAPS data likewise indicate that high CO concentrations were present over Angola, Zaire, and Congo throughout the duration of its flight between October 7 and 14, 1984.

The example described in this section confirms the presence of a significant source of atmospheric emissions from Africa that should result in the formation of large amounts of photochemically generated tropospheric ozone. Quantification of the amount of tropospheric ozone that may be generated from biomass burning will be presented later in this study when the TOMS data are utilized in conjunction with a set of SAGE data.

2.2 SAGE Data

Between October 1979 and September 1980, the SAGE instrument obtained 418 profiles between 15°N and 15°S. Of these, 69 (17 percent) indicated the presence of clouds above 100 mb (Woodbury and McCormick, 1986), and the ozone between cloud-top altitude and 100 mb had to be estimated. For the most part, these clouds penetrated less than 1 km above 100 mb, and therefore we believe that any error associated with these profiles had a negligible impact on the analysis discussed in this study. Of these 418 profiles, TOMS measurements were missing for 30. By examining the distribution of total ozone from the TOMS data on the days in which SAGE profiles were available, but TOMS data were missing, total ozone values were estimated in regions where the data gaps were small and where it appeared that spatial and temporal gradients (which could be estimated by examination of the total ozone distribution both before and after the SAGE profile was obtained) were very weak. Thus, 19 of the 30 missing TOMS total ozone values were estimated in this analysis.

A summary of the location of the SAGE profiles that were used in this study is presented in Figure 2. Different symbols are used to denote each quarter of the year. This figure illustrates the uniform coverage of the SAGE profiles that were used in this study. On the other hand, it also demonstrates the relative paucity of observations over any particular locale, limiting the usefulness of this data set for deriving a climatology or even a pronounced seasonal cycle that may be present at a particular site.

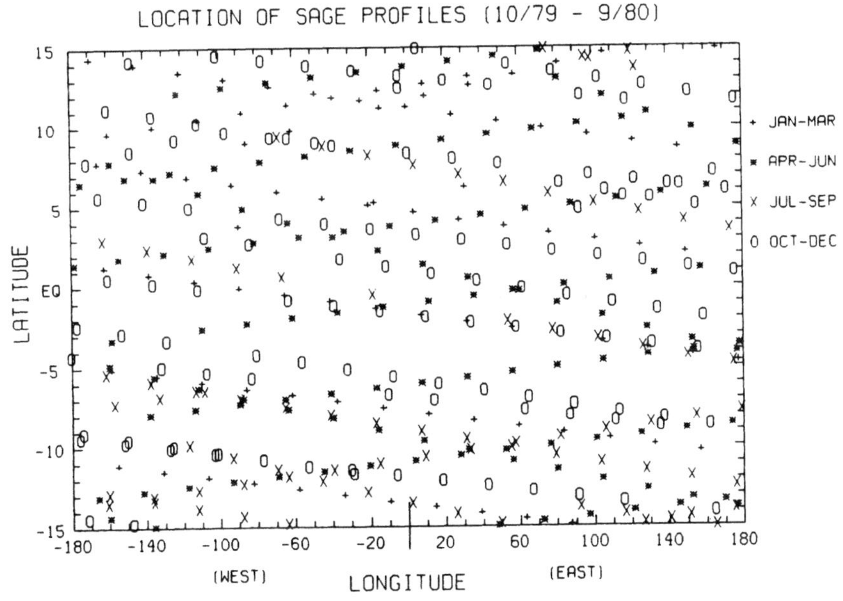

Figure 2. The location of SAGE profiles obtained in the Tropics between October 2, 1979, and September 30, 1980, is shown as a function of latitude and longitude. The various symbols indicate the quarter of the year during which a particular profile was obtained.

3. DERIVATION OF THE TROPOSPHERIC OZONE RESIDUAL

The TOMS data used for this analysis were averaged into 5° latitude by 5° longitude blocks (up to 20 gridded data points) for each day. From this matrix, the total ozone amount for this larger area was used in which each SAGE profile was co-located. Each total ozone value was then multiplied by 1.066 to account for the bias described by Bhartia et al. (1984). To derive an integrated tropospheric ozone amount, the integrated amount of ozone in the stratosphere (i.e., the amount above 100 mb, or above the cloud top) was calculated from each SAGE ozone profile. This quantity was then subtracted from the TOMS total ozone amount. The tropospheric residual values shown in Figure 3 have been summed over the 30°-latitudinal domain (i.e., between 15°N and 15°S) and plotted as a function of longitude. The 5°-longitudinal resolution has been smoothed with a 3-point running filter in which each point is weighted by the number of observations in each 5° interval. Using the running-mean values, each of the 72 points plotted in Figure 3 is comprised of between 7 and 21 concurrent SAGE and TOMS measurements. Without the running mean, each 5°-increment tropospheric residual value would have been obtained from 1 to 13 observations.

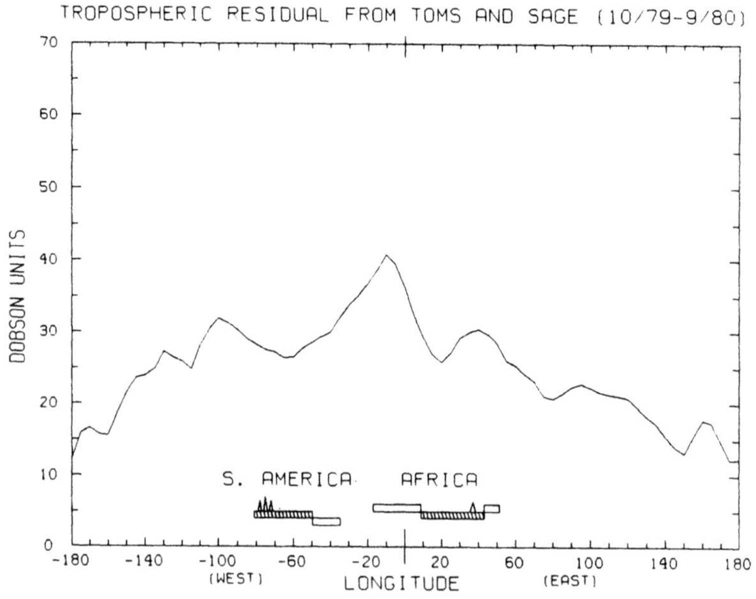

Figure 3. The integrated amount of total ozone found in the troposphere from coincident measurements obtained from the TOMS and SAGE instruments is shown as a function of longitude. See text for details.

The longitudinal domains of both South America and Africa are also shown in the bottom part of Figure 3. The hatched bar depicts where the land masses of each continent straddle the equator; the open bars show where the land mass of each lies between the equator and 15°N (above the hatched bar) and between the equator and 15°S (below the bar).

A distinct longitudinal gradient is seen in Figure 3 with the highest value of 42 D.U. found at 5°W and the lowest values of 12 D.U. at 175°E and 180°E. The second highest value (32 D.U.) is located at 105°W. Assuming that easterly winds prevail at these latitudes, then these two peaks are located slightly downwind of Africa and South America, respectively. In addition, the third highest peak, 31 D.U. at 45°E, coincides with the east coast of Africa.

The lowest residual tropospheric ozone amounts are located in the eastern tropical Pacific Ocean. It should be noted that the MAPS flights in both 1981 (Reichle et al., 1986) and 1984 (Reichle, 1986) measured very low concentrations of CO in this general area. Thus, the strong correlations between total ozone and CO in the Tropics found for the 1981 MAPS data set (Fishman et al., 1986) appear to be similarly valid for 1984. Furthermore, it appears that the relationship between total ozone and CO discussed in Fishman et al. (1986) is likewise valid for the residual tropospheric ozone and CO. It is not, therefore, unexpected that there is a strong positive correlation between total ozone and the residual tropospheric ozone between 15°N and 15°S. The calculated correlation coefficient between these two quantities is +0.835.

4. COMPARISON OF RESIDUAL TROPOSPHERIC OZONE DERIVED FROM TOMS AND SAGE WITH OTHER DATA SETS

In Figure 4, some representative vertical profiles obtained from in situ instrumentation aboard aircraft platforms are presented. Profiles from aircraft observations should not have as many problems associated with them as profiles obtained from ozonesondes at different stations (see discussion by Logan, 1985). Each of these profiles was obtained in the Tropics. The thin solid line (from Fishman et al., 1987) and the measurements indicated by the error bars at 1 and 6 km (Routhier et al., 1980) were obtained over the equatorial Pacific Ocean south of Hawaii. The dashed line (Seiler and Fishman, 1981) is the average of a series of profiles off the west coast of South America, whereas the heavy solid line (Delany et al., 1985; Crutzen et al., 1985) summarizes a set of profiles that are representative of a region influenced by widespread biomass burning. If we make some assumptions about the nature of these profiles between the top of them and 100 mb, we can estimate that the profile over the equatorial Pacific Ocean should have a tropospheric ozone column content of 10-15 D.U. At the other extreme, the composite profile representative of

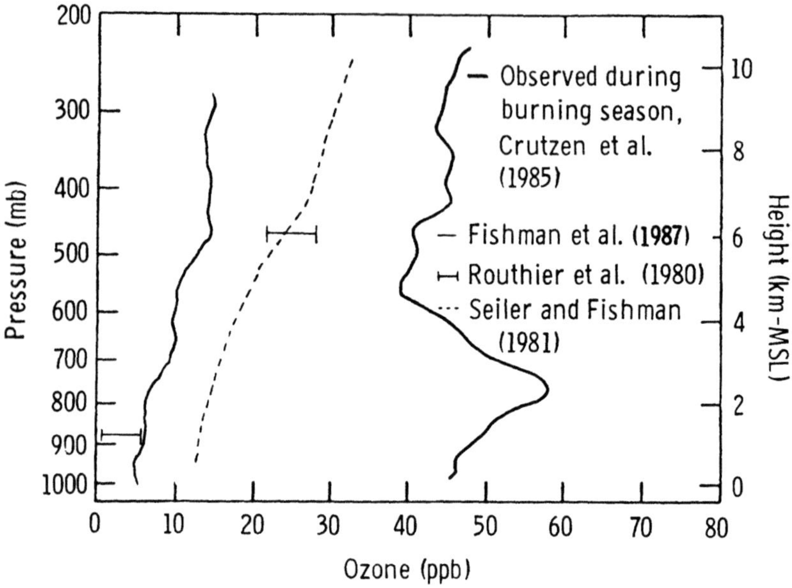

Figure 4. A set of representative tropospheric ozone profiles from the Tropics is shown. The profile depicted by the thin solid line (Fishman et al., 1987) and the concentrations denoted by the error bars at 1 km and 6 km (Routhier et al., 1980) were obtained over the equatorial Pacific Ocean, whereas the solid dashed line (Seiler and Fishman, 1981) is an average of tropical profiles off the Pacific coast of South America. None of these profiles were strongly influenced by the presence of biomass burning. The heavy solid line depicts a set of profiles from Brazil (Crutzen et al., 1985) that had been elevated because of nearby biomass burning.

one which has been influenced by biomass burning should have a tropospheric ozone column content of 40-50 D.U. This range of integrated tropospheric ozone values illustrated by these profiles is consistent with the tropospheric residual ozone range that has been derived from the TOMS and SAGE data sets.

Examination of the integrated amount of ozone between the surface and 100 mb over Natal (Brazil: 6°S, 35°W) shows a range from 25-35 D.U. in February-March to 40-55 D.U. in September-October, when the influence of biomass burning should be most pronounced (Logan and Kirchhoff, 1986). If, from Logan and Kirchhoff's analysis, we assume that an annual average of 35-40 D.U. is representative of the amount of ozone in the troposphere over Natal, then the 31 D.U. at 35°W in Figure 3 appears to be too low. Such a finding, however, is not surprising since the tropospheric residual at 35°W incorporates mostly oceanic area, where the influence of biomass burning should be considerably less than at Natal. In addition, we should expect the tropospheric residual obtained from TOMS data to be underestimated if the model calculations of Klenk et al. (1982) are correct. Nonetheless, despite the caveats involved, the tropospheric ozone residual values that have been computed from these two satellite data sets do appear reasonable in light of the limited amount of in situ data available for comparison. Future studies should be conducted which incorporate other tropical ozone data sets.

5. SUMMARY AND POSSIBLE SIGNIFICANCE

An extension of the study described by Fishman and Larsen (1987) has been presented. We have utilized a set of TOMS total ozone measurements which has been arbitrarily modified to account for the observed bias described by Bhartia et al. (1984) in conjunction with the SAGE data base described in Fishman and Larsen (1987) to derive the longitudinal distribution of integrated tropospheric ozone in the Tropics. The results indicate that a pronounced longitudinal gradient of inferred tropospheric ozone is present with 3-4 times more tropospheric ozone present downwind of Africa than over the pristine regions of the equatorial Pacific Ocean. Furthermore, the integrated amount of residual tropospheric ozone obtained from the use of these two data sets is consistent with the relatively few in situ observations that are available.

It is clear that the feasibility of using the technique described in this study for the purpose of gaining an insight into the global distribution of tropospheric ozone needs to be tested more rigorously. If, however, the results presented herein are verified, then some important implications evolve from these findings. Perhaps the most obvious point is that the source of tropospheric ozone along the African west coast is the largest in the Tropics. This premise is supported by both sets of global CO measurements that are available (Reichle et al., 1986; Reichle, 1986). Comparing the 70 ppbv of ozone at the surface observed by Cros et al. (1987), which very likely is the result of widespread biomass burning in western Africa, with measurements of ozone over the Amazon Basin in Brazil (Andreae et al., 1987; Gregory et al., 1987; Delany et al., 1985; Kirchhoff, et al., 1987; Browell et al., 1987), it may be significant that none of these studies over Brazil observed concentrations as high as those reported by Cros even though widespread biomass burning was observed in Brazil. Furthermore, all of the Brazilian studies indicated that ozone concentrations above the surface were higher than those at the surface. Even the ozonesonde profiles that showed concentrations of more than 100 ppbv in the free troposphere (Logan, 1985; Logan and Kirchhoff, 1986) generally indicated surface ozone concentrations of 25-50 ppbv. Although we recognize the danger involved with comparing such varying data sets in such different regimes, it is possible that the extremely high surface ozone concentrations reported by Cros et al. (1987) may be indicative of the relative source strengths of biomass burning in South America and Africa. Again, however, many more measurements are needed to confirm such speculation.

Another significant implication of the results discussed in this study involves the absolute magnitude of the integrated tropospheric ozone amount that has been derived from the residual values of the TOMS and SAGE data sets. The average value of 25-30 D.U. is comparable to the amount of ozone in the troposphere at northern mid latitudes (Bojkov, 1985; Logan, 1985) even though the concentration of ozone in the northern mid-latitude troposphere is generally higher. The fact that the tropical tropopause is generally located at an altitude of 16-18 km, rather than at an altitude of 8-12 km, as in mid latitudes, compensates for the lower concentrations generally observed at low latitudes. Since stratosphere-troposphere exchange processes occur primarily at mid and high latitudes, other sources of tropospheric ozone must predominate at low latitudes. These sources must be either intra-tropospheric meridional transport from higher latitudes, or in situ photochemical generation within the Tropics. Another consideration is the atmospheric lifetime of tropospheric ozone in the Tropics, compared with its lifetime at other latitudes. Fishman (1985) estimates that the rate of destruction of tropospheric ozone due to deposition is only one-third to one-half the rate of destruction at northern mid latitudes. One the other hand, photochemical calculations indicate that the rate of loss of ozone due to photochemistry is two to three times larger in the Tropics than at northern mid latitudes (Fishman et al., 1979; Logan et al., 1981). The photochemical calculations also suggest that the amount of ozone photochemically destroyed in the troposphere is considerably greater than the amount believed to be lost by deposition. The implication of such calculations is that the sum of the sinks in the tropical troposphere is probably larger than the sum at northern mid latitudes. Therefore, there must be a sizeable in situ photochemical source of tropospheric ozone in the Tropics that may be comparable to, or even larger, than the in situ source at northern mid latitudes. Such a conclusion is likewise consistent with the generalization that the magnitude of the source of ozone-precursor trace gases emitted to the atmosphere by biomass burning is comparable to the magnitude of the source of these gases from fossil fuel combustion (e.g. see Fishman et al., 1986, and references therein).

Lastly, the analysis presented in this study has produced a set of data that is only longitudinally dependent in the Tropics. Because of the relatively few number of concurrent SAGE and TOMS measurements available to us, no attempt has been made to examine the latitudinal or seasonal dependence of the tropospheric residual calculations described here. Future studies should examine the possibility of extracting such information as more satellite data become available.

6. ACKNOWLEDGMENTS

The author thanks J. C. Larsen for his help in processing of the SAGE data and S. D. Johnson for her help in the preparation of the manuscript. This study has benefited from discussions with P. K. Bhartia, A. J. Krueger, and R. D. McPeters of NASA Goddard Space Flight Center. This research has been funded in part by NASA's Global Tropospheric Chemistry Program.

7. REFERENCES

Andreae, M. O. et al., 1987: 'Biomass burning emissions and associated hazo layers over Amazonia.' J. Geophys. Res., (in press).

Bhartia, P. K., K. F. Klenk, C. K. Wong, D. Gordon, and A. J. Fleig, 1984: 'Intercomparison of Nimbus 7 SBUV/TOMS total ozone data sets with Dobson and M83 results.' J. Geophys. Res., **89**, 5227-5238.

Bojkov, R. D., 1985: 'Tropospheric ozone, its changes, and possible radiative effects.' WMO Special Environmental Report, No. 16, World Meteorological Organization, Geneva, pp. 94-127.

Browell, E. V., G. L. Gregory, R. C. Harriss, and V. W. J. H. Kirchhoff, 1987: 'Tropospheric ozone and aerosol distributions across the Amazon Basin.' J. Geophys. Res., (in press).

Connors, V. S., H. G. Reichle, Jr., K. D. Sullivan, and M. R. Helfert, 1986: 'Global observations of biomass burning from space.' Paper presented at Fall 1986 American Geophysical Union Meeting, San Francisco. Abstract in Trans. Amer. Geophys. Union, **67**, 878.

Cros, B., R. Delmas, B. Clairac, J. Loemba-Ndembi, and J. Fontan, 1987: 'Survey of ozone concentration in an equatorial region during the rainy season.' J. Geophys. Res. (in press).

Crutzen, P. J., A. C. Delany, J. Greenburg, P. Haagenson, L. Heidt, R. Lueb, W. Pollock, W. Seiler, A. Wartburg, and P. Zimmerman, 1985: 'Observations of air composition in Brazil between the equator and 20°S during the dry season.' J. Atmos. Chem., **2**, 233-256.

Delany, A. C., P. J. Crutzen, P. Haagenson, S. Walters, and A. F. Wartburg, 1985: 'Photochemically produced ozone in the emission from large-scale tropical vegetation fires.' J. Geophys. Res., **90**, 2425-2429.

Fishman, J., 1985: 'Ozone in the Troposphere.' In Ozone in the Free Atmosphere, R. C. Whitten and S. S. Prasad, eds., Van Nostrand Reinhold, New York, 161-194.

Fishman, J. and J. C. Larsen, 1987: 'Distribution of total ozone and stratospheric ozone in the Tropics: Implications for the distribution of tropospheric ozone.' J. Geophys. Res., (in press).

Fishman, J.,S. Solomon, and P. J. Crutzen, 1979: 'Observational and theoretical evidence in support of a significant in situ photochemical source of tropospheric ozone.' Tellus, **31**, 432-446.

Fishman, J., P. Minnis, and H. G. Reichle, Jr., 1986: 'The use of satellite data to study tropospheric ozone in the tropics.' J. Geophys. Res., **91**, 14,451-14.465.

Fishman, J., G. L. Gregory, G. W. Sachse, S. M. Beck, and G. F. Hill, 1987: 'Vertical profiles of ozone, carbon monoxide, and dew point temperature obtained during GTE/CITE 1, October - November 1983.' J. Geophys. Res., **92**, 2083-2094.

Gregory, G. L., E. V. Browell, and L. S. Gahan, 1987: 'Boundary layer ozone: An airborne survey above the Amazon Basin.' J. Geophys. Res. (in press).

Heath, D. F., A. J. Krueger, H. A. Roeder, and B. D. Henderson, 1975: 'The solar backscatter ultraviolet and total ozone mapping spectrometer (SBUV/TOMS) for Nimbus G.' Opt. Eng., **14**, 323-331.

Kirchhoff, V. W. J. H., E. V. Browell, and G. L. Gregory, 1987: 'Ozone profile measurements in Amazonia.' J. Geophys. Res., (in press).

Klenk, K. F., P. K. Bhartia, A. J. Fleig, V. G. Kaveeshvar, R. D. McPeters, and P. Smith, 1982: 'Total ozone determination from the Backscattered Ultraviolet (BUV) experiment.' J. Appl. Meteorol., **21**, 1672-1684.

Logan, J. A., 1985: 'Tropospheric ozone: Seasonal behavior, trends, and anthropogenic influence.' J. Geophys. Res., **90**, 10,463-10,482.

Logan, J. A., M. J. Prather, S. C. Wofsy, and M. B. McElroy, 1981: 'Tropospheric chemistry: A global perspective.' J. Geophys. Res., 86, 7210-7254.

Logan, J. A. and V. W. J. H. Kirchhoff, 1986: 'Seasonal variations of tropospheric ozone at Natal, Brazil.' J. Geophys. Res., **91**, 7875-7882.

McCormick, M. P., T. J. Swissler, E. Hilsenrath, A. J. Krueger, and M. T. Osborn, 1984: 'Satellite and correlative measurements of stratospheric ozone: Comparison made by SAGE, ECC balloons, chemiluminscent, and optical rocketsondes.' J. Geophys. Res., **89**, 5315-5320.

Reichle, H. G., Jr., 1986: 'The distribution of tropospheric carbon monoxide during October 1984 as measured by a satellite-borne remote sensor.' Paper presented at Fall 1986 American Geophysical Union Meeting, San Francisco. Abstract in Trans. Amer. Geophys. Union, 67, 878.

Reichle, H. G., Jr., V. S. Connors, J. A. Holland, W. D. Hypes, H. A. Wallio, J. C. Casas, B. B. Gormsen, M. S. Saylor, and W. D. Hesketh, 1986: 'Middle and upper tropospheric carbon monoxide mixing ratios measured by a satellite-borne remote sensor during November 1981.' J. Geophys. Res., **91**, 10,865-10,888.

Routhier, F., R. Dennett, D. Davis, E. Danielsen, A. Wartburg, P. Haagenson, and A. C. Delany, 1980: 'Free tropospheric and boundary layer airborne measurements of ozone over the latitude range of 58°S and 70°N.' J. Geophys. Res., **85**, 7307-7321.

Seiler, W. and J. Fishman, 1981: 'The distribution of carbon monoxide and ozone in the free troposphere.' J. Geophys. Res., **86**, 7255-7266.

Woodbury, G. E. and M. P. McCormick, 1986: Zonal and geopotential distributions of cirrus clouds determined from SAGE data.' J. Geophys. Res., **91**, 2775-2786.

STRATOSPHERE-TROPOSPHERE EXCHANGE OF OZONE

G. Vaughan
Physics Dept.
University College of Wales
Aberystwyth
Wales SY23 3BZ

ABSTRACT. The flux of ozone from stratosphere to troposphere is highly episodic, occurring mostly in conjunction with baroclinic instability and the amplification of large-scale troughs in the upper tropospheric flow. Tropopause folding has been studied extensively over the past 30 years, and its salient features are now understood both experimentally and theoretically. Important questions remain, however, about its exact contribution to the cross-tropopause ozone flux. Other exchange mechanisms have been postulated but not studied in detail -e.g. cut-off lows and steady jet streams. Estimates of the global flux of stratospheric ozone into the troposphere derived from radioactivity measurements in the 1960s are in reasonable agreement with recent calculations using general circulation models, but many areas of uncertainty surround both types of estimate.

1. INTRODUCTION

The standard model for the exchange of air between stratosphere and troposphere was first proposed by Brewer (1949), based on observations of extremely low stratospheric humidities. The model calls for an extension of the Hadley circulation through the high, cold tropical tropopause, followed by poleward and downward motion in the subtropical and extratropical stratosphere. The circulation is completed by subsidence in the extratropical stratosphere (particularly in the winter hemisphere) and outflow of stratospheric air into the extratropical troposphere. On the basis of this simple model, the outflow of stratospheric ozone into the troposphere should be greatest during the winter and spring seasons, when the reservoir of ozone in the lower stratosphere is greatest (Dutsch, 1978; Bowman and Krueger, 1985) and the incidence of tropopause folds is also at a maximum (Danielsen, 1968). Confirmation of the basic pattern has come from General Circulation Model (GCM) studies (Mahlman et al, 1980; Kida, 1983; Levy et al, 1985) and from analyses of the removal of radioactive fallout debris (Reiter, 1975; Danielsen, 1968).

I. S. A. Isaksen (ed.), Tropospheric Ozone, 125–135.

The greatest obstacle to an accurate assessment of the flux of ozone from stratosphere to troposphere lies in the episodic nature of the exchange processes. Tropopause folding associated with rapid cyclogenesis was identified by Reed (1955) and Reed and Danielsen (1959), and has since been the subject of many case studies. The dynamics of the folding process, discussed in Section 2, is fairly well understood, but because of the folds' small vertical scale and their association with vigorous small-scale mixing, conventional numerical models (e.g. GCMs or numerical forecast models) have met only limited success in representing them. Thus, extrapolation from a few well-documented case studies to a global total for the mass of ozone delivered to the troposphere in folding events has proved elusive, although some estimates are available (Reiter, 1975; Danielsen and Mohnen, 1977). Recently, attention has bees drawn to cut-off cyclones in the upper troposphere as possible agents of stratosphere-troposphere exchange (Bamber et al, 1984; Hoskins et al, 1985), and there have also been suggestions that the sub-tropical jet streams may play a part (Allam and Tuck, 1984). Quantitative estimates of the ozone flux associated with these mechanisms have not been reported as yet, and it is not known whether they are as important as tropopause folding. Stratospheric and tropospheric air are distinguished by very different values of potential vorticity, P. This scalar quantity is defined as the dot product of the absolute vorticity and the gradient of potential temperature, divided by the density, and is conserved in adiabatic, inviscid flow (Ertel, 1942). An acceptable approximation to P for the scales of length of concern here is (Reiter, 1972, pp45-7):

$$P_\theta = -g \frac{d\theta}{dp} \left(f + \frac{V}{R_\theta} - \left.\frac{dV}{dn}\right|_\theta \right) \qquad \ldots\ldots\ldots\ldots\ldots\ldots \; 1$$

where f is the Coriolis parameter, g the acceleration due to gravity, θ the potential temperature, p the pressure, $\underline{V}$ the wind velocity, R_θ the radius of curvature of the flow and $\underline{n}$ a unit vector perpendicular to $\underline{V}$. The quantity in brackets represents the component of absolute vorticity pependicular to an isentropic surface. Potential vorticity is created in the stratosphere by a vertical gradient in the diabatic heating rate, and destroyed in the lower troposphere. In the lower stratosphere and upper troposphere it is well conserved by individual air parcels and may thus be used as a dynamical tracer. The tropopause corresponds to a marked gradient in P, and indeed may be defined as a value of $P_\theta = 1.6 \times 10^{-6}\, Km^2 kg^{-1} s^{-1}$ (Danielsen and Hipskind, 1980; WMO, 1985, p.153).

Measurements of humidity in the lower midlatitude stratosphere show clearly that this is a transition region between the moist troposphere and the extremely arid stratosphere above about 18km. Cluley and Oliver (1978) and Foot (1984) show a "typical" midlatitude humidity sounding with a smooth decrease in mixing ratio from about 20ppmm at the tropopause (220mb) to about 3ppmm at 140mb. Their data also revealed considerable variability in the lower stratospheric humidity, with several measurements of more than 20ppmm well into the stratosphere. Ozone profiles in this region often display a very marked layered structure (Hilsenrath et al, 1986), suggesting that air of very different origins may be interleaved within it. The high static stability of the stratosphere ensures that vertical mixing is greatly suppressed, but the stratosphere is no more stable than the troposphere to transport and mixing along isentropic surfaces. Large-scale inertia-gravity waves have been identified in the stratosphere (Sidi and Barat, 1986) and have been suggested as a cause for the layered ozone profiles (WMO, 1985, p.218). The laminar structure is not peculiar to the stratosphere: ozone profiles in the troposphere show similar layers (Seiler and Fishman, 1981; Fishman and Seiler, 1983; Fishman et al, 1987), and indeed the tropopause folding process itself is an example of laminar isentropic flow.

These observations of water vapour and ozone demonstrate that tropospheric air must enter the lower midlatitude stratosphere directly, without passing through the Brewer circulation. Data from the LIMS instrument on NIMBUS-7 (WMO, 1985, p.466) confirms this, with clear evidence of elevated water vapour mixing ratios above 100 mb poleward of about 45° in the winter hemisphere. Isentropic transport of tropospheric air into the lower midlatitude stratosphere can occur either from the tropical upper troposphere, probably via one of the subtropical jet streams, or at the exit region of the polar front jet streams (Allam and Tuck, 1984), but very little is known about the extent of such transport. The resulting hybrid nature of lower stratospheric air makes it difficult to use estimates of the mass of air exchanged annually between stratosphere and troposphere (e.g. Reiter, 1975) to deduce the amount of ozone exchanged.

2. TROPOPAUSE FOLDING

The concept of tropopause folding was first introduced by Reed (1955) and Reed and Danielsen (1959), and has been described in detail by Danielsen (1968), Danielsen et al (1970), Danielsen and Mohnen (1977), Danielsen and Hipskind (1980), Danielsen et al (1987) and Shapiro (1974, 1978, 1980). Tropopause folding often accompanies rapid cyclogenesis at the surface and frontogenesis near the tropopause; typically, the process begins when geostrophic confluence of the upper level flow enhances the horizontal temperature gradient. The need to maintain thermal wind balance accelerates the air in the confluence zone and generates a thermally direct ageostrophic circulation across the flow, with descent on the cyclonic side of the jet and ascent on the anticyclonic side. Crucial to the development of the fold is the advective effect of the ageostrophic motions and the sharp gradient in potential vorticity near the tropopause (Hoskins, 1982), which ensure that the descending air is confined to a comparatively small region near the cyclonic side of the jet stream. The process is adiabatic, so the descending stratospheric air must flow southward beneath the jet along the isentropic surfaces which span both the lowest levels of the stratosphere and the troposphere (fig.1). Two-dimensional modelling studies based on the semi-geostrophic equations have succeeded in simulating this phenomenon (Hoskins, 1971, 1972; Keyser and Shapiro, 1986).

The connection between tropopause folding and cyclogenesis arises because large folds occur when frontogenesis occurs upstream of a diffluent trough in the upper level flow. A jet streak, propagating into this trough, develops a downward and southward extension along the

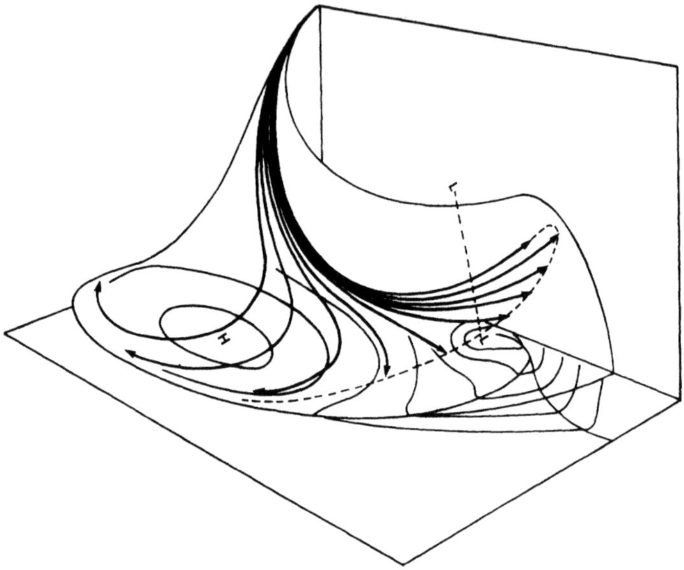

Fig.1. Schematic of trajectories of air extruded in a tropopause fold, relative to the surface pressure pattern. Reproduced from WMO, 1985

boundary of the developing frontal zone. High shear vorticity and high stability characterise the descending stratospheric air in the frontal zone, which must of course maintain its potential vorticity (eqn. 1). Downstream of the trough axis, pronounced divergence occurs in the upper tropospheric flow, both because of the diminishing cyclonic curvature and because of the ageostrophic circulation at the jet exit. This provides the classic conditions for explosive cyclogenesis poleward of the jet axis (Uccellini et al, 1984, 1985; Keyser and Shapiro, 1986).

As the air in the fold descends, mixing occurs with the surrounding tropospheric air at both boundaries (Shapiro, 1978, 1980). Under the influence of frontogenesis, the tendency of stratospheric air to diffuse out of the fold is countered by an inflow of tropospheric air (Danielsen et al, 1987). Thus, as the air descends, it gradually acquires a greater and greater tropospheric component, and the concentrations of tracers of stratospheric origin (ozone and potential vorticity) decrease. This is very well exemplified by the results of Browell et al (1987), who flew an airborne LIDAR instrument perpendicular to the axis of a fold in the South-Western USA. The LIDAR measured profiles of ozone, and of volcanic aerosols of stratospheric origin resulting from the el Chichon eruption, beneath the aircraft's track. The profiles show clearly how the fold descended southward, gradually getting thinner in vertical extent and showing decreasing concentrations of the tracer species. Excellent correspondence was found by Browell et al between the position of this fold and that shown in a potential vorticity cross-section calculated solely from synoptic radiosonde data. Tracer gradients were considerably sharper on the upper (equatorward) side of the fold, showing that the mixing tends to be greatest on the lower, or poleward side, where convection penetrates the stratospheric air.

Because of the inflow of tropospheric air into the fold, and because of the smaller vertical extent of the fold at lower levels, the fold must spread out horizontally. Fig.1, originally presented by Danielsen (1968), shows schematically the trajectories of air parcels during a folding event. The air reaching the lower levels turns anti-cyclonically towards the subtropics, and is not likely to re-enter the stratosphere. However, the air remaining in the fold in the upper troposphere turns cyclonically towards the jet exit region, where the direction of the original ageostrophic circulation is reversed and the air re-enters the lower stratosphere. This is important for two reasons: firstly, it makes for difficulty in assessing accurately the mass of stratospheric air irreversibly transferred into the troposphere in a fold; and secondly, the prospect that some tropospheric air enters the stratosphere at the jet exit means that some of it may re-enter the troposphere in the next folding event downstream (air directly injected into the lowest levels of the midlatitude stratosphere is unlikely to be drawn into the mainstream stratospheric circulation). Thus, although the folding process is well understood both experimentally and theoretically the crucial question of its contribution to the tropospheric ozone budget is not yet solved.

Browell et al (1987) presented a regression analysis between ozone and potential vorticity for the case described above.They found reasonable correlation between the two, (r=0.89) with a slope of about 50 ppbv per $10^{-6} Km^2 kg^{-1} s^{-1}$, and a very small intercept of 0.1 ppbv. This confirms that both potential vorticity and ozone are conserved by small-scale mixing. If these regression parameters prove to be applicable to folds in general, rather than being confined to Browell et al's case study, the opportunity may arise for an accurate assessment of the folds' contribution to the ozone budget; however, the slope is quite different to the 120 ppbv per $10^{-6} Km^2 kg^{-1} s^{-1}$ derived by Danielsen (1968). Numerical forecast models are now approaching the point where they can represent realistically the potential vorticity distributions in a fold; comparisons between aircraft measurements made by the U.K. Meteorological Office beneath a jet stream and a cross-section derived from their fine mesh forecast model were shown in WMO (1985, p.209). The agreement was encouraging, as was a similar comparison with the ECMWF forecast model (D. McKenna, private communication, 1986). Isentropic trajectory analyses within such models could answer several of the questions raised above, such as the fraction of the fold remaining in the troposphere, or the extent of return flow into the stratosphere at the jet exit. A combination of potential vorticity fields and isentropic trajectories, combined with a known relationship between the former and ozone concentrations, would enable considerable progress to be made in calculations of the outflow of stratospheric ozone into the troposphere. A slight note of caution should be sounded: small scale mixing is so intense near jet streams, and near folds in particular, that the assumption of air parcel integrity underlying trajectory calculations may not be sufficiently valid - although considerable success has been achieved with this method in the past (Danielsen, 1968, 1980).

Few case studies are available describing tropopause folds which do not accompany rapid cyclogenesis, although aircraft measurements in upper level frontal zones often encounter dry, ozone-rich air (Briggs and Roach, 1963; Vaughan and Tuck, 1985). A direct ageostrophic circulation does develop at a jet entrance even in the absence of large-scale frontogenesis (Mahlman, 1973), and may distort the tropopause locally so that frontogenesis occurs beneath the jet even though cyclogenesis does not occur downstream (Keyser and Shapiro, 1986). Thus even steady jet streams (such as the subtropical jet streams) could transport ozone into the troposphere. It is not known how reversible such small deformations of the tropopause may be, or how much they contribute to tracer transport.

Tropopause folds, and indeed all deformations of the tropopause,

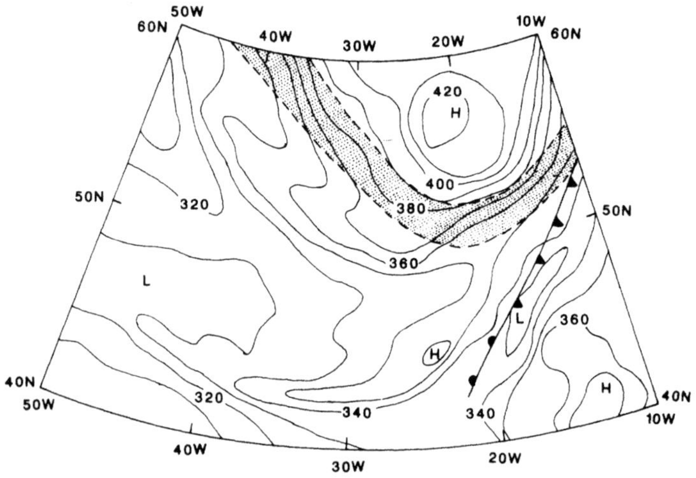

Fig.2. TOMS map for April 16, 1983 over the North Atlantic., showing a tongue of high total ozone curving anticyclonically towards lower latitudes. Contour values are in Dobson units. The 300mb jet stream is shown as the stippled area; its speed exceeded $50ms^{-1}$ on the western side of the trough. The position of the surface front is also shown.

increase the total amount of ozone in a vertical column of atmosphere. These features therefore show up very well on maps of total ozone measured by the TOMS instrument on NIMBUS-7 (Bowman and Krueger, 1985). Typically, a fold shows as a region of enhanced total ozone equatorward and parallel to the jet stream, with a branch curving anticyclonically away from the jet (WMO, 1985, p.229). Fig.2 shows a structure of this kind at the base of a trough in the upper tropospheric flow which was not associated with surface cyclogenesis (although a weak surface front was present).

3. CUT-OFF LOWS

The process by which a trough-ridge system in the upper troposphere intensifies to a blocking pattern has been reviewed by Hoskins et al (1985). Maps of potential vorticity on the 300K isentropic surface show an amplification of the wave, leading to the extrusion of a tongue of high potential vorticity from the polar stratosphere towards lower latitudes.This then develops a cyclonic circulation and becomes detached from the main stratospheric reservoir. On synoptic charts, the vortex is seen to contain cold air and a low tropopause, with stratospheric air constituting a potential vorticity anomaly at pressure levels normally resident in the troposphere at that latitude. This anomaly (together with the cold air over warm sea or land) induces convective instability in the vortex, allowing cumulonimbus clouds to penetrate the stratospheric air and re-establish the tropopause at a higher altitude. Aircraft measurements of hydrocarbons, fluorocarbons and ozone in a cut-off low by Bamber et al (1984) confirmed that the air between 7 and 10 km possessed characteristics intermediate between those of the stratosphere and troposphere, with strong evidence of mixing by convective storms. Hoskins et al (1985) state that cut-off lows can only survive the mixing for about two days, but the example studied by Bamber et al remained a coherent feature for more than two weeks. In addition, TOMS showed smaller-scale patches of high total ozone circulating into the vortex (WMO, 1985, p.205), suggesting an additional exchange mechanism associated with large cut-off lows.

Elevated ozone concentrations at ground-level lasting several days were reported by Derwent et al (1978), and attributed to tropopause folds. It is interesting to note that a cut-off low was present near their measurement site at that time. No explicit studies have been made of the contribution of cut-off lows to stratosphere-troposphere exchange, but they are not uncommon events, and their potential for exchange has been clearly demonstrated.

4. ESTIMATES OF THE CROSS-TROPOPAUSE OZONE FLUX

Early estimates of the ozone flux were based on estimates of the annual mass exchange between stratosphere and troposphere. Atmospheric nuclear explosions between 1958 and 1962 introduced strontium-90 into the stratosphere. This radioactive isotope was a unique tracer for bomb debris, and therefore provided (inadvertently!) an excellent tracer for stratospheric air in the troposphere. Extensive measurements of Sr^{90} in the lower stratosphere and in tropopause folds were combined by Danielsen (1964) with measurements of its surface deposition to derive a mass outflow rate for the northern hemisphere. Danielsen and Mohnen (1977) used this estimate, together with ozonesonde data and the ozone/potential vorticity ratio of Danielsen (1968), to predict an annual outflow of 7.8×10^{10} ozone molecules $cm^{-2}s^{-1}$ in the northern hemisphere, entirely due to tropopause folds. Their estimate, of course, depends on having the correct value for the ratio (see below).

General circulation models have also been used to estimate the transport of ozone from stratosphere to troposphere (Mahlman et al, 1980; Gidel and Shapiro, 1979; Levy et al, 1985). Gidel and Shapiro used model transport of potential vorticity, together with climatological values of zonal mean ozone mixing ratios, to deduce the ozone flux. Effectively, they determined the ratio of potential vorticity to ozone at a number of levels and latitudes and combined these data with potential vorticity fluxes (their data do not support the idea of a constant value for the ratio, but they used model, not observed, potential vorticity). The GFDL group (Mahlman et al, 1981; Levy et al, 1985) represent ozone explicitly in their GCM, but only include chemical processes in the middle stratosphere. Both models predict a greater outflow of ozone in the northern hemisphere than in the southern, by a factor of about 2. Their estimate for the northern hemisphere flux lies between about 4 and 7×10^{10} molecules $cm^{-2}s^{-1}$ - a little smaller than that of Danielsen and Mohnen but in good general agreement. Without explicit representation of tropopause folding and the other exchange processes, the models' success is encouraging, suggesting that the volume of cross-tropopause flow is determined by large-scale processes which are explicitly represented (for instance, the flux of ozone from the middle to the lower stratosphere). However, Levy et al (1985) discuss a number of instances when the model's ozone 'climatology' is quite at variance with the atmosphere. Many of these may be attributed to the lack of chemistry in their model, but they also consider that transport deficiencies play a major role. The apparent agreement in cross-tropopause fluxes calculated by different methods may therefore be illusory, and a more complete picture will have to await a more detailed theoretical treatment.

REFERENCES

Allam, R. J. and A. F. Tuck. 'Transport of water vapour in a stratosphere-troposphere general circulation model. 2. Trajectories' Quart. J. Roy. Meteorol. Soc., **110**, 357-392, 1984.

Bamber, D. J., P. G. Healey, B. M. Jones, S. A. Penkett, A. F. Tuck and G. Vaughan. 'Vertical profiles of tropospheric gases - chemical consequences of stratospheric intrusions' Atmos. Environ., **18**, 1759-1766, 1984.

Bowman, K.,P. and A. J. Krueger. 'A Global Climatology of Total Ozone from the Nimbus 7 Total Ozone Mapping Spectrometer' J. Geophys. Res., **90**, 7967-7976, 1985.
Brewer, A. W. 'Evidence for a world circulation provided by the measurements of helium and water vapour distribution in the stratosphere' Quart. J. Roy. Meteorol. Soc., **75**, 351-363, 1949.
Briggs, J. and W. T. Roach. 'Aircraft observations near jet streams' Quart. J. Roy. Meteorol. Soc., **89**, 225-247, 1963.
Browell, E. V., E. F. Danielsen, S. Ismail, G. L. Gregory and S. M. Bleck. 'Tropopause fold structure determined from airborne lidar and in situ measurements' J. Geophys. Res., **92**, 2112-2120, 1987.
Cluley, A. R. and M. J. Oliver. 'Aircraft measurements of the humidity in the low stratosphere over southern England 1972-1976' Quart. J. Roy. Meteorol. Soc., **104**, 511-526, 1978.
Danielsen, E. F. 'Project Springfield report' Defence Atomic Support Agency 1517, Washington D. C., 1964.
Danielsen, E. F. 'Stratospheric-tropospheric exchange based on radioactivity, ozone and potential vorticity' J. Atmos. Sci., **25**, 502-518, 1968.
Danielsen, E. F. 'Stratospheric source for unexpectedly large values of ozone measured over the Pacific ocean during Gametag, August 1977' J. Geophys. Res., **85**, 401-412, 1980.
Danielsen, E. F., R. Bleck, J. Shedlovsky, A. Wartburg, P. Haagensen and W. Pollock. 'Observed distribution of radioactivity, ozone and potential vorticity associated with tropopause folding' J. Geophys. Res., **75**, 2353-2361, 1970.
Danielsen, E. F. and R. S. Hipskind. 'Stratospheric-tropospheric exchange at polar latitudes in summer' J. Geophys. Res, **85**, 393-400, 1980.
Danielsen, E. F., R. S. Hipskind, S. E. Gaines, G. W. Sachse, G. L. Gregory and G. F. Hill. 'Three-dimensional analysis of potential vorticity associated with tropopause folds and observed variations of ozone and carbon monoxide' J. Geophys. Res., **92**, 2103-2111, 1987.
Danielsen, E. F. and V. A. Mohnen. 'Project Dustorm report: Ozone transport, in situ measurements and meteorological analyses of tropopause folding' J. Geophys. Res, **82**, 5867-5877, 1977.
Derwent, R. G., A. J. Eggleton, M. L. Williams and C. A. Bell. 'Elevated ozone levels from natural sources', Atmos. Environ., 12, 2173-2177, 1978.
Dutsch, H. U. 'Vertical Ozone distribution on a global scale' Pageoph., **116**, 511-529, 1978.
Ertel, H. 'Ein neuer hydrodynamischer wirbelsatz', Meteor. Z., **59**, 277-281, 1942.
Fishman, J., G. L. Gregory, G. W. Sachs, S. M. Beck and G. F. Hill. 'Vertical profiles of ozone, carbon monoxide, and dew-point temperature obtained during GTE/CITE 1, October - November 1983' J. Geophys. Res., **92**, 2083-2094, 1987.
Fishman, J. and W. Seiler. 'Correlative nature of ozone and carbon monoxide in the troposphere: Implications for the tropospheric ozone budget' J. Geophys. Res., **88**, 3662-3670, 1983.
Foot, J. S. 'Aircraft measurements of the humidity in the lower strato-

sphere from 1977 to 1980 between 45N and 65N' Quart. J. Roy. Meteorol. Soc., **110**, 303-320, 1984.
Gidel, L. T. and M. A. Shapiro. 'General circulation model estimates of the net vertical flux of ozone in the lower stratosphere and the implications for the tropospheric ozone budget' J Geophys. Res., **85**, 4049-4058, 1980.
Hilsenrath, E., W. Attmannspacher, A. Bass, W. Evans, R. Hagemeyer, R.A. Barnes, W. Komhyr, K. Mauersberger, J. Mentall, M. Proffitt, D. Robbins, S. Taylor, A. Torres and E. Weinstock. 'Results from the Balloon Ozone Intercomparison Campaign (BOIC)' J. Geophys. Res., **91**, 13137-13152, 1986.
Hoskins, B. J. 'Atmospheric frontogenesis: some solutions' Quart. J. Meteorol. Soc., **97**, 139-153, 1971 .
Hoskins, B. J. 'Non-Boussinesq effects and further development in a model of upper tropospheric frontogenesis' Quart. J. Roy. Meteorol. Soc, **98**, 532-541, 1972.
Hoskins, B. J. 'The mathematical theory of frontogenesis' Ann. Rev. Fluid Mech., **14**, 131-151, 1982.
Hoskins, B. J., M. E. McIntyre and A. W. Robertson. 'On the use and significance of isentropic potential vorticity maps' Quart. J. Roy. Meteorol. Soc., **111**, 887-946, 1985.
Keyser, D. and M. A. Shapiro. 'A review of the structure and dynamics of upper level frontal zones' Mon. Weather Rev., **114**, 452-499, 1986.
Kida, H. 'General circulation of air parcels and transport characteristics derived from a hemispheric GCM. Part 2. Very long-term motion of air parcels in the troposphere and stratosphere' J. Meteorol. Soc. Japan, **61**, 510-523, 1983.
Levy, H. B., J. D. Mahlman, W. J. Moxim and S. Liu. 'Tropospheric ozone: the role of transport' J. Geophys. Res., **90**, 3753-3771, 1985.
Mahlman, J. D. 'On the maintenance of the polar front jet stream' J. Atmos. Sci., **30**, 544-557, 1973.
Mahlman, J. D., H. B. Levy and W. J. Moxim. 'Three-dimensional tracer structure and behaviour as simulated in two ozone precursor experiments', J. Atmos. Sci., 37, 655-685, 1980.
Reed, R. J. 'A study of a characteristic type of upper level frontogenesis' J. Meteorol., 12, 226-237, 1955.
Reed, R. J. and E. F. Danielsen. 'Fronts in the vicinity of the tropopause' Arch. Met. Geophys. Bioklim., 11, 1-17, 1959.
Reiter, E. R. 'Atmospheric transport processes. Part 3: Hydrodynamic tracers', USAEC Report TID-25731, Colorado State University, Fort Collins, Colorado, 1972.
Reiter, E.R. 'Stratospheric-tropospheric exchange processes' Rev. Geophys. Space Phys., **13**, 459-474, 1975.
Seiler, W. and J. Fishman. 'The distribution of carbon monoxide and ozone in the free troposphere' J. Geophys. Res., **86**, 7255-7265, 1981.
Shapiro, M. A. 'A multiple-structured frontal zone - jet stream system as revealed by meteorologically instrumented aircraft' Mon. Weather Rev., **102**, 244-253, 1974.
Shapiro, M. A. 'Further evidence of the mesoscale and turbulent structure of upper jet stream - frontal zone systems' Mon. Weather Rev., **106**, 1011-1111, 1978.

Shapiro, M. A. 'Turbulent mixing within tropopause folds as a mechanism for the exchange of chemical constituents between the stratosphere and troposphere' J. Atmos. Sci., 37, 994-1004, 1980.
Sidi, C. and J. Barat. 'Observational evidence of an inertial wind structure in the stratosphere' J. Geophys. Res., **91**, 1209-1218, 1986.
Uccellini, L. W., P. J. Kocin, R. A. Petersen, C. H. Wash and K. F. Brill. 'The President's day cyclone of 18-19 February 1979: Synoptic overview and analysis of the subtropical jet streak influencing the precyclogenetic period' Mon. Weather Rev., **112**, 31-55, 1984.
Uccellini, L. W., D. Keyser, K. F. Brill and C. H. Wash. 'The President's day cyclone of 18-19 February 1979: Influence of upstream trough amplification and associated tropopause folding on rapid cyclogenesis' Mon. Weather Rev., **113**, 962-988, 1985.
Vaughan, G. and A. F. Tuck. 'Aircraft measurements near jet streams' in Atmospheric Ozone, ed. C. S. Zerefos and A. Ghazi, D. Reidel, 1985.
WMO. 'Atmospheric Ozone 1985' WMO Report no. 16, World Meteorological Organisation, Case Postale No. 5, Geneva.

MEASUREMENTS OF OZONE AT TWO BACKGROUND STATIONS IN FINLAND

S. M. Joffre, H. Lättilä, H. Hakola, P. Taalas and P. Plathan
Finnish Meteorological Institute
Air Quality Department
Sahaajankatu 22E
00810 Helsinki - Finland

ABSTRACT. We present preliminary results of 14 months of ozone observations carried out on a small island in the Baltic Sea. Gross statistics from the data indicate a typical maritime environment with small diurnal amplitude and high average concentrations. Comparison with a few months' observations at a site in Central Finland confirms the importance of different loss mechanisms at the surface. Enhanced ozone concentrations produced by either long-range transport, photochemical processes or stratospheric intrusion were identified.

1. INTRODUCTION

Concern about the effects of ozone in our environment has launched monitoring programmes in many countries. The complexity and variability of the physical/chemical processes acting in the production and destruction of ozone are important obstacles to our understanding of ozone behaviour. Local effects are embedded within larger meso-, synoptical and global scale patterns, making it important to study ozone concentration variations in a variety of environments.

We present here some preliminary results of about one year of ozone measurements at a northern ($\sim 60^\circ N$) maritime site in the Finnish archipelago. They will also be briefly compared to concurrent measurements at a continental site in Central Finland. Interpretation of the data based both on meteorological and a few chemical measurements will be presented for long-term averages and episodic occurrence.

2. MEASUREMENTS: LOCATION AND TECHNIQUES

In July 1985 the Finnish Meteorological Institute started measurements of ozone on the island of Utö ($59^\circ 47'N$, $21^\circ 23'E$, height 7 m), located about 80 km southwest of mainland Finland. The location of this rocky treeless island on the southern margin of the Åland Archipelago is favourable for picking up situations with air masses unaffected by anthropogenic emission for hundreds of kilometres. Utö is a EMEP-station.

I. S. A. Isaksen (ed.), Tropospheric Ozone, 137–146.

Ozone measurements were started in summer 1986 at another EMEP-site at Ähtäri in Central Finland (62°33'N, 24°13'E, 162 m). The area is densely forested but with some cultivated clearings. Measurements of NO_2 were also started at both stations. The measurement programme is shown in Table 1, while the location of the stations with the emission grid for northern Europe is shown in Fig. 1.

Ozone measurements were performed using an ozone monitor 1003-RS manufactured by Environnement SA under DASIBI license. It is based on UV-radiation absorption by ozone at 240 nm. The sensors used at Utö and

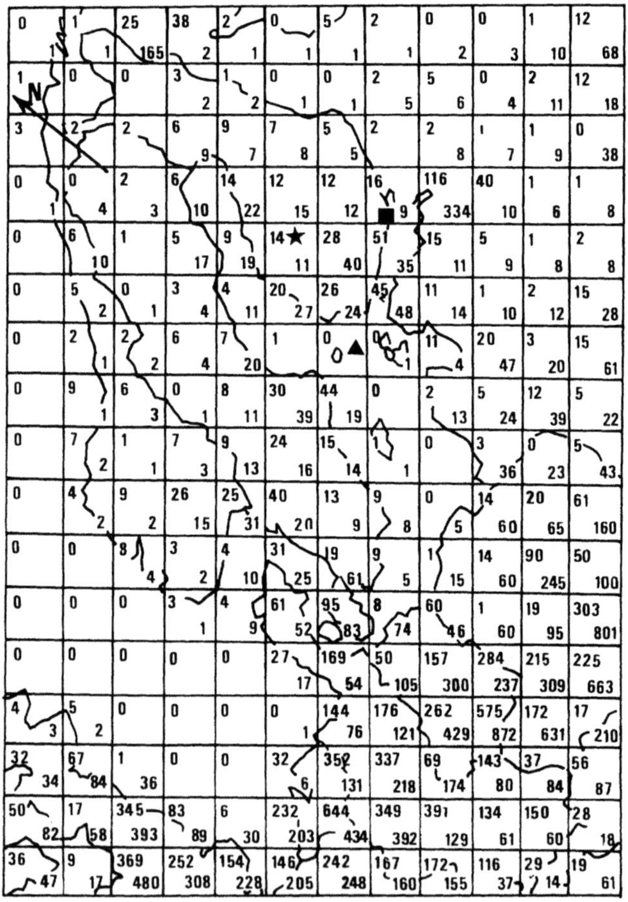

Fig.1. *Location of the Finnish EMEP-stations: (▲) Utö, (★) Ähtäri, and (■) Virolahti. The map is overlayed with the EMEP-grid with NO_x-emission for 1985 (upper-left value) and SO_2-emission for 1980 (lower-right value).*

Table 1. *Measurement programme at the two Finnish stations*

		J F M A M J J A S O N D 85	J F M A M J J A S O N D 86	J F M A M J 87
Utö	O_3	←——→ (Jul–Oct)	←—————— (from Feb)	———————
	NO_2		←——→ (Jul–Dec)	→ (Jan) ←—— (from Mar)
Ähtäri	O_3		←—— (from Aug)	———————
	NO_2			←—— (from Mar)

Ähtäri have been intercalibrated against other sensors used in the Nordic countries so that results are intercomparable. The measurement accuracy is better than 5 μg/m^3. The presence of a diesel generator (on the northern side of the site) has introduced some perturbations into the ozone measurements. These effects were eliminated from the results by constraining the relative standard deviation to be below 14%.

Measurements of NO_2 were performed using an automated version of the Salzman method. The resolution is 0.2 μg/m^3 for one-hour sampling periods. The Nordic intercalibration of NO_2-sensors did not allow firm conclusions due to the large NO_2 concentrations used and the wide scatter between the measuring methods.

3. RESULTS

3.1. Variability of ozone concentration

Ozone concentration variability at Utö and Ähtäri is shown in Fig.2. No definite conclusions can be drawn from this short measurement period - however, we observe a concentration maximum in July and a minimum in December. A secondary maximum in March was due to a strong episode.

Although we do not have measurements for the whole year from the site at Ähtäri, it appears that in autumn ozone concentrations there are significantly smaller than at Utö. This may be due to the slow deposition of ozone to the sea surface around Utö (e.g. a dry deposition velocity of 0.01 cm/s has been reported by Wesely et al., 1981) and, on the other hand, to natural emission of NO from cultivated land around the site at Ähtäri (e.g., Crutzen, 1983). In this connection, it is interesting to notice that ozone concentrations at Ähtäri increase from September to November. This may be due to the decrease in dry deposition as the stomata of plants close, but the decreasing trend of biogenic nitrogen oxide emission as the air temperature decreases (Williams et al., 1987) might also play some rôle.

Due to its photochemical origin and to meteorological conditions, ozone concentrations show a strong daily variation, this pattern being limited to within the atmospheric boundary layer (ABL) (e.g., Garland and Derwent, 1979). In Fig. 3 we show the diurnal cycle at Utö and Ähtäri for the autumn period with classification of the data according

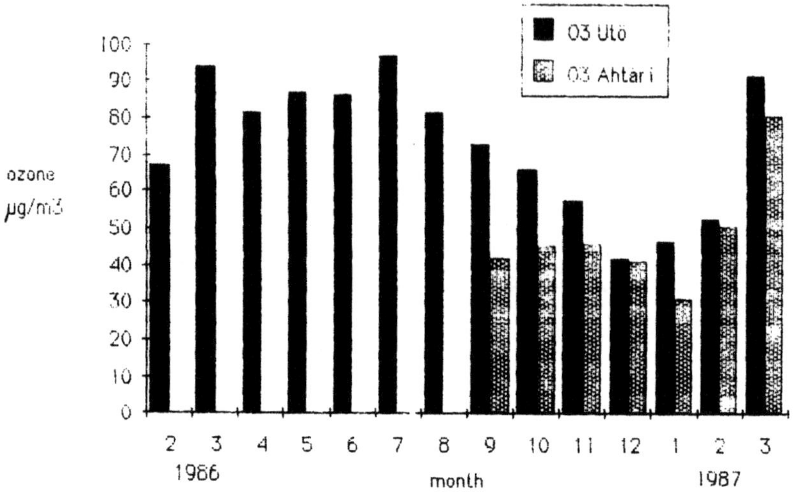

Fig. 2 *Monthly average of ozone concentration at Utö (black columns) and Ähtäri (grey columns) since February 1986. The standard deviations of each column are between 10 and 20* $\mu g/m^3$.

to wind speed. No clear daily trend is apparent at Utö due to very weak diurnal variability in the maritime environment (sea surface temperature and ABL height). The same conclusions were reached by Galbally et al. (1986) from the data of Cape Grim (Tasmania). Typically, in autumn, the marine ABL is convectively mixed. Mean values are rather high due to the low deposition and the probable downward influx of tropospheric ozone into the ABL via entrainment effects. This mixed case displays a maximum in late afternoon.

On the other hand, the diurnal cycle of ozone concentration in the weak wind case has the peculiarity of a maximum at night, although there is a secondary maximum at noon. It is unclear whether this nocturnal maximum is due to a meteorological or chemical mechanism.

During the OECD comparison of summer 1985 (Grennfelt et al., 1987), it was found that the ozone concentration at Utö followed the same diurnal pattern as at Westerland (German Bight), the only other truly maritime station in the comparison. All other stations had quite a different cycle with a marked maximum in the afternoon.

In contrast, for the same autumn period at Ähtäri, the diurnal cycle is much clearer, particularly at low wind speeds (Fig. 3). This is connected with the diurnal variability of the intensity of turbulent fluctuations at continental sites. The frequent nocturnal low-wind situations together with a shallow ABL result in a removal of ozone to vegetation which is quicker than its transport downward from aloft. It has been observed that the wind speed and ozone concentration are well correlated. Obviously, the diurnal cycle is still more marked in summer time, whereas in wintertime it is nearly absent. This latter feature is also enhanced

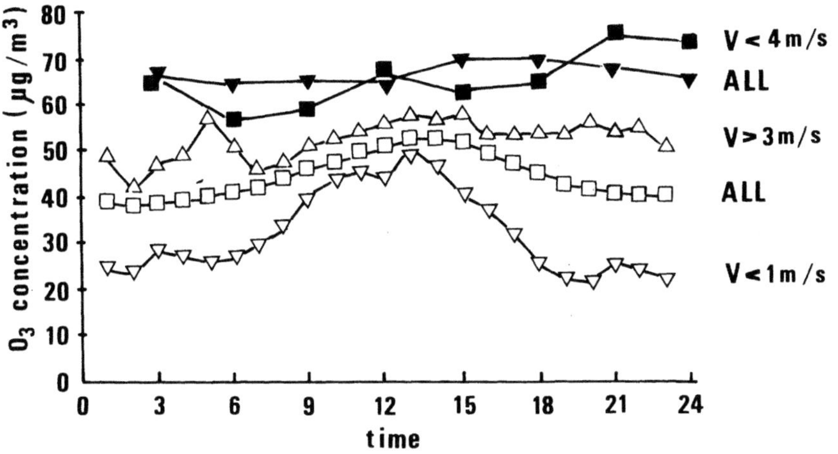

Fig. 3 *Diurnal variation of ozone concentration at Utö (full symbols) and Ähtäri (open symbols) for the period 5.09-1.12. 1986 for different wind speed classes.*

by the snow cover at these latitudes.

Figure 4a shows the monthly average of the hour of occurrence of maximum and minimum ozone concentrations at Utö and Ähtäri. Figure 4b shows the variation of the monthly average of the diurnal cycle at both locations. It will be noticed that the time of the maximum is less variable (with the exception of November and December 1986) than the time of the diurnal minimum. The annual cycle of the diurnal amplitude is also clear.

3.2. Correlations between high ozone concentrations and other parameters

We selected from the whole database sunlit situations with ozone concentrations larger than 100 µg/m^3 and solar elevation angle larger than 15°. It appears that such situations occur preferentially when NO_2 concentrations are 7-9 µg/m^3 (more than 50% of the cases). On the other hand, taking all observations irrespective of ozone episodes, nitrogen dioxide concentrations span over a wider spectrum (80% in the range 1-7 µg/m^3) having a maximum of occurrence at 5 µg/m^3 (20%). Thus, episodic high O_3 concentrations imply increased NO_2 concentrations. This would imply photochemical ozone formation in long-range transported plumes containing nitrogen oxides and hydrocarbons or may reflect the role of nitrogen oxides in the photochemical process.

Next, we compare the frequency of occurrence of high O_3 concentrations (>100 µg/m^3) with respect to meteorological parameters to the corresponding frequency distribution involving all ozone measurements. We find that there is no significant difference for wind speed and relative humidity. On the other hand, high ozone concentrations are favoured when

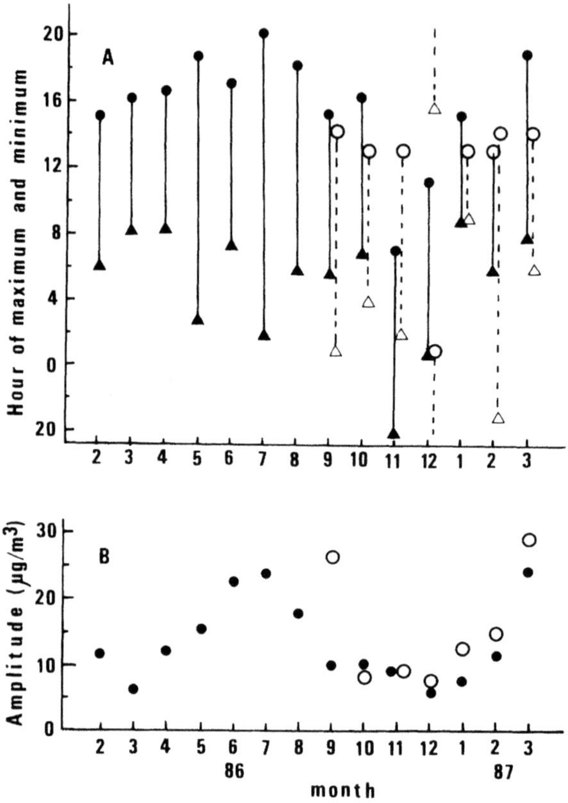

Fig. 4 *(a) monthly average of the time of occurrence of the maximum (circles) and minimum (triangles) at Utö (full symbol) and Ähtäri (open symbol), respectively; and (b) monthly average of the diurnal amplitude in ozone concentration at the same sites.*

the wind direction is in the range 120°-210°, which naturally reflects air masses coming from European industrialized areas. The influence of air temperature on O_3 concentrations appears clearly, with the occurrence of high concentrations at the highest temperatures (> 14°C).

3.3 Ozone episodes

In this section we shall concentrate on three shorter-scale events of O_3 concentration increase involving different mechanisms for ozone production at ground level.

One major source of background tropospheric ozone is stratospheric intrusion (e.g. Singh et al., 1978). Figure 5 shows the time dependence (5-10 May 1986) of O_3 concentration at Utö together with measurements of

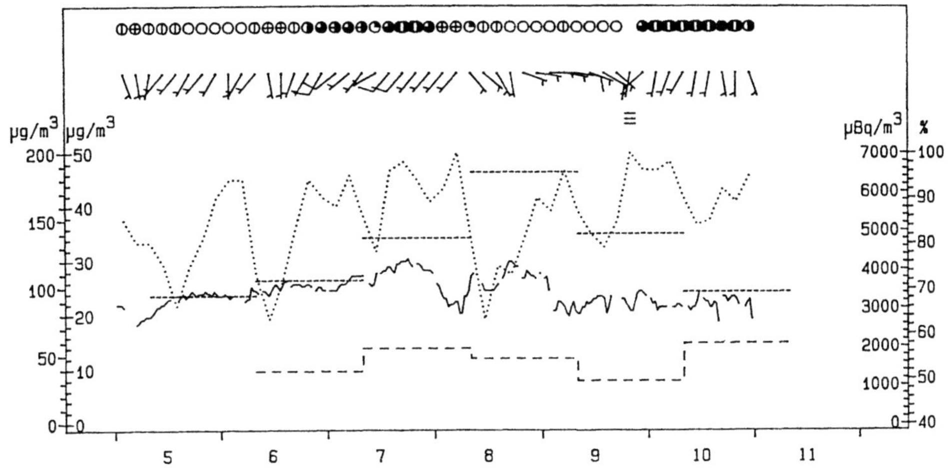

Fig. 5 *Time dependence of ozone concentration (continuous line) during the period 5-11 May 1986, together with NO_3^- + HNO_3 (long dash), ^7Be (short dash) and relative humidity (dotted line). The total cloudiness and the wind vector are shown at the top of the figure.*

a stratospheric tracer ^7Be. We note the slow increase of ozone while the ^7Be concentration is doubled. Concomitant measurements of NO_3^- + HNO_3 show a stationary behaviour, i.e. there is no indication of long-range transport of polluted air. Inspection of radiosonde profiles from Stockholm on May 8th shows a thin, very dry layer centred at 900 mb, which may indicate air of stratospheric origin (e.g. Mukammal et al., 1985). Danielsen and Mohnen (1977) have shown that spring is a period of maximum occurrence of stratospheric intrusions due to enhanced baroclinity. Utö was situated approximately south of an active low pressure system and on the north-west side of a high centre.

A second interesting episode occurred between 21-28 July 1986 when high ozone concentrations at Utö at the beginning of the period, probably due to long-range transport of polluted air from Great Britain, Benelux and Germany, were followed by a decrease (concentrations halved) due to convective activity and wet scavenging of nitrates and nitric acid. Finally a second rise of ozone concentration occurred from July 27 due to photochemical production (see Fig. 6). The first maximum period with $[O_3]$ rising to 126 $\mu g/m^3$ on July 21 occurred during the night so that ozone was probably produced elsewhere, in the plume of western European industrialized areas (NO was converted to NO_2 by other species, mainly hydrocarbons). The same pattern continued until July 23 with early morning ozone minima (due to dark reactions of NO_x and ozone) superimposed on the continuous high level of ozone ($\sim$ 130 $\mu g/m^3$). The sky was mostly overcast at Utö during this period.

On July 24 the warm front and on July 25 the cold front associated

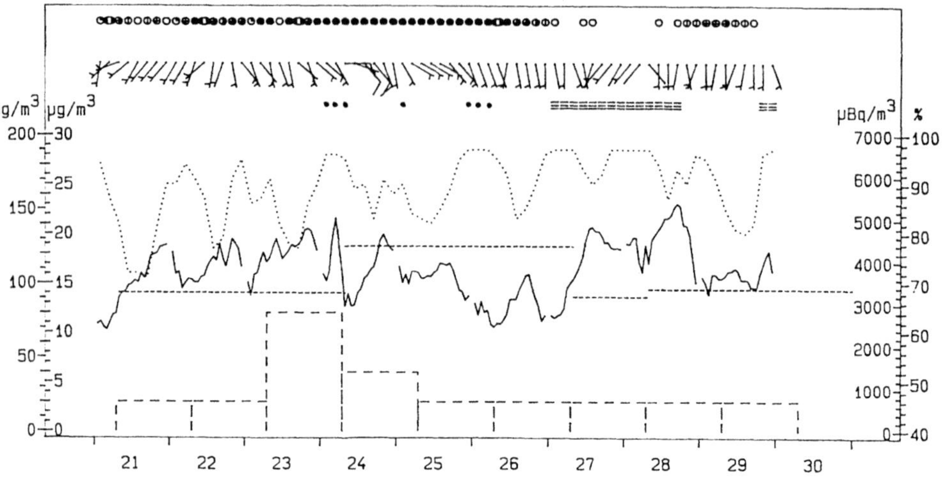

Fig. 6 *Time dependence of ozone concentration during the period 21-30 July 1986. The explanation of the symbols is as in Fig. 5 except $SO_4^=$ concentrations are shown as the long-dashed line (the 0-30 μg/m³ scale).*

with earlier cloudiness swept over Utö and brought rain. The ozone concentration decreased to 62 μg/m³. Thick clouds decreased UV-radiation penetration into the lowest layers of the atmosphere and moreover, cloud activity enhanced vertical mixing and the consequent dilution of O_3 concentration within the ABL. During the next days, July 27 and 28, a high-pressure zone formed over Europe with low cloudiness, favourable for photochemical activity resulting in a new rise of ozone concentrations up to 154 μg/m³.

Finally, we present a sequence of concomitant O_3 and NO_2 measurements at Utö (14-18 December 1986) showing the clear negative correlation between the concentrations of these two gases (Fig. 7). At the beginning of the period (December 14-15) clean air masses arrived from the North Atlantic or Arctic Sea ($[NO_2] \sim 2\text{-}4$ μg/m³ and $[O_3] \sim 55$ μg/m³). On December 16 the air mass had travelled over the Western Soviet Union before reaching Utö. This is reflected in the rising concentration of NO_2 and particulate sulphate. At the same time O_3 concentrations decreased.

4. CONCLUSION

Analysis of 14 months of ozone measurements at a northern maritime site and interpretation of the results have shown that the maritime environment has a marked influence on ozone behaviour when compared to a shorter data sequence from Central Finland.

Detailed meteorological interpretation of the data provides a rather satisfactory qualitative interpretation of ozone variability in spite of

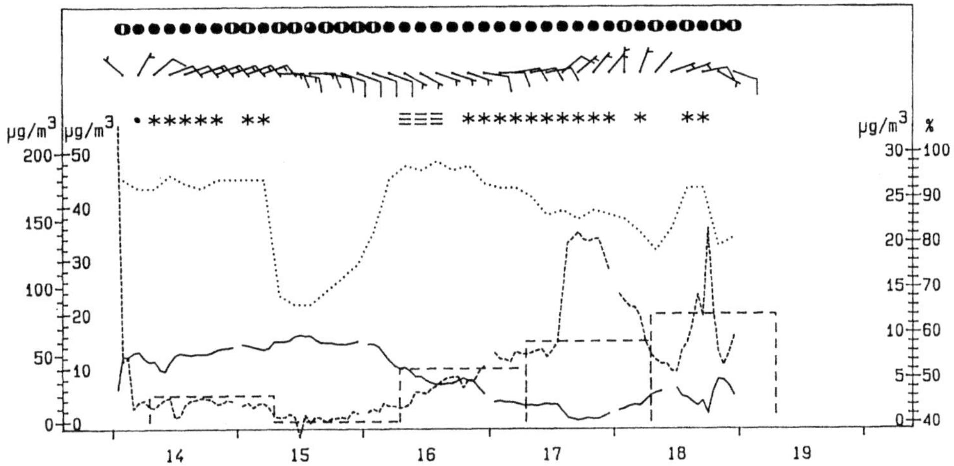

Fig. 7 *Time dependence of O_3 and NO_2 (short-dashed curve) during the period 14-19 December 1986, together with daily concentration of aerosol $SO_4^=$ (long-dashed line), humidity (dotted curve) and weather characteristics.*

our limited chemical measurement programme.

Extension of the chemical species measurements (NO and hydrocarbons) together with improved trajectory calculations (3-D) and a more complete description of the lower atmosphere (ABL height and stability) should enable a still better understanding of O_3 behaviour and provide a potential for predicting adverse ozone concentrations.

ACKNOWLEDGEMENTS

This work was funded by the Acidification Research Project of the Ministry of Agriculture and the Ministry of the Environment.

REFERENCES

Crutzen, P.J., 1983:'Atmospheric interactions - Homogeneous gas reactions of C, N, and S containing compounds'. In *The Major Biogeochemical Cycles and their Interactions* (B. Bolin & R.B. Cook Eds.), John Wiley, New York.

Danielsen, E.F. and V.A. Mohnen, 1977: 'Project Dustorm report: ozone transport, *in-situ* measurements, and meteorological analyses of tropopause folding'. *J. Geophys. Res.* 82, 5867-5877.

Galbally, I.E., A.J. Miller, R.D. Hoy, S. Ahmet, R.C. Joynt and D. Attwood, 1986: 'Surface ozone at rural sites in the Latrobe valley and Cape Grim, Australia'. *Atmos. Environ.* 20, 2403-2422.

Garland, J.A. and R.G. Derwent, 1979: 'Destruction at the ground and the diurnal cycle of ozone and other gases'. *Quart. J. R. Met. Soc.* **105**, 169-183.

Grennfelt, P., J. Saltbones and J. Schjoldager, 1987: 'Oxidant data collection in OECD-Europe 1985-87 (Oxidate)'. NILU-*Report* **22/87**, 109 pp.

Mukammal, E.I., H.H. Neumann and T.R. Nichols, 1985: 'Some features of the ozone climatology of Ontario, Canada and possible contributions of stratospheric ozone to surface concentrations'. *Arch. Met. Geoph. Biocl.* **A34**, 179-211.

Singh, H.B., F.L. Ludwig and W.B. Johnson, 1978: 'Tropospheric ozone: concentrations and variabilities in clean remote atmospheres'. *Atmos. Environ.* **12**, 2185-2196.

Wesely, M.L., D.R. Cook and R.M. Williams, 1981: 'Field measurement of small ozone fluxes to snow, wet bare soil, and lake water'. *Bound.-Layer Meteor.* **20**, 459-471.

Williams, E.J., D.D. Parrish and F.C. Fehsenfeld, 1987: 'Determination of nitrogen oxide emissions from soils: results from a grassland site in Colorado, United States'. *J. Geophys. Res.* **92**, 2173-2179.

OZONE-CLIMATE INTERACTIONS ASSOCIATED WITH INCREASING ATMOSPHERIC TRACE GASES

Wei-Chyung Wang, Nien Dak Sze, Gyula Molnar,
Malcolm Ko and Steve Goldenberg,
Atmospheric & Environmental Research, Inc.
840 Memorial Drive
Cambridge, MA 02139
USA

ABSTRACT. Study of climatic effects of increase in concentration of atmospheric trace gases has so far focused on the direct greenhouse warming. The indirect effect on climate of the seasonal and latitudinal O_3 changes associated with trace gases has not been explored. Here, we use simple 1- and 2-D models to demonstrate that changes in the O_3 vertical distribution, in particular in the middle and upper troposphere, could trigger a response in meridional heat flux with subsequent feedback effect on climate. Discussion on the effects of trace gases and their induced O_3 changes on future climate on the decadal time scale is also presented.

1. INTRODUCTION

Observations in recent years have clearly indicated that concentration of atmospheric trace gases CFCs, CH_4, N_2O, and CO_2 has been increasing. It is anticipated that the trend of increases will continue (cf. DoE, 1985; WMO, 1986). Increases of these radiatively-and chemically-important trace gases may have climate implications both directly through perturbation to the Earth's radiation budget and indirectly through changes of atmospheric O_3 distribution. Calculations from simple 1-D radiative-convective-photochemical models suggest that in the next few decades the combined effects of increases of trace gases and their induced changes of atmospheric O_3 could be comparable in magnitude to that caused by CO_2 increases (see reviews by Wang et al., 1986; Ramanathan et al., 1987). In addition, the stratospheric cooling caused by O_3 depletion resulting from increases of trace gases may be also comparable tc that caused by CO_2 increases.

I. S. A. Isaksen (ed.), Tropospheric Ozone, 147–159.

Recently, 2-D climate models have been used to study the climatic effects associated with increases of trace gases. Wang and Molnar (1985) have used an annual coupled high and low latitude radiative-dynamical model to study the direct greenhouse effect. The model calculated surface warming due to increases of trace gas concentrations is substantially smaller than the value calculated from the 1-D model using the commonly-used 6.5 °C km^{-1} critical lapse rate. The smaller value can be attributed largely to the effects of more realistic simulations of the different temperature (lapse rate) regimes in high and low latitudes. In the low latitudes the lapse rate is governed by the moist-adiabatic processes whereas in the high latitudes it is strongly affected by baroclinic processes. Compared to feedbacks in conventional simulations, the feedbacks in these simulations are more effective in reducing the magnitude of increased downward thermal radiation directed toward the surface due to increases of trace gas concentrations through the increased vertical sensible and latent heat transport. Consequently, these processes provide a stronger negative feedback effect and reduce the magnitude of the surface warming calculated from the 1-D models using 6.5 °C km^{-1} critical lapse rate.

Two dimensional photochemical models have also been used to study the changes of stratospheric O_3 due to trace gases (cf. WMO, 1986). The results indicate that the calculated changes in O_3 distributions due to increases of trace gases are a strong function of latitudes and seasons. The changes of O_3, particularly along the horizontal direction, could yield different effects on surface temperature at different latitudes and seasons.

In this paper, we present results of model studies of O_3-climate interactions associated with increases of trace gases, in particular the CFCs. The role of O_3-climate interactions is demonstrated in the sensitivity calculations by comparing the trace gases' surface warming effects with and without O_3 changes. We also discuss the future climatic effects due to trace gases and their induced O_3 changes.

2. MODELS

To illustrate the importance of O_3-climate interactions associated with trace gases, we use three climate models: (1) the 1-D model, (2) the 2-D annual model of the northern hemisphere, and (3) the 2-D seasonal global model. The interactions will be examined by comparing the surface warming effects between model results with and without including the O_3 changes. The important point here in the model studies is that the three models use the same radiation parameterization so that the major differences between the models are whether or not the latitudinal and seasonal characteristics are incorporated.

The 1-D models have been used widely in trace gases' studies (see Wang et al., 1986 for a review) and the 2-D annual model has been described in Wang and Molnar (1985). Below we described briefly the 2-D seasonal model.

The 2-D seasonal model, shown in Figure 1, is an extension of the annual model and includes three latitudinal zones representing the tropical latitudes (30°S-30°N), southern hemispheric extratropical latitudes (30°S-90°S), or the southern zone and northern hemispheric extratropical latitudes (30°N-90°N), or the northern zone. The model computes the evolution and distribution of temperature from the heat balance for the atmosphere and for the subsurface (ocean/land). Because of the large difference in thermal inertia between land and oceans, which is important in simulating the seasonal cycle, both land and ocean sectors are included within each latitude zone. Ice and snow albedo feedbacks are parameterized for both hemispheres. Empirically derived horizontal (meridional and zonal) heat flux parameterizations, are used while the vertical dynamical heat flux is parameterized through dynamical adjustment processes. The seasonal model uses the same radiation parameterization as the 1- and 2-D annual model, in which the atmospheric gases, clouds, and the surface characteristics are included.

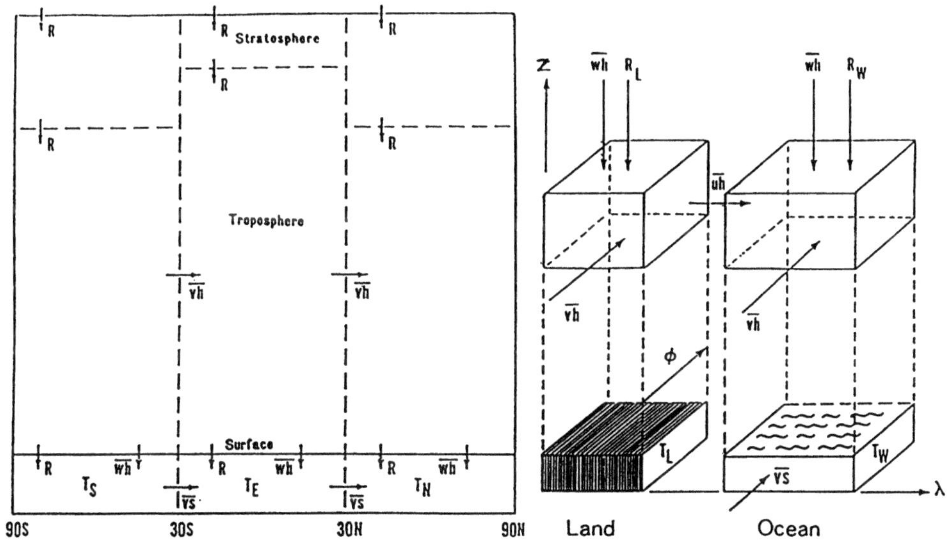

Figure 1: Schematic illustration of the heat balance components of the global and seasonal radiative-dynamical climate model with three latitudinal zones and land/ocean sectors within each latitude zone. The coordinates of ϕ, λ, and z are latitude, longitude and altitude, respectively. The energy components are: $\overline{vh}$ - meridional heat flux, based on averaged ocean/land temperature and used as the same in land and ocean sector; $\overline{uh}$ - land/ocean heat exchange, $K(T_W - T_L)$ with K, T_W and T_L, the diffusion coefficient and the ocean and land surface temperatures, respectively; $\overline{vs}$ - meridional ocean heat transport; $R_{L,W}$ - separate land and ocean radiation heat flux; and $\overline{wh}_{L,W}$ - separate land and ocean vertical heat flux, based on dynamical adjustment.

In order to simulate the current climate, we need to include the seasonal variation of ocean heat transport. However, information about ocean heat transport and its potential changes during climate changes are lacking. We use the simpliest approach. The value of ocean heat transport and storage for the present climate is derived from model surface energy balance with observed seasonal ocean surface temperature. During climate change experiments, the sum of the ocean heat transport and storage is assumed to remain the same as that of the present climate. This approach has been used in general circulation models (see Hansen et al., 1984).

The 2-D model calculated seasonal land surface temperatures of the present climate are in good agreement with observations, although the model simulated amplitude of the seasonal cycle is slightly larger than the observed values. However, a more critical test of the model simulation of the present climate is the seasonal variation of tropospheric lapse rate as shown in Table 1. The close agreement between model and observations indicate that the dynamical adjustment used in the model is capable of simulating the vertical temperature gradient, which is important in simulating the meridional and vertical dynamical heat transport.

Table 1. Seasonal Mean Tropospheric Lapse Rate (°C km^{-1}) of Northern Zone. The observations were calculated between surface and 100 mb.

	Winter	Spring	Summer	Fall
Model	3.90	4.49	5.16	4.92
Observations	3.85	4.44	4.92	4.68

3. RESULTS

Both sensitivity and scenario experiments have been conducted to examine the O_3-climate interactions associated with increases of trace gases. For the sensitivity experiments, we illustrate the importance of radiation-dynamics interactions related to O_3 changes using the three models. However, for the scenario calculations, we rely mainly on the 1-D model because of the amount of computations involved.

3.1 Sensitivity Calculations

We have used the 2-D annual and seasonal models to perform "open-loop" studies of the climatic effects due to O_3 perturbation associated with CFCs' increases. The "open-loop" means that the feedback effect of O_3-induced temperature changes on chemistry in not included. The purpose of the study is to examine the effect on atmospheric temperature of the redistribution of O_3 resulting from the increases of CFCs. To perform the study, we have used the CFCs and O_3 distributions consistently calculated from the 2-D photochemical-dynamical model

(Ko et. al, 1985a,b, and results cited in Chapter 13 of WMO, 1986). In addition to the simulation of the present atmosphere, calculations were carried out using enhanced CFC surface concentrations that correspond to the steady-state conditions for a high flux case, i.e., CFC fluxes based on two times 1980 CFC production rate.

Because of the coarser latitudinal resolution used in the 2-D climate model, we have averaged the CFCs and O_3 calculated from the 2-D photochemical model. The corresponding surface concentration of CFC-11, CFC-12, and column O_3 for the present atmosphere are summarized in Table 2. Because of the relative symmetry between the southern and northern hemispheres, both in latitudinal and seasonal variations after the averaging (see discussion below), we show only the northern hemispheric values in Table 2.

Table 2. 2-D photochemical-dynamical model calculated CFC-11 and CFC-12 concentrations, and column O_3 amount or present and assumed high CFC fluxes (two times 1980 production rate)

	CFC-11 (ppbv)	CFC-12 (ppbv)	Column O_3 (cma)*	ΔO_3(%)
<u>Present</u>				
0-30°N	0.168	0.270	0.265	-
30-90°N	0.179	0.297	0.354	-
0-90°N	0.174	0.284	0.309	-
<u>High Flux</u>				
0-30°N	1.85	4.75	0.236	-10.9
30-90°N	1.97	5.23	0.269	-24.0
0-90°N	1.91	4.99	0.253	-18.1

*Centimeter atmosphere at STP

It can be seen that CFC-11 has northern hemispheric mean concentration of 0.174 and 1.91 ppbv for present and perturbed cases; the latitudinal gradient is relatively weak. For CFC-12, a much larger surface concentration and increases are calculated.

The northern hemispheric mean column O_3 is reduced by 18.1% primarily because a larger (24%) depletion is calculated in the high latitudes. These differences in the latitudinal O_3 responses to CFCs' increases can be clearly seen in Figure 2a. Above 30 km, O_3 decreases in all the latitudes with similar magnitude. However, significant differences exist in the regions below 30 km, especially in the upper troposphere and the lower stratosphere where O_3 perturbation has maximum effect on surface temperature (see Wang and Sze, 1980). In this region,

O_3 decreases/increases may decrease/increase the greenhouse effect and thus lead to a cooling/warming effect. Because of the averaging processes, such a contrast in latitudinal variation of O_3 perturbation cannot be seen in a 1-D model.

The seasonal aspect of the CFCs-induced O_3 perturbations is shown in Figure 2b. The difference between January and July appears to be small in the tropical region. However, at the extra-tropical regions, the difference is quite large -- for example, about a factor of 2 around 20 km.

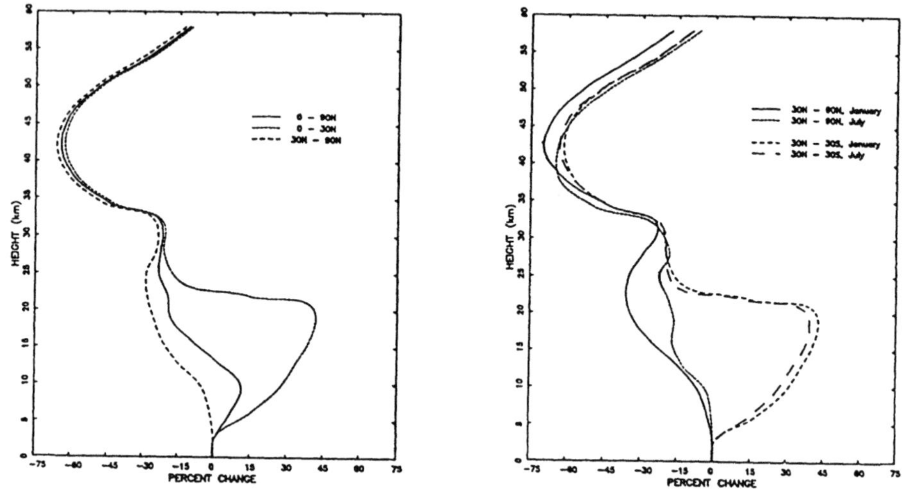

(a) (b)

Figure 2: Calculated percent changes of (a) annual mean and (b) seasonal O_3 distribution caused by increases of CFCs. The changes correspond to the difference between the reference atmosphere with no CFC and the steady state atmosphere based on continuous emissions of CFC at two times 1980 CFC production rates.

In summary, the CFCs-induced O_3 perturbation has a very strong latitudinal variation, especially in the lower stratosphere and upper troposphere, a region critically important to the tropospheric climate. There are some seasonal variations, but the contrast appears not as strong as the latitudinal contrast.

Using the CFCs and O_3 concentration discussed above, the 2-D annual model calculated changes in vertical and meridional temperatures are summarized in Table 3. In this table, we have shown the temperature changes in separated stratosphere, troposphere, and the surface, as well as the changes in northward dynamical heat fluxes. In the calculations, we have used fixed cloud altitude and fixed relative humidity parameterizations and included ice-albedo feedback.

The results suggest that the surface temperature could be warmed by 0.36°C resulting directly from CFCs (case A). The effect of the CFCs-induced O_3 perturbation is to significantly reduce the CFCs direct warming effect to 0.16°C (case B). The effect is particularly large at high latitudes because of the O_3 depletion throughout the atmosphere (see Figure 2a). However, it is surprising to find that the low latitude CFCs-induced warming is also reduced despite the O_3 increases in the troposphere (see Figure 2a).

Table 3. Two-dimensional annual model calculated temperature change ΔT(°C) in stratosphere (S), troposphere (T), and ground (G) due to increases of CFC-11 and CFC-12 and the associated O_3 change. Dynamical heating caused by changes of northward transport of heat ΔD_m(Wm^{-2}) are also indicated.

		0-30°N	30-90°N	0-90°N
A. CFCs				
	S	0.51	0.30	0.41
ΔT(°C)	T	0.64	0.53	0.59
	G	0.34	0.38	0.36
ΔD_m(Wm^{-2})		-0.07	0.07	
B. CFCs + O_3				
	S	-0.61	-1.92	-1.27
ΔT(°C)	T	0.54	0.33	0.44
	G	0.26	0.06	0.16
ΔD_m(Wm^{-2})		0.35	-0.35	

The reason for the low latitude negative feedback can be seen in the dynamical heat flux responses shown in Table 3. Despite the O_3 increases in the troposphere at low latitudes, which has a warming effect on surface temperature, the meridional heat transport increases as a result of the smaller high latitude surface warming (relative to low latitude) caused by increases of CFCs. This results in a cooling at the low latitude. It turns out that the cooling effect more than compensates the warming caused by tropospheric O_3 increases. The results demonstrate that to study the O_3 climatic effects would need at least the use of a 2-D model. It is particularly important that the interaction between the meridional and vertical dynamic heat transport and the radiation heat transport be considered. The results shown in Table 3 also indicate that CFCs will provide a heating effect throughout the atmosphere, while the combined effect of CFCs and O_3 is to cool the stratosphere and to warm the troposphere and the surface.

Because of the importance of the meridional dynamic heat transport in causing the (net) O_3 negative feedback, we have also used the 2-D seasonal model to examine the CFCs and O_3 climatic effects on the seasonal basis. To do this, we have used the seasonal O_3 data shown in **Figure 2b**.

For this particular CFCs + O_3 climate experiment, we have prescribed the sea surface temperature while the land surface temperature is calculated. For the present climate, the sea surface temperature is the climatological values. When the high CFCs case is used, we specified the sea surface temperature to be the value as calculated from the annual model. In so doing, we constrain the response of sea surface temperature to be the same as the annual model and only allow the land surface temperature to change in responding to changes in CFCs and O_3.

The 2-D seasonal model results are summarized in Table 4. For CFCs alone, the whole globe on the annual basis is warmed by about 0.35 °C with relatively weak seasonal variation. However, the effect is much larger at northern hemisphere high latitude winter and fall, and southern hemisphere high latitude spring. The asymmetry in the temperature response between the two hemispheres is mainly caused by the different land/ocean distributions.

Table 4. 2-D Seasonal Model Calculated Surface Warming Effect (°C) due to CFCs and O_3. The seasons refer to northern hemisphere.

		90-30°S	30°S-30°N	30-90°N	Globe
A.	CFCs				
	Winter	0.30	0.33	0.50	0.37
	Spring	0.45	0.30	0.31	0.34
	Summer	0.34	0.31	0.35	0.33
	Fall	0.32	0.30	0.47	0.35
	Annual	0.35	0.31	0.41	0.35
B.	CFCs + O_3				
	Winter	0.10	0.28	0.24	0.23
	Spring	0.27	0.26	-0.05	0.19
	Summer	0.12	0.25	0.08	0.18
	Fall	0.10	0.25	0.15	0.19
	Annual	0.15	0.26	0.11	0.19

The effect of O_3 changes is still calculated to provide a strong negative feedback effect on the direct CFCs' surface warming. The annual- and global-mean temperature change is 0.19 °C (case B), which is about 47 percent smaller than that calculated for CFCs alone (case A). The effects of O_3 changes also exhibits strong seasonal variation. It is particularly interesting that at the northern hemisphere, high-latitude spring time, the net effect of CFCs and O_3 is a cooling rather than a warming.

To summarize the effect of latitudinal and seasonal O_3 variations and their effect on climate, we have compared the 2-D annual and seasonal model calculated global- and annual-mean surface temperature changes caused by CFCs and O_3. We also used a 1-D model to illustrate the importance of consideration of latitudes and seasons in the study of CFCs + O_3 climatic effects. The 1-D study, however, is a "closed-loop" study. The results shown in Table 5 clearly indicate that it is important to include the consideration of latitudinal O_3 variation.

Table 5. Model Calculated CFCs' Surface Warming Effect (in °C) and O_3 Feedback Effect (in %). High CFC fluxes and fixed cloud altitude parameterization are used.

	1-D Radiative-Convective-Photochemical Model	2-D Radiative-Dynamical Model: Annual Version	2-D Radiative-Dynamical Model: Seasonal Version
A. CFCs	0.52	0.36	0.35
B. CFCs + O_3	0.46	0.16	0.19
O_3 Feedback, (B - A)/A	-11.5	-55.6	-47.2

3.2 Scenario Calculations

In this section, we present results from our 1-D model on assessment of O_3 response and changes in surface temperature to trace gas increases.

The increase in CO_2 is expected to continue as fossil fuel power is likely to remain the dominant energy supply in the next few decades. Current estimates are that present level of CO_2 (340 ppmv) would double some time in the next centry (cf. DoE, 1985). Atmospheric measurements showed an increase of 1-2 percent per year in the surface concentration of CH_4 since 1977. However, it is unclear whether the observed increase is due to enhanced emission associated with changes in the biosphere or changes in tropospheric OH. The trend for N_2O is estimated to be between 0.2 to 0.3 percent per year between 1976 and 1980. Projections for future trends are uncertain because of lack of detailed information on the budget of N_2O. The assumed future growth rates for CO_2, N_2O, CH_4 and CFCs for several scenarios are shown in Table 6 while their surface concentrations are shown in Figure 3.

Table 6 Trace Gases Scenarios

A. CO_2, N_2O and CH_4

Species	Growth Rate
CO_2	0.55%/year from 1985
N_2O	0.25%/year from 1955
CH_4	1% from 1960

B. CFCs growth rate Scenarios

	φCFC*	
Scenario	1985-2000	after 20
1	no CFC emissio	
2	constant at 1984 rat	
3	3.0	constant 2008 ra
4	3.0	constant a 1984 rat

* Assumed CFC release rates after year 1985. Prior to 1985, historic release data are used. The emission rate at the year 2008 corresponds to double the present-day CFC production if 3% annual growth were maintained through 1985-2008.

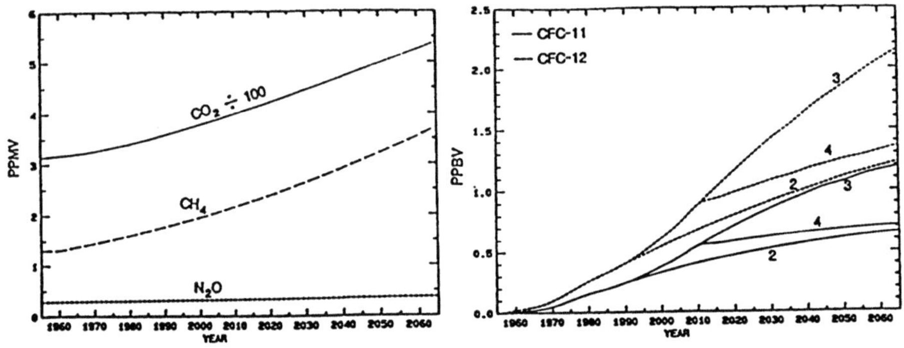

Figure 3: The surface concentrations of assumed N_2O, CH_4 and CO_2 and calculated CFC-11 and CFC-12 for the period 1955-2065.

The 1-D calculated changes in the column abundance of O_3 in the period are shown for different scenarios in Figure 4. In the absence of CFCs (scenario 1), an increase of 1.3% is expected by the year 2060. The calculated O_3 shows a decreasing trend in all other scenarios with CFC emission with the change in O_3 ranging from about -1% (scenario 2) to -3.5% (scenario 3). It is interesting to note that in the cut back scenario (scenario 4) the depletion number obtained in the year 2060 is similar to that for scenario 2 in which no growth in emission is assumed.

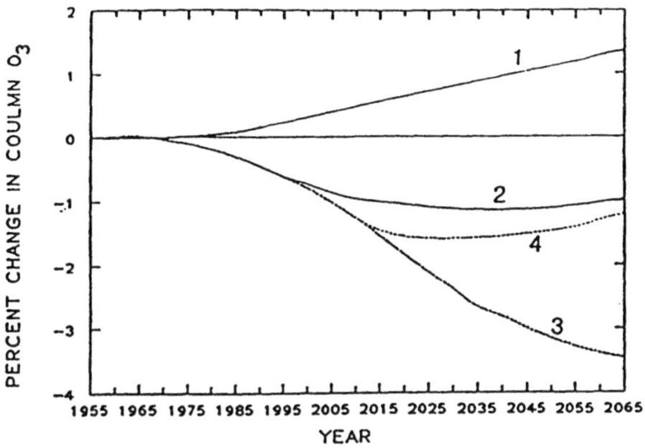

Figure 4: The calculated changes in the column amount of O_3 for the period 1955-2065 for assumed scenarios (see Table 6).

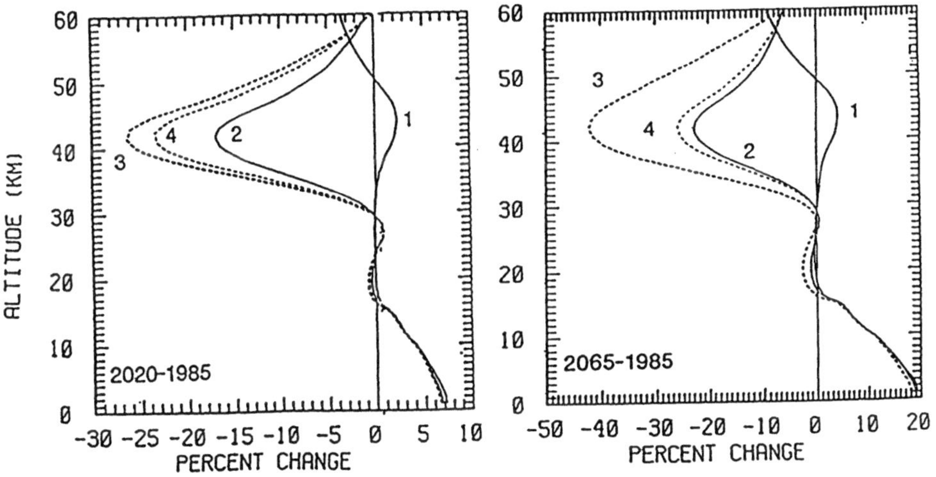

Figure 5: The 1-D calculated changes in the vertical distribution of O_3 concentration for assumed scenarios for the periods of 1985-2020 and 1985-2065.

Figure 5 shows the calculated changes in O_3 as functions of altitude for the periods 1985-2020 and 1985-2065. Basically, CFCs decreases the stratospheric O_3 while CH_4 is responsible for the O_3 increases in the troposphere.

The calculated changes in surface temperature are shown in Figure 6. In addition to the scenarios considered, the changes due to CO_2 only are included for comparison. The calculations are based on 6.5 °C km^{-1} critical lapse rate, fixed cloud altitude, fixed relative humidity, and ice albedo feedback. Note that the first parameterization tends to calculate a larger increase in surface temperature than the use of the moist adiabatic lapse rate, and the second parameterization calculates a smaller surface warming than the fixed cloud temperature. For the period 1955-1985, the model calculates a 0.55°C surface warming, which is somewhat larger but consistent with observations (cf. Jones et al., 1986). The results suggest that CO_2 alone could warm the surface temperature by 1.6°C at 2065, and that the combined effect of N_2O and CH_4 could further enhance the surface warming by 50 percent to 2.4°C at 2065. Scenario 3 gives the largest CFC contribution, and the total surface warming for this scenario is 3°C at 2065.

These scenario calculations suggest that the combined effect of trace gases in the next few decades may substantially increase the greenhouse warming. The contribution from O_3 changes, primarily from tropospheric increases as shown in Figure 5, is calculated to have a small warming effect.

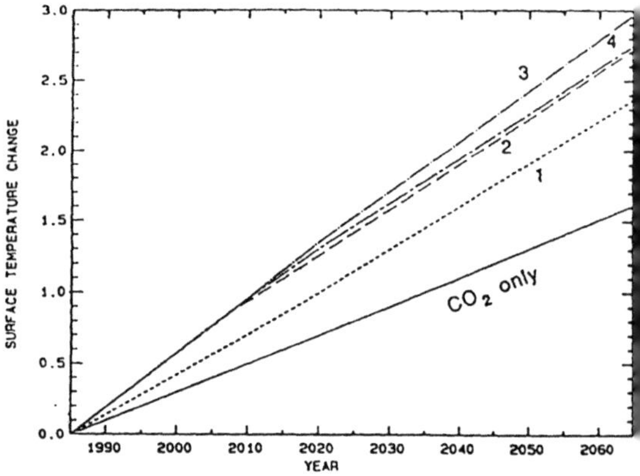

Figure 6: The 1-D calculated change in surface temperature for the period 1985-2065 for assumed scenarios. See Table 6 for description scenarios.

4. SUMMARY AND DISCUSSION

Atmospheric O_3 is a radiatively and chemically important gas. Its latitudinal and seasonal redistribution caused chemically by increases of atmospheric trace gases may lead to perturbation of radiation budget, which may trigger a response in the meridional heat flux with subsequent feedback effect on climate. Our sensitivity calculations indicate that interactions between radiation and dynamical heat flux caused by the CFCs-induced O_3 changes in the middle and upper troposphere may provide a strong negative feedback effect on the CFCs greenhouse effect. However, the strength of the negative feedback depends critically on the contrast in the changes in troposheric O_3 between different latitudes and seasons. Further studies using coupled perturbations of atmospheric trace gases are warranted.

The 1-D scenario calculations of increasing atmospheric trace gases in the next few decades are also performed. The results suggest that the combined effect of N_2O, CH_4, and CFCs and their induced-O_3 changes may substantially augment the CO_2 surface warming. The tropospheric O_3 is calculated to increase slightly and its contribution to the total surface warming is a small warming effect.

Acknowledgement. This research was supported by the Fluorocarbon Program Panel of the Chemical Manufacturers Association (CMA-85-553 & 555), the Climate Dynamics Program of the National Science Foundation (ATM-8519350), and the Carbon Dioxide Research Division of the Department of Energy (DE-AC02-86ER60485)

REFERENCES

DoE (Department of Energy), 1985: Projecting the climatic effects of increasing carbon dioxide. Department of Energy, ER-0237, Washington, D.C.

Hansen, J., A. Lacis, D. Rind, G. Russell, P. Stone, I. Fung, R. Ruedy and J. Lerner, 1984: Climate sensitivity: Analysis of feedback mechanisms. In Climate Processes and Climate Sensitivity, Geophys. Monogr. Ser., vol. 29, J. E. Hansen and T. Takahashi (eds.), AGU, Washington, D.C., 130-163.

Jones, P. D., S. C. B. Raper, R. S. Bradley, H. F. Diaz, P. M. Kelly and T. M. Wigley, 1986: Northern hemisphere surface air temperature variations: 1951-1984. J. Clim. & Appl. Meteorol., 25, 161-179.

Ko, M. K. W., K. K. Tung, D. K. Weisenstein, and N. D. Sze, 1985a: A zonal-mean model of stratospheric tracer transport in isentropic coordinates: Numerical simulations of nitrous oxide and nitric acid. J. Geophys. Res., 90, 2313-2329.

Ko, M. K. W., D. Weisenstein, N. D. Sze, and K. K. Tung, 1985b: Simulation of O_3 distribution using a two-dimensional zonal-mean model in isentropic coordinate. In Atmospheric Ozone, C. S. Zerefos and A. Ghazi (eds.). D. Reidel Publ. Company.

Ramanathan, V., L. Callis, R. Cess, J. Hansen, I. Isaksen, W. Kuhn, A. Lacis, F. Luther, J. Mahlman, R. Reck and M. Schlesinger, 1987: Climate-chemical interactions and effects of changing atmospheric trace gases. Rev. Geophys. (in press).

Wang, W.-C., D. Wuebbles, W. M. Washington, R. G. Isaacs, and G. Molnar, 1986: Potential climatic effects due to perturbations other than CO_2. Rev. Geophys., 24, 110-140.

Wang, W.-C., and G. Molnar, 1985: A model study of the greenhouse effects due to increasing atmospheric CH_4, N_2O, CF_2Cl_2, and $CFCl_3$. J. Geophys. Res., 90, 12971-12980.

Wang, W.-C., and N.D. Sze, 1980: Coupled effects of atmospheric N_2O and O_3 on the Earth's Climate. Nature, 286, 589-590.

WMO (World Meteorological Organization), 1986: Atmospheric Ozone 1985. Global Ozone Research and Monitoring Project, Report No. 16, World Meteorological Organization, Geneva.

OZONE EFFECTS ON VEGETATION

B. Prinz
Landesanstalt für Immissionsschutz des Landes Nordrhein-Westfalen
Wallneyer Str. 6, 4300 Essen
Federal Republic of Germany

ABSTRACT. Ozone is one of the most discussed phytotoxic air pollutant of today. While its importance was first discovered in the late 1960's in California, new impetus in scientific investigations of its phytotoxic effects has been caused by the novel type of forest decline in Central Europe and North America. In the past the emphasis was mainly on acute effects, especially in connection with damage to annual crops. Now chronic effects are the central point of discussion. By this the problems have become much more difficult, because now climatic and edaphic factors are as important as ozone and they very often mimic the ozone effects.

Although there are still many gaps in the knowledge of ozone effects on vegetation, we have reached meanwhile a fairly good picture of the principle modes of actions as well as of the quantitative dose-effects relationships. With regard to the novel type of forest decline in Central Europe a combinatory effect of ozone together with acidic fog and rain and unfavourable soil conditions by lack of nutrients is regarded as the main clue. The climate, in the form of dry warm summers with extensive periods of sunshine is thought to be an important triggering or damage synchronizing factor.

1. INTRODUCTION

While plant damge caused by sulfur dioxide ond other primary pollutants have been investigated for more than 100 years, with admirable scientific results already in the beginning of our century (see for example one of the first reviews prepared by WISLICENUS [1]), the phytotoxic effect of ozone was discovered rather late. The first signs of plant injury were observed in forests in the Los Angeles area in 1944. Ozone as the causative factor was first identified, however, in the late 1960s [6]. Already in the early 1970s there was a flood of published information. Most of it was gathered in the excellent document "Ozone and Other Photochemical Oxidants" by the National Academy of Sciences [2]. Later on much effort was spent to yield dose-response relationships for agricultural crops, especially by the so-called "Na-

I. S. A. Isaksen (ed.), Tropospheric Ozone, 161–184.

tional Crop Loss Assessment Network" (NCLAN) in the USA [3]. Much quantitative information, including those from NCLAN, are presented in the recently published book "Air Pollution by Photochemical Oxidants" [4]. For the east of the USA forest decline as a result of ozone exposure was demonstrated comparatively recently [6]. A new push in the investigation of the effect of ozone on plants was created by a publication of the Landesanstalt für Immissionsschutz (LIS) [5], when the novel type of forest decline in West-Germany was brought into connection to ozone.

Meanwhile a rather complicated picture about the phytotoxic effects of ozone on plants exists. Contrary to former times, the chronic burden of most vulnerable forest ecosystems by long lasting exposure of ozone is the focal point of the discussion. For this reason interactions of ozone with soil, climate as well as other pollutants are automatically of high relevance.

Despite the great success in ozone related investigations, especially during recent years, many important gaps of knowledge still exist. It is hoped that, particularly by intensive investigations in the field of forest decline in Central Europe and North America, some of these gaps will be illuminated in the near future.

In the following the principle mechanisms of the action of ozone on plants as well as the symptomatology is first described. Interactions with soil and climate as well as with other pollutants follow. Some information is then given about the specific sensitivity of individual plant species as well as quantitative dose-response relationships. A last chapter deals with the specific relation between ozone and the novel forest decline.

2. MECHANISMS OF ACTION AND SYMPTOMATOLOGY

The stomata are the principal entrance of ozone to plant tissue. Although ozone itself can change the aperture of the stomata [8], more important is the influence of the genetically and climatically predetermined leaf conductance on the uptake and thus on the effect of this pollutant. So it is understandable that as a general rule each factor, which widens the stomata, automatically increases the sensitivity of the plant [9]. On the other side the responsiveness of the stomata to ozone seems amongst other factors to determine the relative sensitivity of the plant, if comparing for example different varieties of the same plant species [8]. This means that closure of stomata under the influence of ozone works as a protective measure.

Despite many still existing uncertainties there is no question that the predominant toxic reaction of ozone is due to its capability to oxidize organic compounds at different sites of the plant tissue, like specific enzymes and above all the lipid components of the membrane systems. The peroxidation of lipids is deduced from the increase of antioxidants like glutathione, vitamin E and vitamin C under the influence of ozone [20, 21]. This is considered to be a protective measure of the plant. Evidence for the impact of ozone on the plasmalemma can for example be derived from the oberservable loss of turgor shortly appearing after exposure to ozone [10]. Perturbation of the thyllakoid-membrane system is deduced from the distinct reaction of chlorophyll-fluorescence on ozone exposure [11]. Quite a lot of investigations showing ultrastructural changes in chloroplasts, which are partly referenced in [4], lead to the same conclusion.

Although some publications exist which show increased photosynthesis after ozone exposure (for example [12]) the overwhelming investigations demonstrate agreement with the above mentioned destruction of chloroplasts with negative effects on photosynthesis. These results were gained with annual crops [13, 14] as well as with trees [13, 14, 15, 16, 17, 18, 19, 22]. In almost all cases there was a strict relationship between decrease of photosynthesis and yield or growth, even with trees. These effects were found in fumigation experiments where ozone concentrations representative of ambient air were applied. So for example REICH and AMUNDSON [13] state that reductions in net photosynthesis for a great amount of annual and perennial plant species may be occurring over much of the eastern United States. They also stress the point that with regard to forest trees high elevations cause an additional stress due to low fertility, extreme temperatures, lack of soil, high winds, and low availability of water.

The impact of ozone on the photosynthetic apparatus begins very early and with very low concentrations, without any visible injury. Therefore this mode of action is of outstanding importance for chronic exposures as they are typical for perennial plants like forest trees. If the concentrations become higher decay of chlorophyll is quite apparent. This is very often sharply limited to small single groups of epidermal and palisade cells causing flecks or stipples. In a more promoted stage these cells become dead so that chlorosis turns over to necrosis. Very often the affected groups of cells are very tiny but distributed over the whole leaf area. In this case the colour of the leaf changes to a broncing appearance. If the concentrations of ozone are very high, leaves develop dark water-soaked areas within a few hours. For coniferous trees tip dieback of newly elongated needles or mottling over the whole needle length is most characteristic. Ozone symptoms for forest trees are represented by PRINZ [34] and PRINZ et al. [35, 36] as well as by GUDERIAN et al. [32].

Particularly following chronic ozone impacts, premature senescence of leaves and needles, leading to early discolouring and abscission, can appear. This is accompanied by quite a lot of biochemical deviations, as for example increased stress ethylene production [4].

Each leaf passes through phases of different sensitivity. As a general rule, leaves are most sensitive, when unfolded and just developed to normal size. This means that with acute exposures within a single plant the apex as well as the base of the plant very often lacks any injury because this injury occurs predominantly in middle-aged leaves. Conifer needles are most sensitive a couple of weeks after budding, when they are still elongating. The susceptibility decreases when the needles reach full elongation.

Many investigations have dealt with the sensitivity of different varieties of the same plant species. From this it can be derived that the genetic variation in ozone resistance is really great. So there is no wonder that for example in the highly polluted San Bernadino mountains in California one serverely injured tree can be immediately adjacent to an apparently healthy one.

In connection with recent types of forest decline, leaching of nutrients as a result of a combinatory effect of ozone and acidic rain or fog is now intensively discussed. This assumption was first expressed in the LIS-Report No. 28 [5] together with some preliminary experimental results. Meanwhile, this phenomenon could be certified by several experiments [26, 27, 28]. The question is still open whether the enhanced leaching results from the impact of ozone on the cell-membrane system or from ozone-induced cuticular erosion or as a third version from the ozone-induced decay of chlorophyll. The enhanced leaching is principally proved by observation in natural, ozone-exposed forest stands as well as by fumigation experiments in the laboratory [23, 24, 29]. In this connection it is interesting to know that quite recently KRAUSE [33] discovered that ozone effectively increases the content of cytotoxic nitrate in needles of Norway spruce fir as well as in leaves of common oak and common beech. This may be a hint of the oxidation of membrane proteins which are closely associated to membrane lipids. It may, however, also be a result of the impact of ozone on the enzyme nitrate reductase, which would put the most important nitrogen metabolism in a new light.

3. INTERACTIONS WITH SOIL AND CLIMATE

The effect of ozone on vegetation seems to be more dependent on edaphic and climatic conditions than that of any other air pollutant . Thus in one experiment a specific symptom can be provoked, while in another experiment with the same ozone regime but slightly differing climatic conditions this fails. Optimum moisture conditions in soil and high relative humidity in air make the plant most sensitive, while on the other hand water stress induces closure of the stomata and so decreases stomatal conductance for the uptake of ozone [30, 31]. This is in principle a very old statement which helped to explain that the trees on the ridges of the San Bernadino Mountains exhibited a greater response to ozone than the trees at lower elevations at comparable ozone concentrations.

Many investigations have dealt with the interaction between the impact of ozone and the nutrient status in soil (see for example [2, 4]). Some of these results are contradictory, however, and no consistent model for interpreting these interactions are at hand so far. From more recent investigations it is rather interesting in the light of the novel type of forest decline that some effects of ozone on trees only became apparent if the plants were cultivated in soil with a low nutrient supply. So GUDERIAN et al. [32] found that instead of the well known mottling of needles of Norway spruce a complete yellowing could be reached, if the plants suffered from magnesium and calcium deficiency. In agreement with some other authors, for some unknown reasons the ozone-exposed plants exhibited greater contents of magnesium and calcium in the needles than the control plants. On the other hand the leaching of these elements was distinctly increased by ozone exposure, when the needles were washed with an acidic solution after fumigation.

In experiments with Norway spruces, carried out by SELINGER et al. [22], photosynthesis was decreased to a great amount when the plants were cultivated under nutrient deficient conditions and treated with ozone and acidic rain. When the plants were fertilized the effect of ozone still existed but it was not so obvious. Treatment with acidic rain alone increased even the photosynthesis of the fertilized plants in comparison to the control. So the combination of ozone and low nutrient supply in the soil seems to be one of the crucial points for possible injury in forest ecosystems.

Light intensity generally enhances the effect of ozone. This is quite obvious from cases where leaves shadow themselves und symptoms of ozone injury are absent at those sites which are not directly exposed to sunlight [34, 35, 36]. If sunlight, in combination with high temperature, causes wilting then of course light intensity interacts negatively with ozone injury. This may be the reason that the sensitivity of the well kwown bioindicator tobacco Bel W_3 is distinctly increased when the plants are shadowed during exposure.

Experimental results about the coergistic influence of temperature on ozone injury are not so unequivocal that a general statement could be derived. Clearly, as in the case of relative humidity and light, this influence is realised mainly by triggering the aperture of the stomata.

4. POLLUTANT INTERACTIONS

Quantitative aspects of pollutant interactions are rather complicated. Normally over-additive effects are called synergistic, while less than additive effects are called antagonistic. This depends very much on the chosen scale, however. If, for example, the relative reduction in growth or yield is used, this measure can never exceed 100 % so that already additive effects in the high range of this scale are quite obviously impossible. A synergistic effect is evident if a certain symptom only appears if two or more pollutants are applied, while with each of these components in single application this symptom fails to appear. In a classic experiment MENSER and HEGGESTADT [37] could show as a typical synergistic effect that brief exposures of the tobacco Bel W_3 to sulfur dioxide and ozone produced visible injury, while exposure to individual pollutants did not. The literature has grown since that time immensely so that a representative overview about this topic is impossible. For the evaluation of the problem it has to be borne in mind, however, that ozone is very often negatatively correlated to other pollutants. So by quite simple reactions in the atmosphere the synchronous occurrence of high concentrations of ozone and nitrogen monoxide is impossible. But very often also sulfur dioxide and nitrogen dioxide are very low in concentration just in those areas which are typical for high ozone exposure, like in the higher altitudes of the South German subalpine mountains.

In addition to the overviews presented by GUDERIAN et al. [4] and the NAS document [2] the following recent results are of some interest. ASHMORE and ÖNAL [38] combined in a fumigation experiment with spring barley for the first time a comparatively high concentration of ozone (0.18 ppm) with a comparatively low concentration of sulfur dioxide (0.065 ppm), while in all other cases the sulfur dioxide concentration exceeded the ozone concentration. He observed that the plants developed significantly more leaf injury after exposure to ozone alone than after exposure to ozone and sulfur dioxide together. Sulfur dioxide was applied in a concentration which, when applied alone, had no adverse effect on the plants.

In an experiment described by PRINZ et al. [35] the air pollutant regimes typical for industrial zones and alternatively for subalpine mountains were simulated and tested for their effect on common beech. In the first case 150 $\mu g\ O_3/m^3$ for 5 hours per day during daylight and 100 $\mu g\ SO_2/m^3$ continuously were applied, while the "clean air" regime was simulated by 150 $\mu g\ O_3/m^3$ continuously and 60 $\mu g\ SO_2/m^3$ continuously with 500 $\mu g\ SO_2/m^3$ for 2 hours each fortnight. This experiment showed that sulfur dioxide alone had almost no influence on premature leaf fall, compared with the control, while ozone alone and in combination with sulfur dioxide significantly accelerated the senescence. More interesting is the fact, however, that the prolongation of the ozone exposure over the 5 hours within the "low burden regime" had, despite the lower sulfur dioxide concentration, a greater effect on premature leaf fall than the "high burden regime". This means also that the night exposure has a distinct influence on the toxic effect of ozone despite the reduced photosynthetic activity.

SMIDT [39] compared the effect of ozone and sulfur dioxide, alone and in combination, on Norway spruce. He showed that with respect to necrosis as well as to increased peroxidase activity the combination of the pollutants had higher effects than one of these pollutants alone. GUDERIAN et al. [32] fumigated Norway spruce trees with ozone, sulfur dioxide, and nitrogen dioxide, cultivated on nutrient deficient soil. They found that needle chlorosis was especially pronounced when nitrogen dioxide together with ozone and sulfur dioxide was involved.

5. SENSITIVITY OF INDIVIDUAL PLANT SPECIES AND DOSE-RESPONSE RELATIONSHIPS

The great genetic variability of ozone sensitivity has already been mentioned. So there is no wonder that different plant species react in a quite different manner to the influence of ozone. It happens that different species of one genus, for example pine, belong to the most sensitive as well as to the most tolerant groups of plants. In many cases the relative sensitivity or relative tolerance is derived from short-term fumigation experiments. Particularly with respect to long living forest trees these species may react, however, in a very different way to these results, if exposed chronically. A second point, which limits the value of such classifications, results from the fact that in most cases visible injury was taken as a scale of relative sensitivity. If, for example, leaching of nutrients under the influence of ozone in combination with certain climatic and edaphic conditions are the crucial point of plant damage, then these classifications may even mislead the user.

A first comprehensive list of plant species grouped into the three classes of sensitive species, intermediate and less sensitive species was assembled by DAVIS and WILHOUR [40]. The following shorter list (Table I) is taken from PRINZ and BRANDT [41] and considers also other authors than DAVIS and WILHOUR. It has to be read with the caution and restriction which has been mentioned above.

TABLE I. Sensitivity of selected plant to ozone

very sensitive	sensitive	less sensitive
	crops	
spinach	onion	lettuce
oat	beet	carrot
rye	barley	grape
bean	wheat	
potato	corn	
tomato	alfalfa	
tobacco		
	ornamentals	
petunia	chrysanthemum	sultana
	coleus	begonia
	carnation	geranium
		gladiolus
		fuchsia
	trees	
white pine	Ponderosa pine	Scotch pine
ash	apple	sycamore
	maple	birch
	oak	walnut
	poplar	beech
		locust
		yew

Spruce, especially Norway spruce, and silver fir are also assigned in the older literature to the class of less sensitive plants. This is absolutely true if one considers that mottling of the needles can be provoked only at concentrations of more than 500 $\mu g\ O_3/m^3$ applied over several weeks. Related to more complex effects coniferous trees seem to be extremely sensitive, however.

Quantitative dose-effects relationships are also focused on visible injury and yield or growth and this mainly under short-term conditions of exposure. A second lack of the available informations results from the fact that a vast variety of different time patterns of ozone exposure are cited. The cause for this phenomenon is the pronounced diurnal variation of the ozone concentration, at least in the vicinity of the emission sources of the precursors. This diurnal variation often disappears, however, if one goes to the highly burdened areas in the subalpine mountains. Figures 1 to 3 give some examples of the ozone regime in the mountains like Egge-mountains and Eifel in Northrhine-Westphalia compared to the Rhine-Ruhr area as well as for the Black Forest compared to the city of Freiburg and Cologne. For comparison the yearly variation of sulfur dioxide, nitrogen monoxide and nitrogen dioxide as peak values and monthly averages in mountainous forest areas in Northrhine-Westphalia as well as in the Rhine-Ruhr area is also represented (Figure 4 to 6). The main characteristic is that contrary to other components ozone does not so much differ in the peak values but in the average values between the "source"- and "remote"-areas. Since many quantitative dose-effect relationships were gained by the National Crop Loss Assessment Network (NCLAN) [42] which operate with open-top-chambers with filtered and unfiltered airas well as with proportionally added ozone in comparison to ambient air, the rather complicated statistical description of the ozone regime is quite evident. But also in fumigation experiments these complicated time-patterns of concentrations are normally applied in a most different way, to simulate the ambient air situation, so that it is extremely difficult to compare the individual results.

An extensive review of the literature was made for critical information on dose and response by LINZON et al. [43]. Curves of concentration-response and time-response were developed for over 100 species and varieties obtained from a literature review. From these curves, a scatter diagram for a 5 % response was drawn (Figure 7). The values from the scatter diagram were used to generate the dose equation which was based on a modification of the O'Gara equation for sulfur dioxide:

$$C = a + b/T$$

where C = concentration in pphm,
T = time in hours,
a = threshold parameter, and
b = constant.

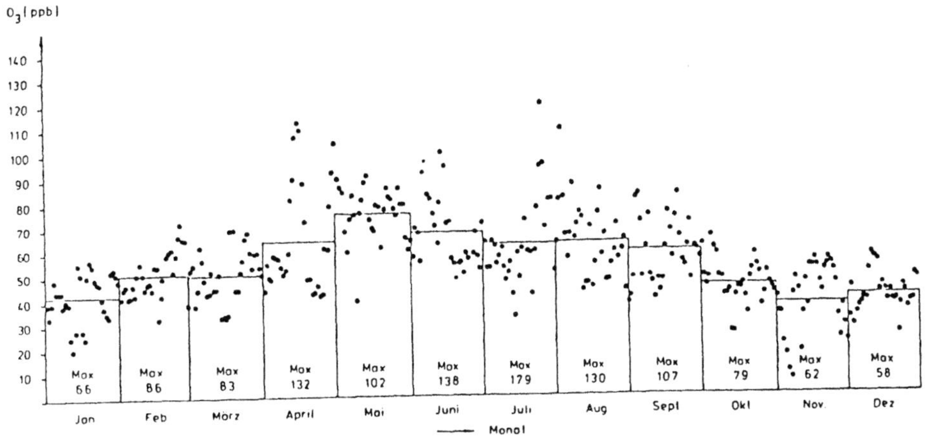

Figure 1. Results of ozone monitoring at the station Schauinsland/ South Black Forest in 1980. Represented are monthly averages (bars), monthly averages (spots), and maximum hourly averages per month (numbers). [5]

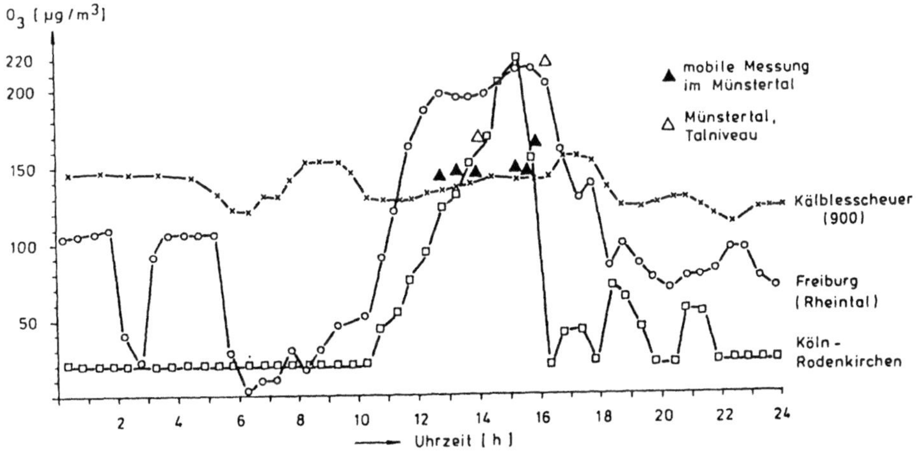

Figure 2. Diurnal variation of hour and half-hour averages at the station Kälbelesscheuer (South Black Forest, 900 m), Freiburg (Rhine valley close to Black Forest), and Cologne on one single day (18. 9. 1982). The triangles represent random samples. [5]

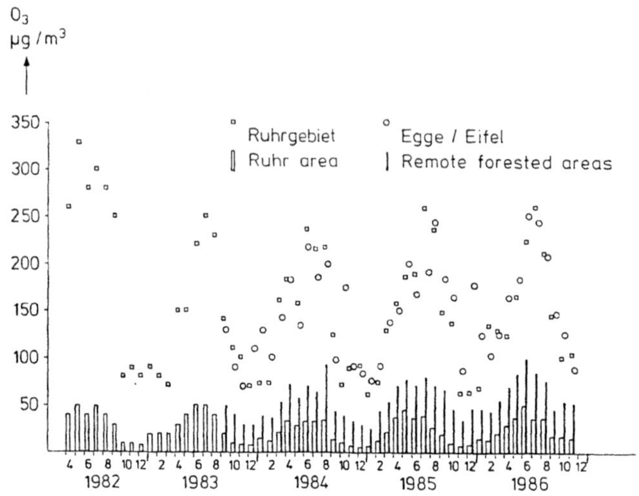

Figure 3. Monthly averages and maximum half hour averages of ozone in the mountainous stations Eggegebirge and Eifel as well as in the Ruhr area (spatial averages of several single stations). [51]

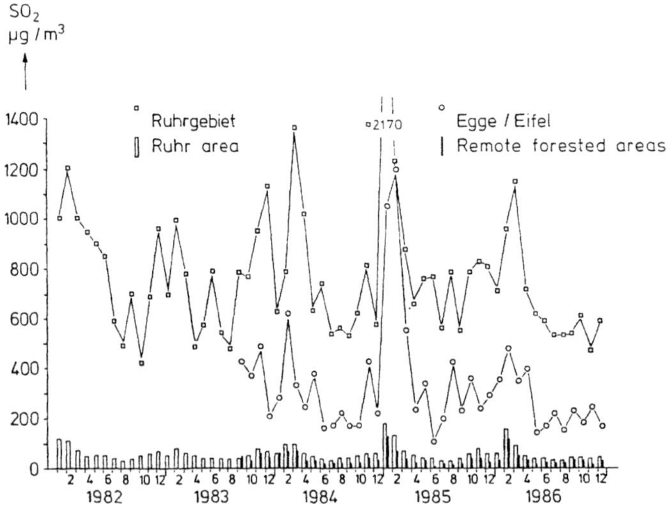

Figure 4. Monthly averages and maximum half hour averages of sulfur dioxide in the mountainous stations Eggegebirge and Eifel as well as in the Ruhr area (spatial averages of several single stations). [51]

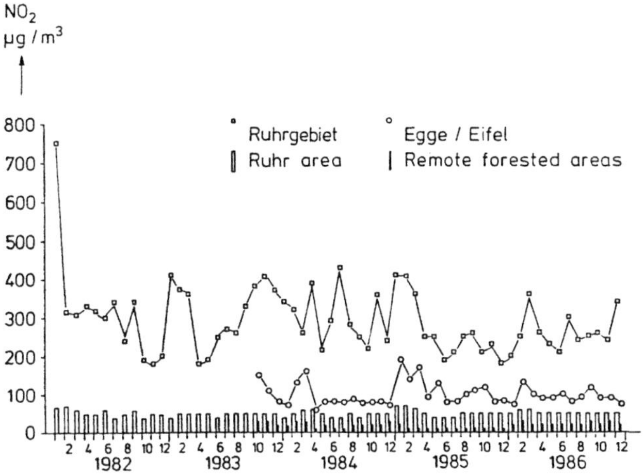

Figure 5. Monthly averages and maximum half hour averages of nitrogen dioxide in the mountainous stations Eggegebirge and Eifel as well as in the Ruhr area (spatial averages of several single stations). [51]

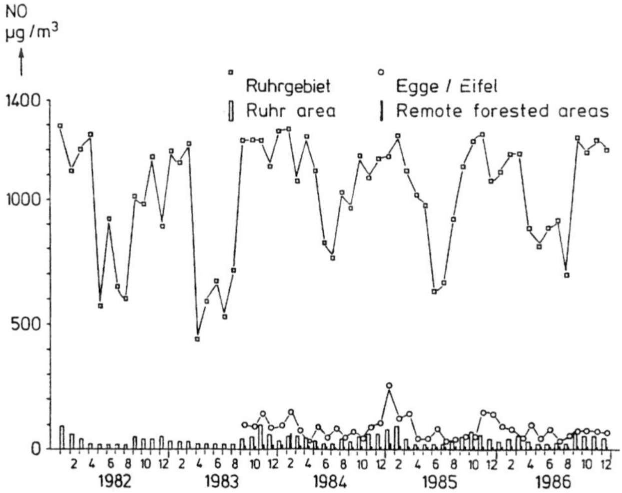

Figure 6. Monthly averages and maximum half hour averages of nitrogen monoxide in the mountainous stations Eggegebirge and Eifel as well as in the Ruhr area (spatial averages of several single stations). [51]

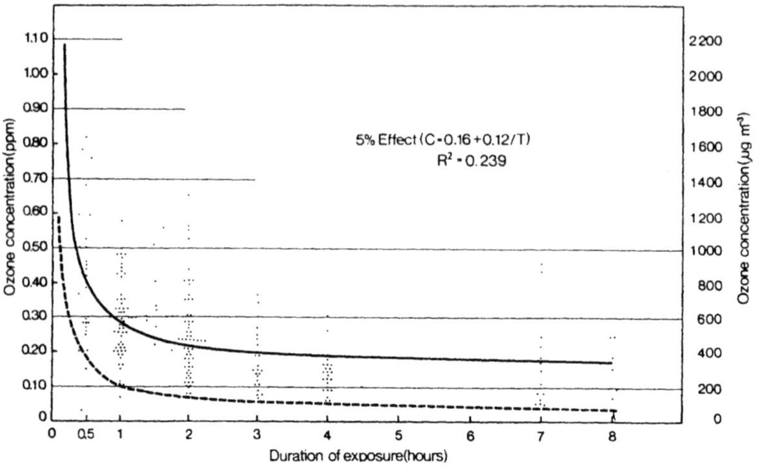

Figure 7. Ozone concentration time scatter diagram for a 5 per cent plant response from over 100 species and varieties tested. [43]

The solid curve shown in Figure 7 represents an average over many species and varieties of plants. The lower broken curve is the minimum or threshold response shown by sensitive plant species grown under favourable conditions.

From JACOBSON [44], in a similar way to LINZON et al., a combination of concentration vs. exposure time was derived (Figures 8), which should protect trees and shrubs as well as agricultural crops. Unfortunately the diagrams from LINZON et al. and from JACOBSON end with a couple of hours as exposure time so that long lasting exposures can not be evaluated by these representations.

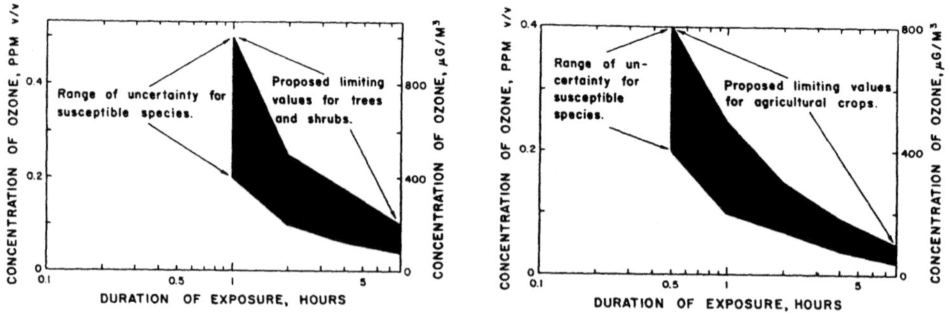

Figure 8. Proposed limiting values for foliar injury to trees and shrubs (left) as well as agricultural crops (right) by ozone. [44]

PRINZ and BRANDT [41] derived from these and other information the following proposals of concentrations limits for the protection of vegetation from the effects of atmospheric ozone (in $\mu g/m^3$) (Table II). It must again be emphasized, however, that these figures say

Table II. Proposed concentration limits for ozone to protect vegetation

Sensitivity	Average over 30 min.	Average 1 hour	Average 4 hours
Very sensitive	300	150	50
Sensitive	500	300	150
Less sensitive	1000	600	300

nothing about the risk of long living plants like forest trees.

The evaluation of the NCLAN data were done by application of the Weibull function and different approaches of exposure description. One possibility for this description was the seasonal mean of the daily 7-h (0900 to 1600 Standard Time) mean O_3 concentrations. The Weibull function is

$$y = \alpha \exp[-(\chi/\sigma)^c] + \epsilon$$

where y is the observed yield and χ is the O_3 concentration. The other variables are parameters to be estimated from the experimental results.

The parameters of the Weibull function are documented in extensive tables but unfortunately difficult to interpret. For the aim of greater transperancy the dose-response curves from an older publication of NACLAN data [46] is used in Figure 9. This figure shows that, in comparison to a "background"-concentration of 0.025 ppm, the yield very soon declines with increasing ozone concentrations. These data are rather meaningful since they represent seasonal 7-h/day mean O_3 concentrations unlike the data discussed previously.

Quantitative relationships between chronic ozone exposure and injury to forest trees are very rare. One conclusion can be drawn from the ozone concentration, which has been measured between 1968 and 1977 in the Sky Forest in the San Bernadino mountains (Table III), where forest decline by ozone was first detected. The data are taken from MCBRIDE and MILLER [47]. In the neighbouring city of Bernadino, 1349 m lower in altitude than the Sky Forest, the concentration was rather more than half. So the critical values have to be considered somewhere between these two points.

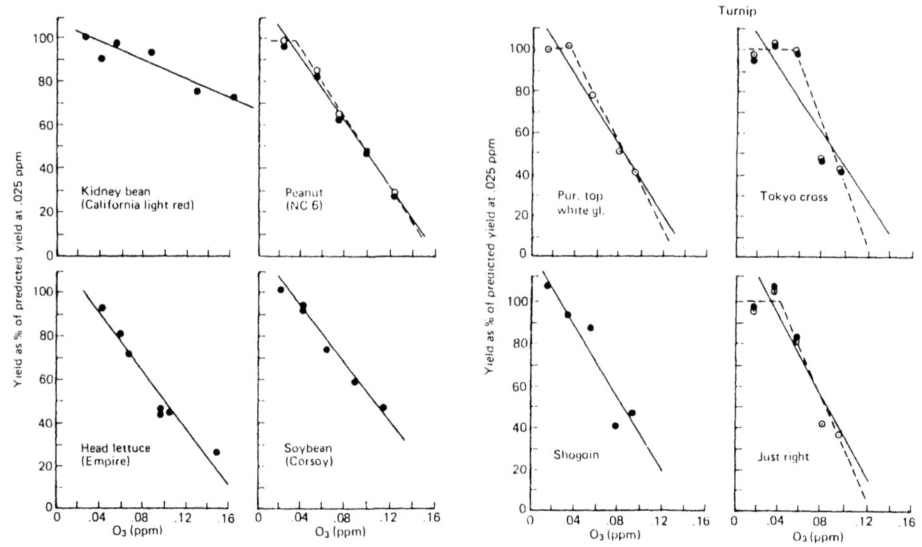

Figure 9: Dose response curves for yield of five plant species as a function of the seasonal 7-h/day mean Ozone concentration. Yield is given as the percent of the yield predicted at 0.025 ppm O_3 by a simple linear model (●) and a plateau-liner model where applicable (o). The linear model is shown as a solid line and the plateau-liner as a dashed line. [46]

Table III. Daylight (6000 to 2000) monthly average concentrations at Sky forest (1709 m above sea level) in ppb

Month	I	II	III	IV	V	VI	VII	VIII	IX	X	XI	XII
Conc.	5	20	26	44	71	91	105	100	64	38	17	8

REICH and LASSOIE [16] fumigated hybrid poplar plants with 0.025, 0.050, 0.085 and 0.125 ppm ozone for 10 weeks with 5.5 hours daily. First effects could not be observed before 6 weeks treatment, which again stresses the importance of chronic exposure. By the end of the study ozone exposure had resulted in decreased plant height and diameter and number of leaves per plant, and in decreased leaf, stem, root and total dry mass per plant as well as increased leaf senescence. With reagard to dry mass the reduction was between 10 and 20 % at a concentration of 0.085 ppm related to 0.025 ppm as a reference concentration. Similar results were gained by REICH et al. [15], when fumigating sugar maple and northern red oak trees.

In a Swedish report about the effects of air pollutants on the vegetation from 1982 [50] the following standards for ozone were proposed (Table IV). A distinction is made between "zero-effect" standard and "realistic" standard. The latter means that economic damage is involved if this standard is exceeded.

Table IV. Swedish proposals for air quality standards for ozone in $\mu g/m^3$

Exposure time	"zero-effect" st.	"realistic" effect
1 hour	100 - 200	200 - 300
7 - 8 hours	100	120
1 month	80	100
3 - 5 months	70	90

6. OZONE AND THE PHENOMENON OF NOVEL FOREST DECLINE IN WEST GERMANY

The problem of the novel type of forest decline in West Germany shall be discussed here only in a very rough manner. More detailed descriptions of this problem are given elsewhere by the author [34, 35, 36, 48, 49].

Different species of trees (coniferous and deciduous ones) are involved. It appears as if silver fir has started first with apparent symptoms of decline (in the mid-1970s) followed by Norway spruce (1980 in Southern Germany, 1982 elsewhere) and some deciduous species such as common beech and oak (1984/1985). With respect to the severity of symptoms and distribution of damage the decline of Norway spruce seems to be the most important aspect.

For Norway spruce the symptoms concentrate more or less on yellowing of needles, closely correlated with the elevation of the stand, as well as to thinning of crown, occuring more in the lower altitudes and in the plain. With respect to specificity, intensity and distribution, the type "needle chlorosis in the higher altitudes of the subalpine mountains" seem to be most important. This damage is found in areas where the concentration of sulfur dioxide in ambient air is at least partly extremely low. Therefore just these areas were called in former times clean air areas. Moreover for the type of needle chlorosis the most scientific information is available. Therefore the discussion shall be restricted in the following to this type of damage.

The yellowing of needles can clearly be associated to deficiency of nutrients, especially of magnesium , calcium , zinc, potassium and manganese. Depending on the specific site condition, i. e. supply of theses nutrients from the soil, the magenesium deficiency type and to a lesser degree the potassium-manganese deficiency type are predominant. Only the sunexposed needles turn yellow and these light induced chlorosis, at least for the magnesium deficiency type, concentrate on the last years's and older needles, where the magnsium content in plant tissue uses to be lowest (Figure 10), while the youngest needles

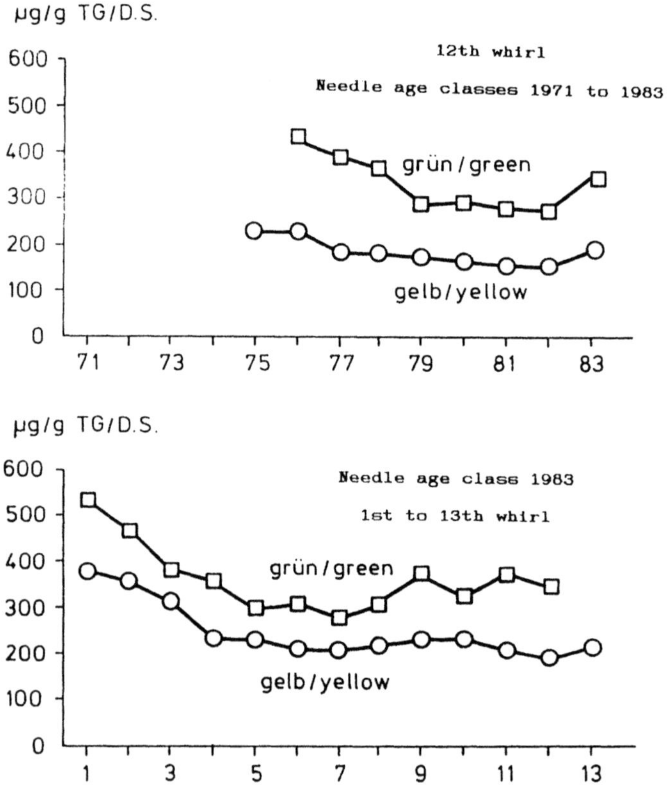

Figure 10. Variation of magnesium content in needles of Norway spruce within different needle age classes of one whirl as well as within different whirls of one needle age class. The youngest whirls and youngest needles are because of the mobility of magnesium best supplied. Due to accumulation during their longer life time the oldest needles contain also a relative high concentration, but obviously physiologically not so effective, i. e. stored at other sites than in the chloroplasts. [25]

with the best supply of magnesium use to be green. Older trees are more affected than younger ones and multiaged and mixed stands with rough canopy are more affected than even-aged and uniform stands with closed canopies.

On the side of potential causal factors it has principally to be taken into account that apart from the abnormal frequency of dry and warm summmers in the past (see Figure 11) the ozone concentration has increased during the last decades, especially at the sites remote from the centres of industrial and urban emission. Moreover the concentration of nitrate in rain has shown during the last years an upward trend, while the concentration of sulfate has decreased. It has to be considered that the most damaged sites in the higher alitudes are characterised by long lasting fog exposures. There are also some indices that the forest soils, due to the influence of acidic deposition have undergone meanwhile a loss of nutrients and a decrease in buffer capacity. Related to this assumption the aluminium concentration in the soil solution, which is principally toxic to plant roots, may also be higher now than in former times, but uncomparable to the ozone concentration the clear spatial correspondence of the aluminium concentration to the intensity of damage is lacking.

As the most probable explanation for the causes of the novel forest decline, which in outline was first presented by PRINZ et al.

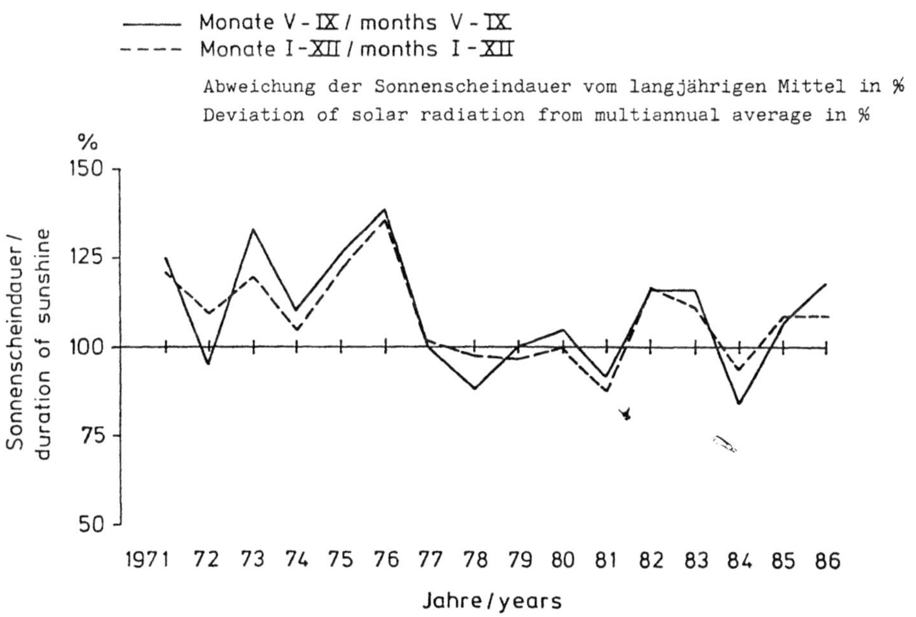

Figure 11. Variation of solar radiation, temperature, and rainfall within the last 15 years at the meteorological station in Lüdenscheid (Sauerland mountains in Northrhine-Westphalia)

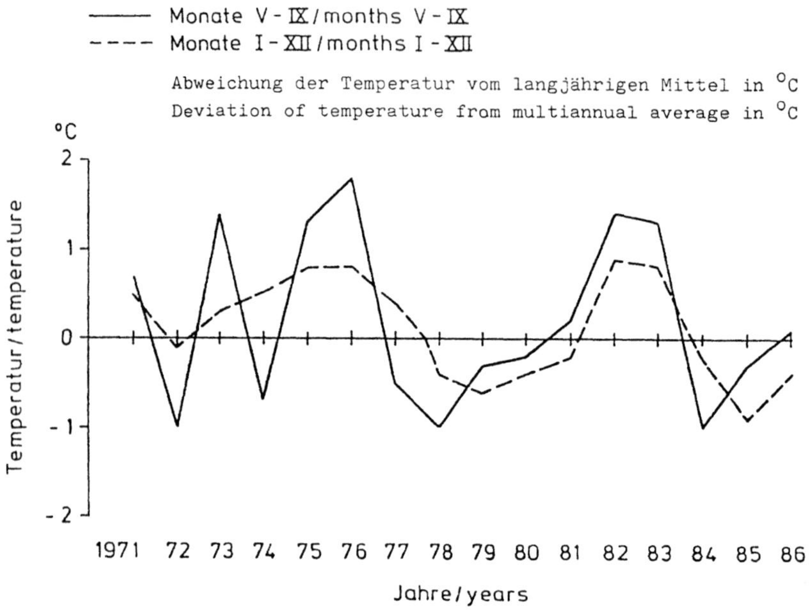

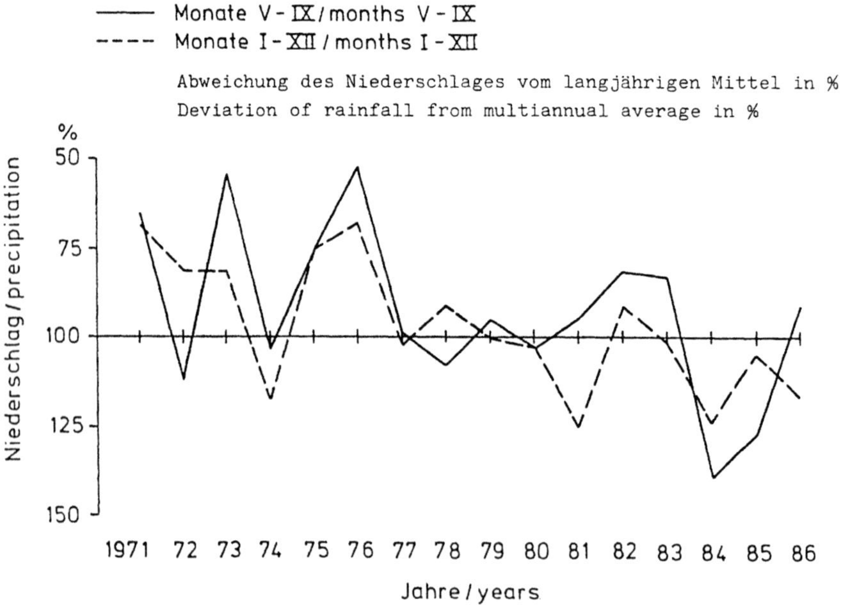

Figure 11. (Contin.)

[5], it is assumed that according to Figure 12 ozone affects the energetic metabolism in a direct way and furthermore weakens the cell

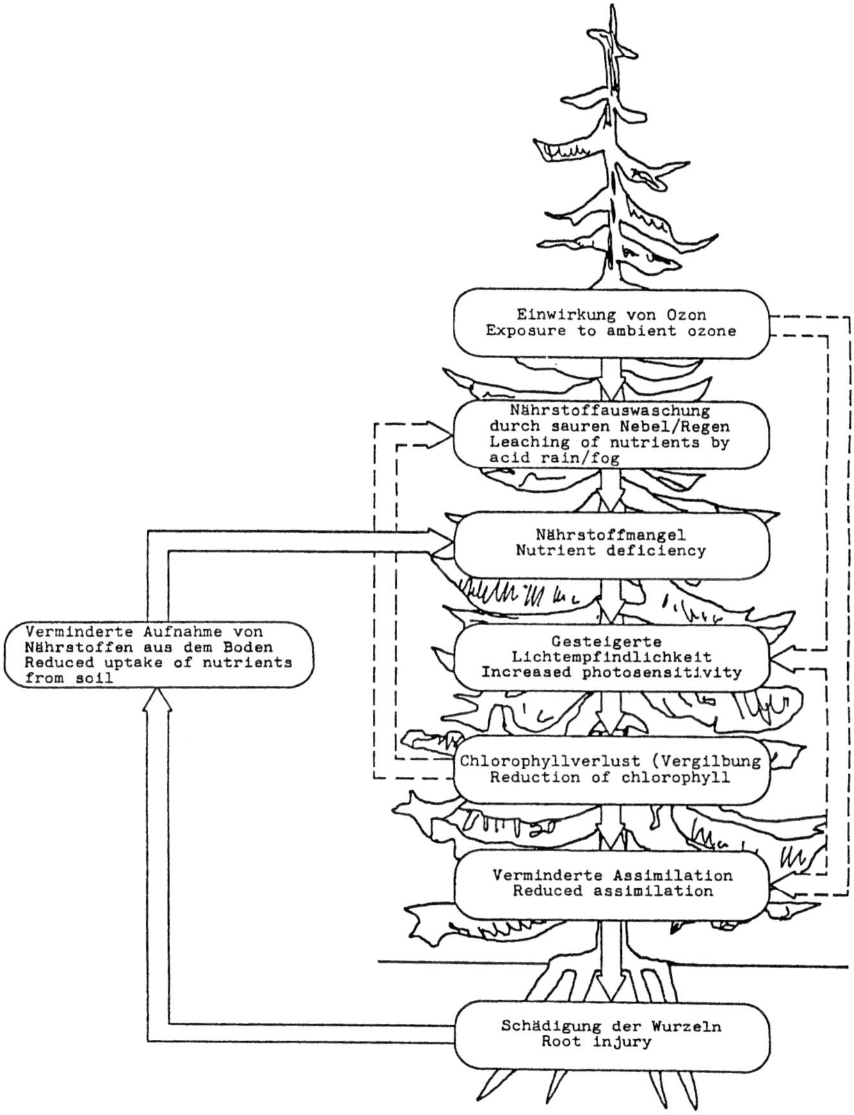

Figure 12. Impact of ozone and acidic fog and rain on coniferous trees. [34]

membrane system so that in combination with acidic rain and fog the leaching of essential nutrients is enhanced. As a further result the root system as the heterotrophous part of the plant is affected and the uptake of nutrients therefore is disturbed which means that from both sides the nutrient supply of the needles is hindered. It is quite evident from this assumption that without any doubt the soil is an important accompanying factor, which helps to explain the microspatial difference (geogenic influence) as well as the long term development of damage (anthropogenic influence). Last not least the climate has to be taken into consideration as a most important trigger factor. This means that the more or less sudden appearance of the symptoms of the novel forest decline at so many sites can only be understood if the climate has had a really strong influence on the outbreak of these symptoms. Summing up these points the following approach explaining the causes of needle chlorosis in the higher altitudes may be used.

Cause for temporal development:
- upward trend of ozone concentration during the last decades
- continuous loss of nutrients in soil by acidic deposition.

Cause for spatial distribution:
- increase of ozone concentration (as a daily average) with increasing altitude
- geogenic differences in nutrient supply.

Cause for short term appearance (and recovery):
- Climate as a triggering or synchronizing factor

ACKNOWLEDGEMENT

The author thanks Dr. G. Vaughan for the most helpful revision of the English manuscript.

REFERENCES

[1] WISCLICENUS, H. (Ed.): 'Sammlung von Abhandlungen über Abgase und Rauchschäden'. Paul Parey, Berlin 1907. Reprint VDI-Verlag Düsseldorf 1985.

[2] 'Ozone and Other Photochemical Oxidants'. Ed. National Academy of Sciences. Washington D.C. 1977.

[3] HECK, W. W. , O. C. Taylor, R. Adams, G. Bingham. J. Miller, E. Preston, and L. Weinstein: 'Assessment of crop loss from ozone' *JAPCA* **32** (1982), 353 - 361.

[4] 'Air Pollution by Photochemical Oxidants'. Ed. R. Guderian. Springer-Verlag Berlin Heidelberg New York Tokyo 1985.

[5] PRINZ, B., G. H. M. KRAUSE und H. STRATMANN: 'Waldschäden in der Bundesrepublik Deutschland'. *LIS-Berichte Nr. 28*. Essen 1982.

[6] PAREMTER, J. R. and P. R. MILLER: 'Studies relating to the cause of decline and death of Ponderosa pine in Southern California'. *Plant Disease Reporter* **52** (1968), 707-711.

[7] DUCHELLE, St. F., J. M. SKELLY, and B. I. CHEVONE: 'Oxidants effects on forest tree seedling growth in the Appalachian Mountains'. *Water, Air, and Soil Pollution* **8** (1982), 363-373.

[8] AMIRO, B. D. and T. J. GILLESPIE: 'Leaf conductance response of Phaseolus vulgaris to ozone flux density'. *Atmospheric Environment* **19** (1985), 807-810.

[9] JENSEN, K. F. and B. R. ROBERTS: 'Changes in yellow poplar stomatal resistance with SO_2 and O_3 fumigation'. *Environ. Pollut. Ser. A* **41** (1986), 235 - 245.

[10] KEITEL, A. and U. ARNDT: 'Ozoninduzierte Turgeszenzverluste bei Tabak (Nicotiana tabacum var. Bel W 3) - ein Hinweis auf schnelle Permeabilitätsveränderungen der Zellmembranen'. *Angew. Botanik* **57** (1983), 193 - 204.

[11] RENGER, G.:' Wirkung von Schadstoffen auf die Primärprozesse der Photosynthese'. Statusseminar "Ursachenforschung zu Waldschäden". Jülich, 30 March - 3 April 1987. (Proc. in prep.)

[12] SCHAUB, H., I. ZWOCH, and R. JURAT: 'Untersuchungen über den Einfluß schwefelhaltiger Verbindungen und Ozon (einzeln und auch in Kombination) auf Wachstum und Entwicklung von Pflanzen'. Statusseminar "Ursachenforschung zu Waldschäden". Jülich, 30 March - 3 April 1987. (Proc. in prep.)

[13] REICH, P. B. and R. G. AMUNDSON: 'Ambient levels of ozone reduce net photosynthesis in tree and crop species'. *Science* **230** (1985), 566 - 230.

[14] REICH. P. B., A. W. SCHOETTLE, R. M. RABE, and R. G. AMUNDSON: 'Response of soybean to low concentrations of ozone: I. Reductions in leaf and whole plant net photosynthesis and leaf chlorophyll content'. *J. Environ. Qual.* **15** (1986), 31 - 36.

[15] REICH, P. B., A. W. SCHOETTLE, and R. G. AMUNDSON: 'Effects of O_3 and acidic rain on photosynthesis and growth in sugar maple and northern red oak seedlings'. *Environ. Pollut. Ser. A* **40** (1986), 1 - 15.

[16] REICH, P. B. and J. P. LASSOIE: 'Influence of low concentrations of ozone on growth, biomass partitioning and leaf senescence in young hybrid poplar plants'. *Environ. Pollut. Ser. A* **39** (1985), 39 - 51.

[17] YANG, Y.-S., J. M. SKELLY, B. I. CHEVONE, and J. B. BIRCH: 'Effects of long-term ozone exposure on photosynthesis and dark respiration of eastern white pine'. *Environ. Sci. Technol.* **17** (1983) 371 - 373.

[18] JENSEN, K.: 'Response of yellow poplar seedlings to intermittent fumigation'. *Environ. Pollut. Ser. A* **38** (1985), 183 - 191.

[19] ARNDT, U. and M. KAUFMANN: 'Wirkungen von Ozon auf die apparente Photosynthese von Tanne und Buche'. *AFZ* (1985), 19 - 20.

[20] KUNERT, K. J. and G. HOFER: 'Lipidperoxidation als phytotoxische Folge atmosphärischer Schadstoffwirkungen'. Statuskolloquium des PEF, Karlsruhe, 10 - 20 March 1987 (Proc. in prep.)

[21] MEHLHORN, H. G. SEUFERT, A. SCHMIDT, and K. J. KUNERT: 'Effects of SO_2 and O_3 on Production of antioxidants in conifers'. *Plant Physiol.* **82** (1986), 336 - 338.

[22] SELINGER, H., D. KNOPPIK, and A. ZIEGLER-JONES: 'Einfluß von Mineralstoffernährung, Ozon und saurem Nebel auf Photsynthese-Parameter und stomatäre Leitfähigkeit von Picea abies (L.) Karst'. *Forstw. Cbl.* **105** (1986), 239 - 242.

[23] KARHU, M. and S. HUTTUNEN: 'Erosion effects of air pollution on needle surfaces'. *Water, Air, and Soil Pollution* **31** (1986), 417 - 423.

[24] MAGEL, E. und H. ZIEGLER: 'Einfluß von Ozon und saurem Nebel auf die Struktur der stomatären Wachspfropfen in den Nadeln von Picea abies (L.) Karst.'. *Forstw. Cbl.* **105** (1986), 234 - 238.

[25] KRAUSE, G. H. M. and B. PRINZ: *LIS-Berichte* (in prep.).

[26] KRAUSE, G. H. M. and B. PRINZ: 'Zur Wirkung von Ozon und saurem Nebel (einzeln und in Kombination) auf phänomenologische und physiologische Parameter an Nadel- und Laubgehölzen im kombinierten Begasungsexperiment'. Statusseminar "Wirkungen von Luftverunreinigungen auf Waldbäume und Waldboden". Proc. ed. by KFA Jülich 1986, 208 - 221.

[27] KRAUSE, G. H. M., K.-D. JUNG, and B. PRINZ: 'Experimentelle Untersuchungen zur Aufklärung der neuartigen Waldschäden in der Bundesrepublik Deutschland'. *VDI-Berichte 560: Waldschäden*. Düsseldorf, Oktober 1985, 627 - 656.

[28] BOSCH, Chr., E. PFANNKUCH, K. E. REHFUESS, P. SCHRAMEL, and M. SENSER: 'Einfluß einer Düngung mit Magnesium und Calcium, von Ozon und saurem Nebel auf Frosthärte, Ernährungszustand und Biomasseproduktion junger Fichten (Picea abies (L.) Karst.)'. *Forstw. Cbl.* **105** (1986), 218 - 229.

[29] SCHMITT, U., M. RUETZE, and W. LIESE: 'Rasterelektronenmikroskopische Untersuchungen an Stomata von Fichten- und Tannennadeln nach Begasung und saurer Beregnung'. *Eur. J. For. Path.* **17** (1987), 118 - 124.

[30] KOBRIGER, J. M. and Th. W. TIBBITTS: 'Effect of relative humidity prior to and during exposure on response of peas to ozone and sulfur dioxide'. *J. Amer. Hort. Sci.* **110** (1985), 21 - 24.

[31] TEMPLE, P. J., O. C, TAYLORL, AND L. F. BENOIT: 'Cotton yield responses to ozone as mediated by soil moisture and evatransporation'. *J. Envirn. Qual.* **14** (1985), 55 - 60.

[32] GUDERIAN, R., K. KÜPPERS, and R. SIX: 'Wirkungen von Ozon, Schwefeldioxid und Stickstoffdioxid auf Fichte und Pappel bei unterschiedlicher Versorgung mit Magensium und Kalzium sowie auf die Blattflechte Hypogymnia physodes'. VDI Berichte 560 "Waldschäden". Düsseldorf 1985, 657 - 702.

[33] KRAUSE, G. H. M.: 'Zur Wirkung von Ozon auf die Nitratbildung in Nadeln und Blättern von Picea abies (L.) Karst., Fagus sylvatica (L.) und Quercus robur (L.)'. Statuskolloquium "Luftverunreinigungen und Waldschäden" Düsseldorf, Oktober 1986 (Proc. in press)

[34] PRINZ, B.: 'Waldschäden in den USA und in der Bundesrepublik Deutschland - Betrachtungen und Ursachen'. *VGB Kraftwerkstechnik* **65** (1985), 930 - 938.

[35] PRINZ, B., G. H. M. KRAUSE, and K.-D. JUNG: 'Untersuchungen der LIS zur Problematik der Waldschäden'. In: *Waldschäden - Theorie und Praxis auf der Suche nach Antworten.* R. Oldenbourg Verlag München - Wien 1985, 143 - 194.

[36] PRINZ, B. und G. H. M. KRAUSE: 'Waldschäden in der Bundesrepublik Deutschland'. *Staub* **47** (1987), 94 - 100.

[37] MENSER, H. A. and H. E. HEGGESTADT: 'Ozone and sulfur dioxide synergism: injury to tobacco plants'. *Science* **153** (1966), 424 -425.

[38] ASHMORE, M. R.: 'Modification by sulphur dioxide of the responses of Hordeum vulgare to ozone'. *Environ. Pollut. Ser. A.* **36** (1984), 31 - 43.

[39] SMIDT, S.: 'Begasungsversuche mit SO_2 und Ozon an jungen Fichten'. *Eur. J. For. Path.* **14** (1984), 214 - 248.

[40] DAVIS, D. D. and R. G. WILHOUR: 'Susceptibility of woody plants to sulfur dioxide and photochemical oxidants. A literature review'. EPA-660/3-76-102.

[41] PRINZ, B. and C. F. BRANDT: 'Study on the impact of the principal atmospheric pollutants on the vegetation'. Ed. CEC 1980, EUR 6644 EN.

[42] HECK, W. W., W. W. CURE, J. O. RAWLINGS, L. J. ZARAGOZA, A. S. HEAGLE, H. E. HEGGESTGAD, R. J. KOHUT, L. W. KRESS, and P. J. TEMPLE: 'Assessing impacts of ozone on agricultural crops: I. Overview'. *JAPCA* **34** (1984), 729 - 817.

[43] LINZON, S. N., W. W. HECK, and F. D. H. MACDOWELL: 'Effects of photochemical oxidants on vegetation'. In: *Photochemical air pollution - formation, transport, effects*. Nat. Res. Counc., Canada 1975.

[44] JACOBSON, J. S.: 'The effects of photochemical oxidants on vegetation'. VDI-Berichte 270, Düsseldorf 1977, 191 - 196.

[45] HECK, W. W., W. W. CURE, J. O. RAWLINGS, L. J. ZARAGOZA, A. S. HEAGLE, H. H. HEGGESTAD, R. J. KOHUT, L. W. KRESS, and P. J. TEMPLE: 'Assessing impacts of ozone on agrucultural crops: II. Crop yield functions and alternative exposure statistics'. *JAPCA* **34** (1984), 810 - 817.

[46] HECK, W. W., O. C. TAYLOR, R. ADAMS, G. BINGHAM, J. MILLER, E. PRESTON, and L. WEINSTEIN: 'Assessment of crop loss from ozone'. *JAPCA* **32** (1982), 353 - 361.

[47] MCBRIDE, J. R. and P. R. MILLER: 'Responses of American Forests to photochemical oxidants'. Proc. of the NATO Advanced Research Workshop "Effects of Atmospheric Pollutants on Forests, Wetlands and Agricultural Ecosystems", Toronto, May 12 - 17, 1985. Springer-Verlag (in press).

[48] PRINZ, B.: 'Effects of air pollution on forests - Critical review discussion papers'. *JAPCA* **35** (1985), 913 - 915 (Suppl. in JAPCA 36 (1986), 15).

[49] PRINZ, B.: 'Major hypotheses on forest damage causes in Europe and North America'. *Environment* (in prep.)

[50] SKÄRBY, L.: 'Effekter av luftföroreninger pa vegetatiom - Fotokemiska oxidanter'. *Naturvardsverket Rapport* snv pm 1562. Solna 1982.

[51] KRAUSE, G. H. M.: 'Impact of air pollutants on above-ground plant parts of forest trees'. European Communities Symposium on Effects of Air Pollution on Terrestrial and Aquatic Ecosystems. Grenoble, France, 18. - 22. 5. 87 (Proc. in prep.).

THE MEASUREMENT OF NO_x IN THE NON-URBAN TROPOSPHERE

F. C. Fehsenfeld, D. D. Parrish, and D.W. Fahey
Aeronomy Laboratory
Environmental Research Laboratories
National Oceanic and Atmospheric Administration
325 Broadway
Boulder, CO 80303

ABSTRACT. The measurements of NO_x that have been made in the non-urban troposphere are reviewed. The NO_x levels in the clean remote troposphere are very low: 0.01 ppbv at the surface in the Mid-Pacific, but increasing with altitude to approximately 0.2 ppbv in the upper troposphere. The upper tropospheric NO_x levels are maintained by lightning, stratospheric injection, high flying aircraft and convective transport of surface pollution. Surface NO_x mixing ratios over large regions of the industrial Northern Hemisphere exceed 1 ppbv. At the surface in more isolated continental areas the NO_x mixing ratio will be intermediate between those found in industrial regions and remote maritime locations. The NO_x mixing ratio in the clean troposphere may be strongly influenced by other reactive nitrogen oxides such as PAN. PAN and PAN-like compounds may be the dominate reactive nitrogen oxide species in many regions of the troposphere. This complicates the estimation of the tropospheric NO_x distribution and presents substantial difficulties in the measurement of tropospheric NO_x levels.

INTRODUCTION

Present estimates indicate that most of the ozone in the troposphere is photochemically produced there (Chameides and Walker, 1973; Fishman et al., 1979; Hov, 1983; Fishman et al., 1985; Logan et al. 1981; Liu et al., 1987). It is generally believed that this production of ozone is NO_x (NO_x = $NO + NO_2$) limited throughout most of the troposphere (Liu et al., 1987). Thus, in order to estimate the ozone production potential of the atmosphere and to determine to what extent the ozone production in the atmosphere is altered by man's activities, it is necessary to know the atmospheric NO_x content.

A predictive capability for tropospheric ozone requires detailed information concerning the NO_x distribution for two reasons. First, the efficiency of ozone production by NO_x decreases with increasing NO_x

I. S. A. Isaksen (ed.), Tropospheric Ozone, 185–215.

level (Liu et al., 1987). The efficiency near urban and heavily industrialized areas which have relatively large NO_x mixing ratios is less than in remote regions which typically have very low NO_x ratios. Second, solar UV intensity, water vapor concentration and non-methane hydrocarbon concentration vary with location, altitude, season and time of day. This variability yields large variations in the concentrations of the oxidizing free radicals. These photochemically produced species interact with NO_x to produce O_3 while in the absence of NO_x they destroy O_3. Thus, knowledge of the variation of NO_x with location and altitude and of its seasonal and diurnal cycles is required.

A complete determination of the NO_x distribution is difficult because it is highly variable for several reasons. First, several natural and anthropogenic sources inject NO_x into the atmosphere. These sources are highly non-uniform, both spatially and temporally. Second, the lifetime of NO_x in the atmosphere is short enough that the non-uniformity of the sources is not strongly attenuated by atmospheric transport and mixing. Furthermore, the lifetime varies strongly with latitude and season. Third, the processes that transport NO_x in the atmosphere are strongly coupled with the NO_x source locales as well as with atmospheric photochemistry. Several aspects of these three points will be discussed further in the interpretation of reported measurements. In particular, experimental results that help to define the NO_x lifetime will be reviewed. These three features combine to yield variability over at least four orders of magnitude of NO_x levels even excluding urban areas.

Much effort has been directed toward measurements of NO_x, but only during the past ten years have techniques been available with sufficient sensitivity and dynamic range to measure NO_x outside urban environments. However, the critical evaluations of NO_x measurement techniques that have recently been undertaken indicate that many of these measurements, particularly the measurement of NO_2, may have suffered from significant interferences (c.f. Fehsenfeld et al., 1987). Nevertheless, our knowledge of the potential interferences allows the information obtained to yield meaningful constraints on the NO_x concentrations at a variety of locations.

In this article, the available NO_x measurement methods and their limitations will be briefly reviewed. The NO_x measurements made using these techniques will be presented and evaluated . The results from surface studies will be divided according to broad location classifications to begin ellucidating the spatial distribution of NO_x in the planetary boundary layer (PBL). The aircraft measurements will be summarized to give an indication of the altitude profile of NO_x. The small amount of information concerning seasonal and diurnal cycles will also be summarized. Finally, since the distribution of NO_x is a strong function of its transport, particularly from anthropogenic sources, the lifetime of NO_x, and therefore its potential for transport, will be discussed.

MEASUREMENT TECHNIQUES

At present a variety of techniques are emerging that are capable of measuring NO, NO_2, NO_X and the sum of the reactive nitrogen oxides (NO_y = NO + NO_2 + PAN + HNO_3 +...) at mixing ratios well below one part in 10^9 by volume (1 ppbv).

The capability of currently available instrumentation to measure NO mixing ratios in the atmosphere at these low levels has now been established rigorously. Two fundamentally different methods have been intercompared: chemiluminescence (NO-O_3 chemical reaction and emission of radiation from the NO_2 product) and laser induced fluorescence (absorption of radiation by NO and then re-radiation at different wavelengths by the excited NO). Over two month-long campaigns, two chemiluminescence instruments and a laser induced fluorescence apparatus made simultaneous ambient measurements from an eastern U.S. rural surface site and from an aircraft in the western U.S. (Hoell et al., 1985; Hoell et al., 1987). The data agreed within 30% in all of these different environments which spanned a mixing-ratio range of 0.0005 to 0.2 ppbv. The conclusion from these intercomparisons that chemiluminescence instruments were capable of reliable NO measurements at low mixing ratios was of great significance since this type of detector has been used for a large majority of the NO measurements that have been made in the rural and remote troposphere.

Most atmospheric measurements of NO_2 and NO_X have used conversion devices to convert NO_2 to NO followed by chemiluminescence detection. Recently, two of these conversion techniques, photolysis and surface conversion of NO_2 to NO on ferrous sulfate ($FeSO_4 \cdot 7H_2O$), have been intercompared (Fehsenfeld et al., 1987). This key result is given in Figure 1. The levels ranged from 0.07 to 100 ppbv during the sampling period. NO_2 was the dominant component of that sum. The surface and photolytic converters agreed well at NO_X levels of 1 ppbv and greater. However, the surface converter systematically reported higher values at lower NO_X levels, and reached a factor of two higher at 0.1 ppbv. Spiking test showed that the ferrous sulfate surface converter was also converting peroxyacetyl nitrate (PAN) to NO. This intercomparison thereby revealed a significant bias in the measurement of NO_2 using this surface converter.

The intercomparison results suggest that all surface converters that are sufficiently robust to convert NO_2 to NO also can be expected to convert other reactive nitrogen oxide species such as PAN. An interference from these compounds in the measurement of NO_2 in chemiluminescence detectors employing surface converters must be considered in the evaluation of the reported results and will enter significantly into the present discussion of these results.

The implication of this interference for interpretation of NO_X measurements made using surface converter/chemiluminescence detectors

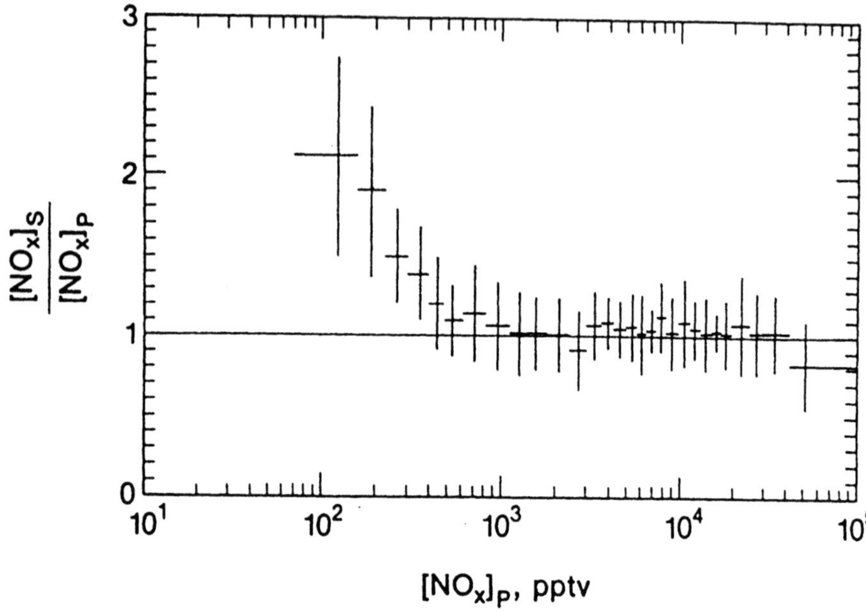

Fig. 1. Plot of $[NO_x]_S/[NO_x]_P$ adapted from Fehsenfeld et al. (1987), where S represents data measured using a ferrous sulfate converter for NO_2 and P indicates a photolytic converter for NO_2, vs $[NO_x]_P$. Each symbol contains the average of 50 separate, simultaneous measurements. For each set, the vertical bar represents the standard deviation of the Measurements while the horizontal bar is the span of the $[NO_x]_P$ mixing ratio for each set of measurements.

can be surmised from the data presented in Figure 2. Shown here is the ratio of PAN to NO_x as a function of the NO_x level (c.f. Singh et al., 1985). In these studies NO_x was measured using a photolysis/chemiluminescence detector that has very low sensitivity to PAN (PAN response <6% that of NO_2). At this surface site located in the western United States the PAN to NO_x ratio is less than 0.5 for NO_x levels greater than 0.5 ppbv. Presumably at these elevated NO_x levels at this site there is not sufficient photochemical aging of NO_2 to produce larger PAN levels. Errors in the measurement of NO_x by detectors that have a sensitivity to PAN would not be important under these circumstances. On the other hand, for NO_x levels less than 0.1 ppbv the PAN to NO_x ratio is 1 or larger. Thus under these conditions NO_x measurements made using surface converters for NO_2 would be subject to errors of a factor of two or more. Because PAN is photochemically produced but thermally unstable, the potential error will depend strongly on location, season and altitude. Thus, all measurements made with surface

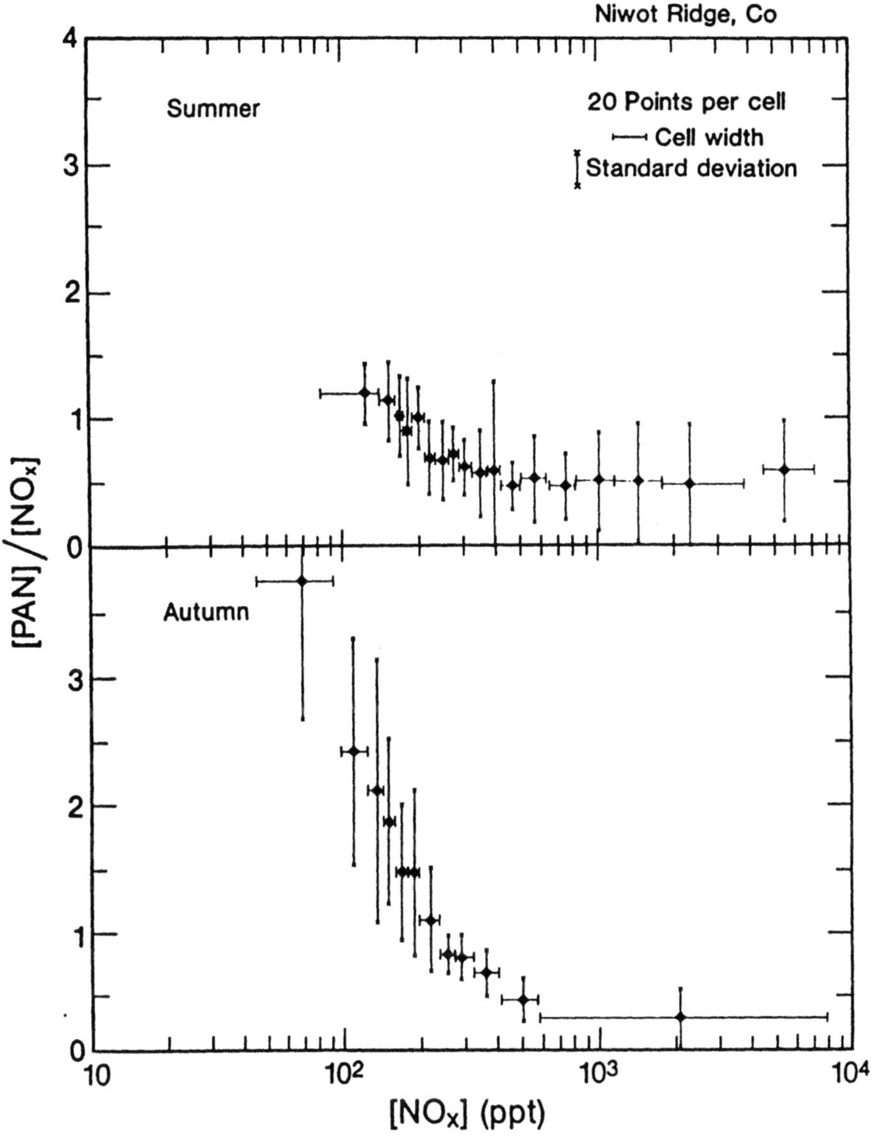

Fig. 2. [PAN]/[NO_x] Ratio as a function of [NO_x] for the summer and autumn of 1984 at Niwot Ridge from Singh et al. (1985). Each symbol gives the average of 20 measurements. The vertical bars indicate the standard deviation of the measurements while the horizontal bars show the range of [NO_x] included in each average.

converters must realistically be considered to be upper limits for NO_2 or NO_x and will be indicated with an asterisk in later discussion.

SURFACE NO_x MEASUREMENTS

Extensive measurements of NO_x have been underway for over twenty years. However, most of the measurement have been made in urban areas and, principally, in the United States. Since this review is aimed at establishing the systematics of the NO_x distribution that is characteristic of the larger regions of the troposphere, the NO_x distribution in urban areas will be specifically excluded from this discussion. The surface measurements will be divided into four principal site types characteristic of the site location relative to NO_x sources. The site types are: (1) rural sites in industrial regions in the Northern Hemisphere, (2) isolated inland sites, (3) coastal inflow sites, and (4) remote maritime sites.

The results obtained at rural sites located in industrial regions of the Northern Hemisphere are listed in Table 1. The NO_x measurements made at many of these locations were recently reviewed by Altshuller (1986) and much of his work is included here. The sites listed in Table 1, which are principally concentrated in the Eastern United States and Western Europe, are embedded in a complex matrix of anthropogenic NO_x emissions. For this reason they are never sufficiently remote from anthropogenic NO_x sources to have atmospheric concentrations that reflect the natural background NO_x. Almost all these measurements were made with chemiluminescence detectors with surface converters for NO_2. However, since the NO_x levels are so large with average NO_x mixing ratios ranging from 3 ppbv to 20 ppbv, the uncertainty contributed by conversion of NO_y species other than NO_2 may not be great. Another source of error in the reported results is the problem of maintenance of quality in these long term studies. At these sites, there is great variability in the levels on a short term basis. This variability indicates the influence of the local meteorology in transporting NO_x from local sources to these sites.

Measurements have also been carried out at more isolated sites in North America. A resume of these results are shown in Table 2A. These sites were chosen to elucidate the mechanisms that control the photochemical production of O_3 as well as to establish the background NO_x levels. The NO_x levels at these sites are also dominated by anthropogenic NO_x under certain meteorological conditions. The reported levels, although much lower than the less isolated sites listed in Table 1 are substantially greater than levels expected for remote locations. Even so, the levels are sufficiently low that NO_x mixing ratios determined with chemiluminescence detectors using surface converters may suffer significant interferences from NO_y compounds such as PAN.

The factors that shape the annual NO_x distribution at these isolated rural surface locations are clearly shown in Figure 3. The data were taken at Niwot Ridge, Colorado from October, 1980 through Septem-

Table 1. Average NO, NO_2, and/or NO_x mixing ratios measured at non-urban monitoring locations. (*Upper limit for NO_2 and NO_x)

Reference	Location	[NO] ppbv	[NO_2] ppbv	[NO_x] ppbv	Measurement Period
	A. <u>USA</u>				
Research Triangle Institute, 1975	McHenry, MD		6*		June-Aug., 1974
	DuBois, PA		10*		
	McConnelsville, OH		6*		
	Wilmington, OH		6*		
	Wooster, OH		6*		
Decker et al., 1976	Bradford, PA	2	3*		June-Sept., 1975
	Creston, IA	4	2*		
	Deridder, LA	1	3*		June-Oct., 1975
					1975
Martinez and Singh, 1979	Montague, MA	3	7*		Aug.-Dec., 1977
	Scranton, PA	3	11*		
	Indian River, DE	3	5*		1977
	Res. Tr. Pk., NC	10	13*		
	Lewisburg, WV	1	5*		
	Duncan Falls, OH	1	8*		
	Fort Wayne, IN	3	7*		
	Rockport, IN	3	10*		
	Giles Co., TN	5	11*		
	Jetmore, KA	1	4*		Apr.-May, 1978
Pratt et al., 1983	Lamoure Co., ND	2.4	1.7*		1978
		4.8	1.5*		1979
		3.3	2.8*		1980
		2.7	2.1*		1981
	Wright Co., MN	3.2	5.4*		1978
		3.0	6.7*		1979
		3.5	5.8*		1980
		2.9	4.7*		1981

Table 1. Average NO, NO_2 and/or NO_x mixing ratios measured at non-urban monitoring locations (Cont.) (*Upper limit for NO_2 and NO_x)

Reference	Location	[NO] ppbv	[NO_2] ppbv	[NO_x] ppbv	Measurement Period
Pratt et al., 1983	Traverse Co. MN	3.6	3.7*		1978
		4.8	3.6*		1979
		4.0	2.9*		1980
		2.0	2.2*		1981
Parrish et al., 1986	Scotia, PA			3.0 (2.1)[1]	June-July, 1986
	B. WESTERN EUROPE				
Harrison and McCartney, 1980	Heysham Peninsula near Lancaster, UK		10*		May-Sept., 1977
Martin and Barber, 1981	Near Bottesford, Leicestershire, UK	6-10	9-10*		1978
Messina, 1985	Deuselbach FRG		19*		Jan
			20*		Feb
			12*		March
			10*		April
			6*		May
			6*		Aug
Broll et al., 1984	Deuselbach FRG	0.8	5.2*	6.0*	Sept, 1980
		1.4	15.6*	17.0*	Oct
		1.3	17.3*	18.7*	Nov
		1.9	17.1*	19.0*	Dec
		1.0	7.7*	8.7*	March, 1981
		2.1	11.2*	13.3*	April
		0.5	4.6*	5.1*	May
		0.3	4.0*	4.3*	June
		0.6	3.9*	4.5*	July
		0.8	4.9*	5.7*	Aug

[1]Median shown in parenthesis

Table 2. Average NO, NO_2, and/or NO_x mixing ratios measured at isolated inland sites, and coastal inflow sites. (*Upper limit for NO_2 and NO_x)

Reference	Location	[NO] ppbv	[NO_2] ppbv	[NO_x] ppbv	Measurement Period
	A. ISOLATED INLAND				
Kelly et al., 1980	Niwot Ridge, CO			0.2*	Jan-April, 1979
Kelly et al., 1982	40 km WNW Pierre, SD		1.2*		July-Sept.,1978
Kelly et al., 1984	Schaeffer Obs. Whiteface Mt., NY	$\leq$0.2 [NO_x]		1.1*	July, 1982
Bollinger, et al., 1984	Niwot Ridge, CO			0.80 (0.29)[1]	1980,1981,1983, 1984
	B. COASTAL INFLOW				
Cox, 1977	Adrigole, CO Cork, Eire	<0.2	0.34 *		Aug-Sept., 1974
Platt and Perner, 1980	Loop Head Lighthouse, Eire		$\leq$0.18		April, 1979
Helas and Warneck, 1981	Loop Head Lighthouse, Eire	$\leq$0.1 [NO_2]	0.10 *		June, 1979
Parrish et al., 1985	Point Arena, CA			0.37 (0.26)[1]	April-May, 1985

[1]Median shown in Parenthesis

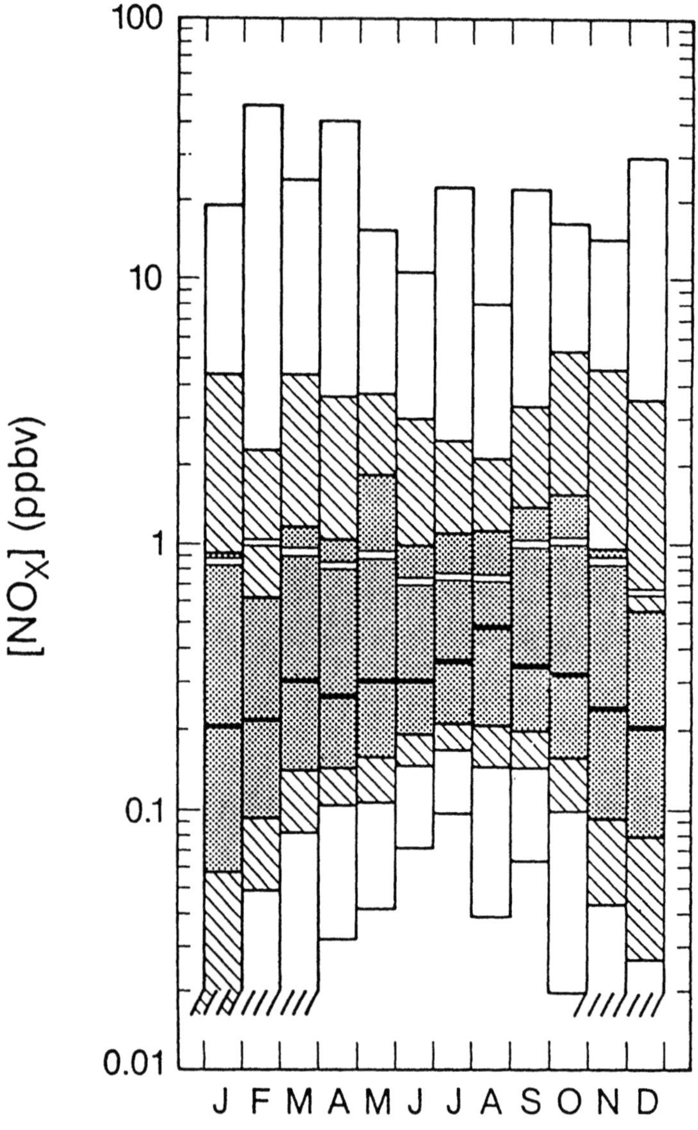

Fig. 3. Histogram of NO_x measurements made at Niwot Ridge, CO from October, 1980 through September, 1981. Each bar contains all the NO_x measurements made at this site for each month. The striped region in each bar indicates the central 90% of the measurements, and the shaded region the central two-thirds of the distribution. The white bar is the average while the black bar is the median.

ber, 1981. Figure 3 shows a histogram of the NO_x measurements. The sampling station used in this study is at 3.05 km elevation approximately 50 km northwest of the Denver metropolitan area and approximately 1.7 km higher in elevation. The prevailing winds at the site are westerly and transport relatively clean continental air to this site. These conditions account for the lowest NO_x levels. Interruptions of this air flow pattern by easterly winds occasionally occur and transport air that is anthropogenically enriched in NO_x to the site. These easterly winds produce the highest observed NO_x levels.

Both the highest and lowest NO_x levels at this site are observed in the winter. During the winter NO_x levels at this continental site can drop to 0.02 ppbv or less for extended periods. The meteorological factors that contribute to these low levels are discussed by Bollinger et al. (1984). In the summer the range of NO_x is less. Convective mixing of pollution during easterly flow conditions dilutes the anthropogenic NO_x. The NO_x during the summer is further reduced by the enhanced photochemistry. In contrast, the lowest NO_x levels in summer are much higher than in winter. This is attributed to increased natural sources of NO_x, lightning and soil emissions, during the summer, which can influence NO_x in clean continental air. In addition, the site is less isolated from anthropogenic sources to the east by a reduced persistence in the westerly winds. For this reason the averages shown in Figure 3, which are biased by limited periods of very high NO_x concentrations are almost independent of season, while the medians, which more nearly reflect typical conditions, are lowest in winter and peak in the late summer.

An isolated site at Whiteface Mt. in the eastern United States was studied by Kelly, et al. (1984). The levels of NO_x measured were higher than those reported in the western United States, and the authors note that their converter was sensitive to PAN and other organic nitrates. The small NO to NO_x ratios observed indicate that such species made substantial contributions and thus the average NO_x level is probably below 1 ppbv.

Surface measurements have been made at three coastal inflow sites. A summary of these measurements is given in Table 2B. The NO_x levels observed at these sites located on the coasts of California and Ireland may be expected to reflect maritime air with very low levels of NO_x that are carried inland by the prevailing westerly wind. However, depending on both local and synoptic air movement, these sites can be affected by anthropogenic pollution from on shore. In addition, off shore, near-coastal maritime activity can significantly perturb the natural background. Nevertheless, the NO_x levels recorded at these sites are typically less than that at surface sites located inland.

The relative differences in the NO_x levels at continental rural sites and the coastal inflow sites are shown in Figure 4. In this bar graph the NO_x levels are presented for Point Arena, CA during the spring of 1985, Niwot Ridge, CO during the summer of 1984 and Scotia, PA during the summer of 1986. At Point Arena the average NO_x level was

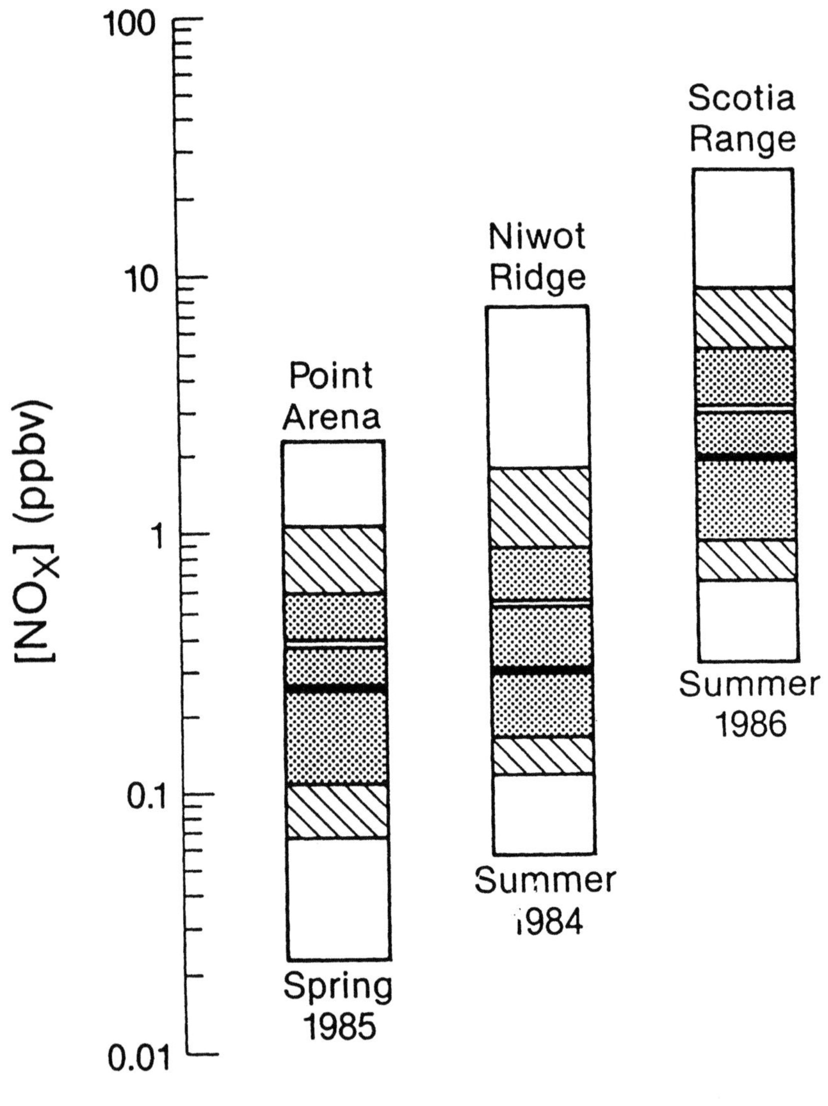

Fig. 4. Histogram of NO_x measurements made at Pt. Arena, CA, Niwot Ridge, CO and Scotia Range, PA (see Tables 1 and 2). Each bar contains all the NO_x measurements at these sites for the specified periods. The striped region indicates the central 90% of the measurements, the shaded region the central 67% of the measurements. The white bar indicates the average, the black bar the median.

0.37 ppbv, at Niwot Ridge the average was 0.55 ppbv and at Scotia the average was 3 ppbv. The progressive increase in the NO_x level from coastal inflow to isolated rural to Eastern U.S. rural is as expected.

The data in Figure 4 represent sites which are strongly influenced by anthropogenic NO_x sources. Even so, very low levels of NO_x are observed: (e.g., 0.022 ppbv at Point Arena). Since these measurements were made during seasons with the largest natural emissions of NO_x, the low levels of NO_x that were recorded suggest that in remote locations NO_x levels should be exceedingly small.

The measurements at Point Arena can be compared with those made at the coastal inflow site at the western edge of the European continent at Loop Head Lighthouse, Ireland. The average mixing ratio for NO_2 obtained from the measurements of Helas and Warneck (1981) at this site in June, 1979 was 0.10 ± 0.087 ppbv. However, NO_2 mixing ratios in air masses of marine origin gave a minimum that occurred during the nighttime of 0.037 ± 0.006 ppbv and a 24 hour average of 0.084 ± 0.049 ppbv. It was noted by Helas and Warneck (1981) that these measurements made with a ferrous sulfate converter could have interferences from PAN and thus represented upper limits. Confirmation of this concern is provided by the NO measurements that were made during this study. These levels never exceeded 10% of the NO_2 levels. Such large imbalances in the photochemical stationary state are not expected. It is thus reasonable to conclude that the measured NO_2 contains a substantial interference, probably from PAN. However, it is interesting to note that Platt and Perner (1980) using Long-Path Optical Absorption, presumably with no PAN interference, measured comparable NO_2 levels during April, 1979 at this site.

Ground level measurements have also been carried out at more remote surface locations, which may represent natural background NO_x levels. These are summarized in Table 3. Galbally (1977) found NO_x concentrations at Cape Grim, Tasmania to lie between 0.05 ppbv and 0.3 ppbv when meteorological conditions discriminated against anthropogenic pollution. When the meteorology at this site was more favorable for contamination of the sampled air masses by anthropogenic pollution, the NO_x levels ranged from 0.05 ppbv to 3 ppbv. It should be noted that these measurements were made using a molybdenum converter. Therefore, as noted by Galbally (1977), the instrument was probably sensitive to other NO_y compounds, such as PAN and HNO_3.

A similar result was obtained somewhat later by Stedman and McEwan (1983) at an isolated site: Mt. John, New Zealand (43°59'S). The studies used a chemiluminescence detector with ferrous sulfate and molybdenum converters. For most of the sampling period the NO level was indistinguishable from the background implying an NO mixing ratio of 0.007 ppbv or less. However, the surface converter/chemiluminescence detector yielded a median value of the mixing ratio reported as NO_y of 0.127 ppbv. The very small ratios of NO to NO_y indicate that the NO_y signal was dominated by compounds other than NO and NO_2. However, tests using a nylon wool plug to remove HNO_3, indicated that HNO_3 was

Table 3. Average NO, NO_2, and/or NO_x mixing ratios measured at remote locations. (*Upper limit for NO_2 and NO_x)

Reference	Location	[NO] ppbv	[NO_2] ppbv	[NO_x] ppbv	Measurement Period
Galbally, 1977	Cape Grim, Tasmania, Australia			0.05-0.3*	1977, 1978
McFarland et al., 1979	Mid-Pacific	0.002-0.006 Noon-time values		(0.01)[2]	1978
Broll et al., 1984	Mid-Atlantic				Oct.-Dec., 1981
	a) 40° N		0.07*		
	b) Equator		0.02*		
	c) 20° S		0.04*		
Stedman and McEwan, 1983	Mt. St. John, New Zealand	<0.007		(0.127*)[1]	Feb.-May, 1981
Bottenheim et al., 1986	Alert, N.W.T., Canada		(0.018-.097)[3]		March, 1985

[1]Median shown in parenthesis
[2]Inferred from photochemical stationary state
[3]Range of daily averages

not present in significant concentrations. By elimination, PAN would seem to be a major contributor to the NO_y signal.

Shipboard measurements reported by Broll et al. (1984) showed NO_2 levels over the Atlantic Ocean as a function of latitude. These mid-ocean measurements indicated peak NO_2 mixing ratios averaging 0.07 ppbv about 40°N, a secondary maximum of approximately 0.04 ppbv at 40°S and minimum levels of approximately 0.02 ppbv near the equator. Presumably the maximum levels were recorded during periods that the ship traversed remnants of the continental anthropogenic plume. However, the very low mixing ratios observed by Broll et al. (1984) were obtained using a chemiluminescence detector with a surface converter to reduce NO_2 to NO. As a consequence, the measurements should be regarded as upper limits for NO_x.

Surface measurements of NO_2, HNO_3, and PAN in the Arctic were made by Bottenheim, et al. (1986) in March, 1985. A luminol-NO_2 chemiluminescence technique was used to measure NO_2; this technique has not been shown to reliably measure NO_2 in field intercomparisons. The measurements were made near Alert, N.W.T., a small settlement in the Canadian Arctic. The NO_2 mixing ratio showed considerable variation with the daily average varying from 0.018 ppbv to 0.097 ppbv with most NO_2 attributed to local pollution. On the other hand, PAN levels were approximately constant at 0.2 ppbv. These measurements indicate that PAN is the predominant reactive nitrogen oxide species at this northern location at the onset of spring while NO_x is exceedingly small.

Finally, McFarland et al. (1979) on a ship in the mid-Pacific measured noontime mixing ratios of NO that ranged between 2 pptv and 6 pptv. From these data, Kley et al. (1981) derived an NO_x mixing ratio of 10 pptv by using a NO_x/NO ratio from a simple photochemical equilibrium calculation.

AIRCRAFT MEASUREMENTS

Early airborne measurements of the oxides of nitrogen in the troposphere were made by Schiff et al. (1979) as part of the Global Atmospheric Measurement Experiment of Tropospheric Aerosols and Gases (Gametag). These measurements were made using a chemiluminescence detector. Upper limit values for NO in maritime air were $\leq$ 0.03 - 0.04 ppbv. These measurements clearly showed that the NO_x level is very low in the free troposphere over remote maritime regions.

Subsequently, Kley et al. (1981) measured NO_x in flights over the western U.S. These measurements indicate that NO_x mixing ratios decrease with increasing altitude. However, in these flights the elevated NO_x levels at lower elevation were attributed to anthropogenic pollution from surface sources. By combining the measurement made at the highest elevation during those flights, which were assumed to be uncontaminated, with the available data on NO and NO_x distributions in the clean troposphere and lower stratosphere, Kley et al. (1981) deduced an NO_x profile for the clean troposphere. According to this picture the NO_x mixing ratio increases from 0.01 ppbv in surface air (McFarland et al., 1979) to 0.2 ppbv at the tropopause. This distribution also agreed with the low upper limits of NO established by Schiff, et al. (1979) and with the tropospheric NO_2 column density measured by Noxon (1978) employing optical absorption using the sun as a light source.

Several aircraft studies have been carried out since the summary of Kley, et al. (1981). The parameters of each study are summarized in Table 4. Most of these studies have measured only NO, and the altitude profile of this species will be discussed most fully.

Drummond and Volz (1985) and Drummond, et al. (1987) reported NO values obtained during the STRATOZ II flights that were carried out in June, 1984. The NO mixing ratio was measured from 67°N to 60°S over an altitude range from 0 to 12 km. High NO mixing ratios up to several ppbv were observed in the planetary boundary layer (PBL). In the middle troposphere, between the PBL and 7-8 km (the average altitude of the cloud tops), NO rarely exceeded 0.03 ppbv and often was well below 0.01 ppbv. In the upper troposphere NO was higher in both hemispheres, particularly over the European continent. The increase in the NO mixing ratio in the upper troposphere was attributed to stratospheric injection (c.f., Kley, et al. 1981), high flying aircraft (c.f., Ehhalt and Drummond, 1982), and lightning (c.f., Borucki and Chameidies, 1984), with lightning representing the largest source. In addition, there are indications that convective transport of pollution from the boundary layer, particularly in the northern hemisphere is a source for NO_x in the upper troposphere (c.f., Chatfield and Crutzen, 1984).

Dickerson (1984) measured NO and NO_2 on flights from Frankfurt, FRG to Abu Dhabi, United Arab Emirate (June, 1982) and from Frankfurt to Sao Paulo, Brazil (December, 1982). The measurements were made the northern hemisphere at 9 km were about 200 pptv. Enhanced levels of reactive nitrogen oxides were measured in the proximity of thunderstorms. This enhancement was attributed to convective transport by the storms of polluted boundary layer air to the upper troposphere.

The role of thunderstorms as a mechanism in the transport of air pollutants was discussed further by Dickerson et al. (1987). In these studies enhanced levels of NO_x, CO and several NMHC (non-methane hydrocarbons) were observed well above the PBL. For these short lived photochemically active compounds, the decrease in the mixing ratio was much less than predicted from their PBL mixing ratios assuming eddy diffusion transport and photochemical destruction. In addition, lightning was observed to be the principal source of NO_x in the upper cloud. On the other hand, Pickering et al. (1987) found that convective transport of PBL air by cumuli could be blocked by cold front capping of the PBL. From this it was concluded that a regional estimate of the contribution of convective storms to vertical tropospheric mixing will require detailed information concerning storm distribution, frequency and intensity, PBL NO_x distribution and synoptic structure of the storm system.

Table 4. Summary of aircraft measurements of NO_x in clean troposphere

Reference	Lat (°)	Alt (km)	NO (ppbv)	NO_x (ppbv)	Measurement Period
Drummond and Volz, 1985 Drummond et al., 1987	67N-60S	0-12			June, 1984
Dickerson, 1984	20S-50N	0-10	≤0.1	0.2-0.6*	Dec., 1982
Dickerson, 1985	51-87N	0.2-12	<0.01	<0.6*	March, 1983
Kondo et al., 1987	30-35N	3-8	0.015-0.035	0.2*	Feb., 1983
	40N			1.0*	Jan.-Feb., 1984
Torres and Buchan, 1987	3S-36N	5 km	0.014-0.2		July, Aug., 1985
		PBL	0.01-0.06		
		free trop.	0.013		
Ridley et al., 1987	15-42N	PBL	0.001		Oct., Nov., 1983
		1.2-3.1	0.006		
		4.9-6.4	0.01		
		7.6-10.1	0.017		
Davis et al., 1987	15-42N	≤1.8	0.004		Oct., Nov., 1983
		6	0.02		
		9	0.025-0.035		
Fahey et al., 1987	37-38N	8-15	0.04-0.25		June, 1987

Reactive nitrogen compounds also were measured in the Arctic by Dickerson (1985) using the techniques indicated above. The measurements were made in March 1983 at elevations between 0.2 km and 12 km on flights between 87°N and 51°N. Large mixing ratios of NO_x were reported in the boundary layer (NO_x ~ .6 ppbv) which decreased sharply with altitude. Since these measurement were made using a ferrous sulfate converter, the apparent NO_x levels that were reported also reflect detection of organic nitrates such as PAN. Although the sun was low (10° above the horizon), the NO levels were also very low (< 0.01 ppbv) and may suggest high levels of PAN in the air sampled at these high latitudes (c.f., Bottenheim et al., 1986). The measurements support the contention that the Arctic region may serve as a reservoir for the reactive nitrogen compounds during the polar winter.

Kondo et al. (1987) reported measurements of the oxides of nitrogen over the sea surrounding the Japanese Islands (30-43°N). The measurements were made during the winter of 1983 and 1984 at altitudes between 3 and 8 km. A chemiluminescence detector was used to measure NO. The compounds detected by the ferrous sulfate converter/chemiluminescence detector were interpreted to constitute a measure of the sum of NO, NO_2 and other unstable oxides of nitrogen that were converted to NO by ferrous sulfate. Over the Pacific between 30°-35°N, which is south of Japan, the observed NO_x mixing ratio was 0.2 ppbv while the NO mixing ratio increased with altitude from 0.015 ppbv at 3 km to 0.035 ppbv at 7 km. Over the sea of Japan, which is west of Japan, NO_x increased with latitude and reached 1.0 ppbv at 40°N. These high levels were interpreted to result largely from anthropogenic NO_x sources located on the Asian continent.

Torres and Buchan (1987) measured horizontal and vertical distributions of nitric oxide using a chemiluminescence detector over the Amazon Basin as a part of the Amazon Boundary Layer Experiment (NASA ABLE-2A). In the flight between the Virginia Coast and Manaus, Brazil, NO was typically about 0.014 ppbv at 5 km altitude with mixing ratios as large as 0.2 ppbv in electrically active clouds. Over the Amazon Basin, NO in the lower PBL were 0.025 - 0.06 ppbv over the central basin decreasing to 0.010-0.012 ppbv over coastal regions. The NO mixing ratio decreased with increasing height with values in the free troposphere averaging 0.013 ppbv in regions not influenced by biomass burning.

Ridley et al. (1987), using a chemiluminescence detector, and Davis et al. (1987), using the two-photon laser-induced fluorescence technique, measured NO mixing ratios over the Pacific Ocean in the autumn of 1983 as part of the NASA Global Tropospheric Experiment (GTE). These measurements were made between 15°N and 42°N and from the coast of California to west of the Hawaiian Islands. The lowest NO mixing ratios were obvserved in the PBL with median NO mixing ratios of 0.001 ppbv. The median NO mixing ratios were observed to increase with height. The chemiluminescence method gave 0.006 ppbv between 1.2 km and 3.1 km; 0.01 ppbv between 4.9 km and 6.4 km and 0.017 ppbv between

7.6 km and 10.1 km. The LIF method gave nearly identical results: 0.004 ppbv at $\leq$ 1.8 km, 0.02 ppbv at 6 km, 0.025-0.035 ppbv at 9 km. During high altitude flights, enhanced NO mixing ratios that correlated with ozone gave evidence of stratospheric influence, and enhanced NO mixing ratios were observed in electrically active clouds.

It should be noted that the very good agreement that was obtained in the measurement of NO using a chemiluminescence detector (Ridley et al., 1987) and the Laser-Induced Fluorescence technique (Davis et al., 1987) provides reasonable evidence that the two sets of measurements are accurate determinations of ambient NO at exceptionally small mixing ratios.

Fahey et al. (1987) have recently measured NO using a chemiluminescence detector onboard the NASA ER-2 High Altitude aircraft. The NO instrument was successfully combined with a gold catalytic converter to measure NO_y in the NASA Stratosphere-Troposphere Exchange Project (Russell et al., 1985; Fahey et al., 1985). The data included here represent a single flight over California (38°N) in June 1987 in which the converter was removed in order to measure only NO. The NO values uniformly increase with increasing altitude from <0.03 ppbv in the mid-troposphere to 0.10 ppbv near the tropopause when the solar zenith angle $\geq$83°. The values increase above the tropopause with a gradient of ~0.30 ppbv/km.

Figure 5 presents a comparison of recent aircraft NO measurements with the NO_x profile (dashed diagonal line) of Kley et al. (1981). Measurements that showed strong surface influence are excluded here. The altitude of the Kley et al. (1981) profile was referenced to the tropopause, and that convention is continued in this figure. The rationale for this convention is three-fold. First, Kley et al. (1981) found that there was less scatter between measurements from different groups and the tropospheric data connected smoothly with lower stratospheric data; second, major natural sources in the remote troposphere produce NO_x in the upper troposphere, and third, the tropopause represents a physical boundary for the troposphere. The tropopause height was estimated for each reported measurement from a climatological summary (Roe, 1981).

The data of Drummond et al. (1987) are represented by the circles in Figure 5. Measurements from the PBL, from their flight number 1 over the European continent, and from all times when the solar zenith angle was greater than 75° have been excluded. The remaining data are divided into two sets: northern hemisphere plus the tropics to 20° south latitude (open circles) and the southern hemisphere below 20° south latitude (closed circles). These sets, then, represent the tropics and the northern hemisphere summer, and the southern hemisphere winter, respectively. All data points from Figures 5-14 of Drummond et al. (1987) not excluded by the criteria above are averaged in 1 km intervals relative to the estimated tropopause to give the cirles in

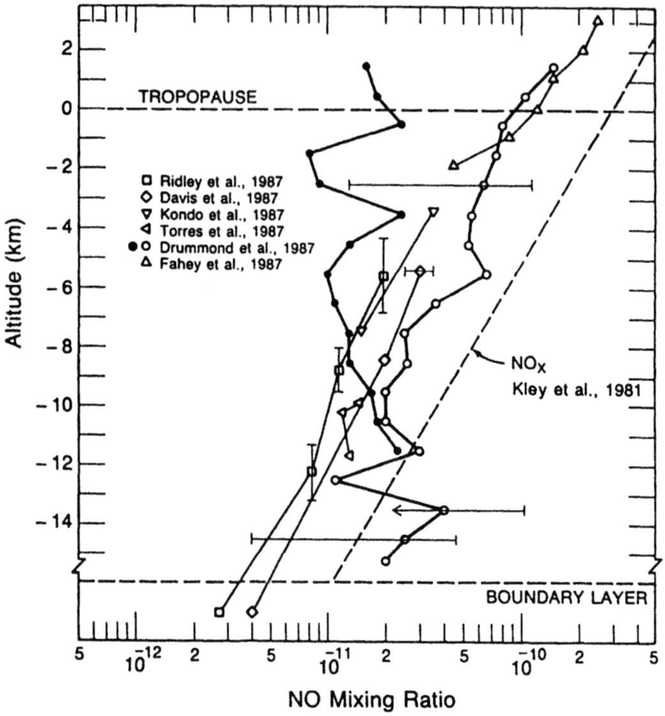

Fig. 5. Altitude profiles of average NO mixing ratios referenced to the tropopause height. Open symbols indicate data from the tropics and northern hemisphere, while the closed circles indicate the one study from the southern hemisphere. The horizontal bars on the circles indicate one standard deviation about the mean. The horizontal bar on the upper diamond indicates the higher average obtained including measurements in the vicinity of electrified clouds and the lower average resulting from exclusion of those values. The vertical bars on the squares indicates the altitude interval included in each average. Measurements from the planetary boundary layer are included separately at the bottom of the figure. Included for comparison is the NO_x mixing ratio profile derived by Kley et al. (1981).

Figure 5 here. Several horizontal error bars are included to indicate one standard deviation about the mean. Often the standard deviation was larger than the mean. These large standard deviations indicate the large variability of the observed NO mixing ratios.

Other data included in Figure 5 are all from the northern hemisphere or tropics. They include the late fall measurements of Ridley et al. (1987) and Davis et al. (1987), the winter measurements of Kondo et al. (1987), and the summer measurements of Torres and Buchan (1987) and Fahey et al. (1987). The vertical bars on the data of Ridley et al. (1987) indicate the altitude range for each average. The horizontal bar on the upper point of Davis et al. (1987) indicates the higher value obtained when data from near electrified clouds were included and the lower value from when those data were excluded.

The data of Drummond et al. (1987) for altitudes more than 10 km below the tropopause are higher than the other measurements. These results are probably due to transport of higher NO_x levels from the PBL. Indeed, Drummond et al. (1987) often observed "C" shaped profiles with minimum values occurring between 2 and 8 km. The Drummond, et al. (1987) flights were generally along continental coastlines while the other studies were generally over oceans.

The southern hemisphere results of Drummond et al. (1987) show a nearly constant altitude profile without a significant increase in NO as the tropopause is approached and crossed (closed circles in Fig. 5). Whether this is a hemispherical or a seasonal difference is not clear, but Kasting and Singh (1986) have predicted a large wintertime decrease in upper tropopause NO_x due to the photochemistry of non-methane hydrocarbons. In contrast, the late fall and winter northern hemisphere NO measurements do show the increase with altitude, but they do not extend as near to the tropopause as might be required to show a similar seasonal difference in the northern hemisphere.

The data presented in Fig. 5 for the NO altitude profiles from the northern hemisphere and the tropics (shown as the open symbols in Figure 5) are surprisingly consistent. The results are generally parallel to, and a factor of 2 to 3 times smaller than, the NO_x profile of Kley et al. (1981). Such consistency is surprising because even for a common NO_x profile, considerable variation in NO would be expected because NO and NO_2 are in rapid photochemical equilibrium that is a strong function of ozone concentration, solar zenith angle, temperature and perhaps other variables. In this connection it should be noted that the data of Fahey et al. (1987) were taken with simultaneous measurements of O_3 and temperature. Combining these data with model calculated NO_2 photodissociation rates (Tallamraju, 1987) and simple photostationary state considerations yields NO_x values within 30% of the Kley et al. (1981) profile. However, it must be recognized that the assumption of the photochemical stationary between NO and NO_2 even in the remote troposphere is open to question. Significant, deviations from the photochemical stationary state near the surface in a rural location has been observed (Parrish et al., 1986b). For several reasons a determination of the balance of the photochemical stationary state as a function of altitude and/or latitude in the remote troposphere is of critical importance.

In summary, the aircraft measurements present a reasonably consistent picture for the tropics and northern hemsiphere. The NO_x levels in the remote troposhere increase with increasing altitude. The Kley et al. (1981) NO_x profile probably reasonably represents the average upper troposphere and the lower maritime troposphere, but the lower continental troposphere is likely to have higher NO_x levels. However, given the large variability observed about the averages, any particular profile may well deviate from the average profile by a large factor. In addition, there is a strong indication that in the wintertime, at least in the southern hemisphere, the NO_x levels remain low throughout the tropospheric height.

The sources of upper troposphere NO_x are associated with lightning, high flying aircraft, anthropogenic pollution convected from the surface, and stratospheric injection. Over industrial regions the profiles are inverted, at least near the surface, due to upward mixing of the high NO_x concentrations in the PBL that are attributable to anthropogenic pollution.

SEASONAL AND DIURNAL CYCLES

Sources and photochemical transformations of NO_x as well as atmospheric mixing and transport have strong seasonal and diurnal variation. Consequently, NO_x levels are expected to have corresponding cycles that are highly specific to a particular location. Because of the limited data, seasonal and diurnal patterns for free tropospheric NO_x have not been deduced, although a possible seasonal cycle is discussed above.

Two studies have been carried out over periods extended enough to reveal seasonal cycles in the PBL. The data from the isolated inland site at Niwot Ridge, CO are shown in Figure 3 and have been discussed earlier. Here the median values peak in the summer. In contrast the results from the non-urban site at Dueselbach, FRG (Messina, 1985; Broll, et al., 1984) summarized in Table 1 reveal a maximum in the winter.

Several studies in rural and isolated locations have shown afternoon peaks in NO_x (Kelly, et al., 1980; Helas and Warneck, 1981; Stedman and McEwan, 1983; Kelly, et al., 1984). In all these cases the interpretion of an afternoon peak is unclear because of the detectors' sensitivity to compounds other than NO_x. The data from Niwot Ridge in Figure 3 made using a detector that is thought to detect only NO and NO_2 also yield an afternoon peak. At this latter site, the afternoon peak is due to meteorological effects which cause transport of anthropogenic NO_x to the site to occur preferentially in the afternoon. In contrast, the NO_x levels at Scotia Range, PA peak shortly after sunrise. The photochemical conversions of NO_x illustrated in Figure 6 would lead to the expectation of an early morning maximum and an early evening minimum in NO_x if the NO_y levels are constant through the day.

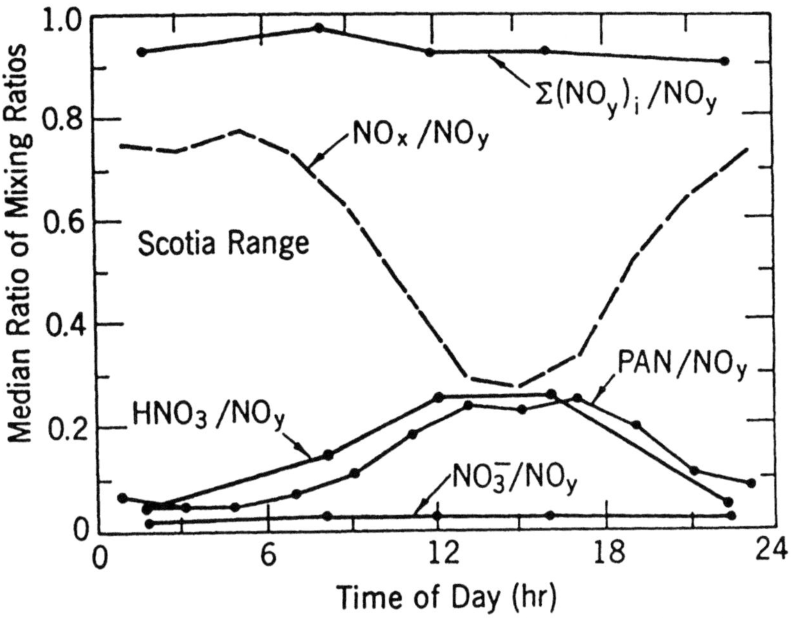

Fig. 6. Diurnal cycle of the ratios of [NO_x], [PAN], [HNO_3] and [NO_3^-] to [NO_y] as measured at Scotia Range, PA during the summer of 1987. $\Sigma(NO_y)_i$ is defined as the sum of the component reactive odd-nitrogen species NO, NO_2, HNO_3, PAN, and particulate nitrate NO_3^-.

Large data bases of NO_x, NO_y and meteorological measurements would be required to distinguish between contributions to the observed seasonal and diurnal cycles from modulation of sources, photochemical transformation, and atmospheric transport.

LIFETIME OF NO_x

The principal factors that determine NO_x lifetimes in the atmosphere are deposition of NO_x and chemical conversion of NO_2 to HNO_3 or particulate nitrate followed by heterogeneous removal. The chemical conversion involves the association during the day of NO_2 with the hydroxyl radical (OH) to form HNO_3, and the nighttime reaction of NO_2 with O_3 to form NO_3 and N_2O_5 followed by gas phase and/or heterogeneous reactions of NO_3 or N_2O_5 to form HNO_3.

During the summer, the NO_x lifetime is probably controlled by the reaction of NO_2 with OH. Model calculations (Liu, et al., 1987) suggest that this lifetime is of the order of 0.5 days for NO_x levels of approximately 1 ppbv. These lifetimes will change with NO_x level and will depend on the ratio of NO_x to NMHC (Liu et al., 1987).

Figure 6 shows the variation in the mixing ratios of the principal components of the reactive odd-nitrogen family (NO_x, PAN, HNO_3, and nitrate particulates (NO_3^-) relative to the total mixing ratio of this family (NO_y). These measurements were made at a surface site in western, Pennsylvania (USA) in June and July of 1986 (Grandrud et al., 1986; Parrish et al., 1986). During the nighttime hours NO_x constitutes the principal component of the NO_y family and accounts for approximately 75% of NO_y with PAN and HNO_3 representing small residual components. During the day the NO_x mixing ratio relative to NO_y falls and reaches a minimum of approximately 28% of NO_y in the afternoon after the photochemically active period of the day. At this time the relative concentrations of PAN and HNO_3 are at maxima, with each species accounting for approximately 25% of the measured NO_y. It should be noted that at this site throughout the day NO_x, PAN and HNO_3 combined account for approximately 90% of NO_y. These data show the daytime conversion of NO_x to HNO_3 and PAN yields a photochemical lifetime for NO_x of approximately eight hours which is consistent with the model result.

The removal rates of NO_x from the atmosphere have been estimated from the decrease in NO_x in an aging urban plume (Spicer, 1982; Spicer et al., 1982). In these studies carried out in August 1978, the decline in NO_x in the Boston urban plume as a function of distance from the urban source yielded lifetimes ranging from 3 hours to 7.5 hours during sunny summertime conditions. Since the dry deposition rate of NO_x to the oceans is expected to be small (Hicks, 1984), this NO_x loss is attributed to chemical destruction of NO_x.

During the winter the lifetime of NO_x is far less certain. In this season the formation of HNO_3 through NO_3 chemistry could be predominant. If NO_3 at night in the winter is directly converted to HNO_3 (or deposited as NO_3^-) NO_x would have a lifetime of about two days (Liu et al, 1987). However, if the NO_3 combines with NO_2 to form N_2O_5 which is subsequently removed as HNO_3 or NO_3^-, the NO_x lifetime could be one day (Liu et al, 1987). Considerable uncertainty surrounds the NO_3 chemistry, however, particularly the products of the reaction between NO_3 and NMHC. At present no measurements have been used to infer the lifetime of NO_x in the winter.

It should be noted that atmospheric NO_x levels in the rural and remote troposphere can be strongly influenced by PAN. The extended lifetime of PAN in the colder regions of the upper atmosphere can allow PAN to act as an effective agent in transferring NO_x from source regions to the remote atmosphere.

CONCLUSIONS

The NO_x levels in urban and industrial areas of the world are determined largely by anthropogenic sources. This leads to near surface and PBL NO_x mixing ratios in rural areas of the eastern United States and Western Europe typically in the range between 1 ppbv and 10 ppbv. In less populated and coastal inflow regions, the NO_x levels depend on the prevailing meteorology and the proximity and distribution of urban sources. In these locations the NO_x mixing ratios typically range between 0.1 ppbv and 1.0 ppbv.

Surface measurements at remote locations reflect the influence of natural sources such as soil emissions and lightning. These are principally terrestrial sources that are seasonal. These sources may provide concentrations of NO_x in the PBL that typically range between 0.02 ppbv and 0.1 ppbv. It should be noted that at most locations on land, particularly in the mid-latitudes, the NO_x levels in ambient air are influenced by anthropogenic activities. In remote maritime air and in the polar regions that are not influenced by anthropogenic activities, the NO_x concentrations are exceedingly small, typically 0.001 ppbv to 0.01 ppbv, and are associated with the downward mixing of NO_x from NO_x reservoirs in the upper troposphere.

The free tropospheric burden of NO_x is also strongly influenced by anthropogenic NO_x, particularly in the northern hemisphere. Considerations involving the atmospheric transport and photochemical lifetime and the contribution of aircraft emissions of NO_x and natural sources such as lightning and stratospheric subsidence, increases the range of NO_x variability in the free troposphere. Here the range of NO_x mixing ratios may vary from 0.02 ppbv in remote regions to 5 ppbv over populated areas. In general, throughout the free troposphere the NO_x mixing ratio increases with height. The atmospheric NO_x profile suggested by Kley et al. (1981) seems valid for most of the free troposphere and for the PBL over remote marine locations. However, the sources of this NO_x are probably associated with a mixture of natural and anthropogenic NO_x. For this reason the upper tropospheric NO_x level that anchors this profile will vary with latitude and season.

Continued progress toward the goal of fully understanding the origins of NO_x in the troposphere and the role of NO_x in tropospheric photochemistry will require more complete measurements of NO_x throughout the troposphere. The majority of the studies reviewed here cannot be used directly to meet this goal since the results provide only an upper limit for NO_x. For the future, it is important that NO_x measurements be made that are free of known interferents or that account for the interferents with separate measurements.

ACKNOWLEDGEMENTS:

This work was supported by the National Acid Precipitation Assessment Program.

REFERENCES:

Altshuller, A. P., The Role of Nitrogen Oxides in Nonurban Ozone Formation in the Planetary Boundary Layer Over N. America, W. Europe and Adjacent Areas of Ocean, Atmos. Environ., 20, 245-268, 1986.

Bollinger, M. J., C. J. Hahn, D. D. Parrish, P. C. Murphy, D. L. Albritton, and F. C. Fehsenfeld, NO_x Measurements in Clean Continental Air and Analysis of the Contributing Meteorology, J. Geophys. Res. 89, 9623-9631, 1984.

Borucki, W. J., and W. L. Chameides, Lightning: Estimates of the Rates of Energy Dissipation and Nitrogen Fixation, Rev. Geophys. Space Phys., 22, 363-372, 1984.

Bottenheim, J. W., A. G. Gallant and K. A. Brice, Measurement of NO_y Species and O_3 at 82°N Latitude, Geophys. Res. Letters, 13, 113-117,1986.

Broll, A., G. Helas, K. J. Rumpel and P. Warneck, NO_x Background Mixing Ratios in Surface Air Over Europe and the Atlantic Ocean, Third European Symposium on Physico-Chemical Behavior of Atmospheric Pollutants, Varese, Italy, 1984.

Chameides, W. L., and J. C. G. Walker, A Photochemical Theory of Troposphere Ozone, J. Geophys. Res., 78, 8751-8760, 1973.

Chatfield, R. B. and P. J. Crutzen, Sulfur Dioxide in Remote Oceanic Air: Cloud Transport of Reactive Precursors, J. Geophys. Res. 89, 7111-7132, 1984.

Cox, R. A., Some Measurements of Ground Level NO, NO_2, and O_3 Concentrations at An Unpolluted Maritime Site, Tellus, 29, 356-363, 1977.

Davis, D. D., J. D. Bradshaw, M. O. Rogers, S. T. Sandholm and S. KeSheng, Free Tropospheric and Boundary Layer Measurements of NO Over the Central and Eastern North Pacific Ocean, J. Geophys. Res., 92, 2049-2070, 1987.

Decker, C. E., L. A. Ripperton, J. J. Worth, F. M. Vukovich, W. D. Bach, J. B. Tommerdahl, F. Smith, and D. E. Wagoner, Formation and Transport of Oxidant Along Gulf Coast and in the Northern U.S., EPA-450/3-76-033, 1976.

Dickerson, R. R., Measurements of Reactive Nitrogen Compounds in the Free Troposphere, Atmos. Environ., 18, 2585-2593, 1984.

Dickerson, R. R., Reactive Nitrogen Compounds in the Arctic, J. Geophys. Res., 90, 10,739-10,743, 1985.

Dickerson, R. R., G. J. Houfman, W. T. Luke, L. J. Nunnermacker, K. E. Pickering, A.C.D. Leslie, C. G. Lindsey, W. G. N. Slinn, T. J. Kelly, P. H. Daum, A. C. Delany, J. P. Greenberg, P. R. Zimmerman, J. F. Boatman, J. D. Ray and D. H. Stedman, Thunderstorms: An Important Mechanism in the Transport of Air Pollutants, Science, 235, 460-465, 1987.

Drummond, J. W. and A. Volz, A Summary of the Nitric Oxide (NO) Measurements Obtained During Stratoz III, 0-12km, 67°N-60°S: Evidence of Air Pollution in the Upper Trophosphere, "Pollutant Cycles and Transport: Modeling and Field Experiments" Eds. F.A.A.M. DeLeeuv, N. D. Van Edmond, Bilthoven, 1985.

Drummond, J. W., D. H. Ehhalt and A. Volz, Measurements of Nitric Oxide Between 0-12 km Altitude and 67°N - 60°S Latitude Obtained During STRATOZ III., J. Geophys. Res., submitted, 1987.

Ehhalt, D. H., and J. W. Drummond, The Tropospheric Cycle of NO_x, in: Chemistry of the Unpolluted and Polluted Troposphere, ed. by H.W. Georgii and W. Jaeschke, D. Reidel, Hingham, MA, 1982.

Fahey, D.W., C.S. Eubank, G. Hübler, and F.C. Fehsenfeld, Evaluation of a Catalytic Reduction Technique for the Measurement of Total Reactive Odd-Nitrogen NO_y in the Atmosphere, J. Atmos. Chem., 3, 435-468, 1985.

Fahey, D.W., D.M. Murphy, C.S. Eubank and M.H. Proffitt, National Oceanic and Atmospheric Administration, Boulder, CO; K.R. Chan, S.G. Scott, and S.W. Bowen, NASA Ames Research Center, Moffett Field, CA, private communication, 1987.

Fehsenfeld, F. C., R. R. Dickerson, G. Hübler, W. T. Luke, L. J. Nunnermacker, E. J. Williams, J. M. Roberts, J. G. Calvert, C. M. Curran, A. C. Delany, C. S. Eubank, D. W. Fahey, A. Fried, B. W. Gandrud, A. O. Langford, P. C. Murphy, R. B. Norton, K. E. Pickering, B. A. Ridley, A Ground-Based Intercomparison of NO, NO_x, NO_y Measuring Techniques, J. Geophys. Res., submitted, 1987.

Fishman, J., S. Solomon and P. J. Crutzen, Observational and Theoretical Evidence in Support of a Significant In-Situ Photochemical Source of Tropospheric Ozone, Tellus, 31 432-446, 1979.

Fishman, J., F.M. Vukovich and E.V. Browell, The Photochemistry of Synoptic-Scale Ozone Synthesis: Implications for the GLobal Tropospheric Ozone Budget, J. Atmos. Chem., 3, 299-320, 1985.

Galbally, I. E., Air Pollution Measurement Techniques, Spec. Environ. Rep. 10, 10, World Meteorol. Organ., Geneva, 1977.

Gandrud, B.W., J.D. Shetter, B.A. Ridley, D.D. Parrish, E.J. Williams, M.P. Buhr, R.B. Norton, F.C. Fehsenfeld, H.H. Westberg, J.C. Farmer, B.K. Lamb and E.J. Alwine, Measurements of Peroxyacetylnitrate at a Rural Eastern U.S. Site, EOS, 67, 884, 1986.

Harrison, R. M. and H. A. McCartney, Ambient Air Quality at a Coastal Site in Rural North-West England, Atmos. Environ., 14, 233-244, 1980.

Helas, G., and P. Warneck, Background NO_x Mixing Ratios in Air Masses over the North Atlantic Ocean, J. Geophys. Res., 86, 7283-7290, 1981.

Hicks, B. B. Dry Deposition Processes. in The Acidic Deposition Phenomenon and its Effects: Critical Assessment Review Papers. Vol. 1, edited by A. P. Altshuller, Chapter A-7, EPA-600/8-83-016AF, 1984.

Hoell, J. M., G. L. Gregory, D. S. McDougal, M. A. Carroll, M. McFarland, B. A. Ridley, D. D. Davis, J. Bradshaw, M. O. Rodgers, and A. L. Torres, An Intercomparison of Nitric Oxide Measurement Techniques, J. Geophys. Res., 90, 12,843-12,851, 1985.

Hoell, J. M., G. L. Gregory, D. S. McDougal, A. L. Torres, D. D. Davis, J. Bradshaw, M. O. Rodgers, B. A. Ridley, and M. A. Carroll, Airborne Intercomparison of Nitric Oxide Measurement Techniques, J. Geophys. Res., 92, 1995-2008, 1987.

Hov, O., One-Dimensional Vertical Model for Ozone and Other Gases in the Atmospheric Boundary Layer, Atmos. Environ., 17, 535-549, 1983.

Kasting, J.F., and H.B. Singh, Nonmethane Hydrocarbons in the Troposphere: Impact on the Odd Hydrogen and Odd Nitrogen Chemistry, J. Geophys. Res., 91, 13,239-13,256, 1986.

Kelly, T. J., D. H. Stedman, J. A. Ritter and R. B. Harvey, Measurements of Oxides of Nitrogen and Nitric Acid in Clean Air, J. Geophys. Res., 85, 7417-7425, 1980.

Kelly, N. A. G. T. Wolf, and M. A. Ferman, Background Pollution Measurements in Air Masses Affecting the Eastern Half of the United States, Atmos. Eviron., 16, 1077-1088, 1982.

Kelly, T. J., R. L. Tanner, L. Newman, P. J. Galvin and J. A. Kadlecek, Trace Gas and Aerosol Measurements at a Remote Site in the Northeast U.S., Atmos. Environ., 18, 2565-2576, 1984.

Kley, D., J. W. Drummond, M. McFarland and S. C. Liu, Tropospheric Profiles of NO_X, J. Geophys. Res., 86, 3153-3161, 1981.

Kondo, Y., W. A. Matthews, A. Iwata, Y. Morita, and M. Takagi, Aircraft Measurements of Oxides of Nitrogen Along the Eastern Rim of the Asian Continent: Winter Observations, J. Atmos. Chem., 5, 37-58, 1987.

Liu, S. C., M. Trainer, F. C. Fehsenfeld, D. D. Parrish, E. J. Williams, D. W. Fahey, G. Hübler and P. C. Murphy, Ozone Production in the Rural Troposphere and the Implications for Regional and Global Ozone Distributions, J. Geophys. Res., 92, 4191-4207, 1987.

Logan, J. A., M. J. Prather, S. C. Wofsy, and M. B. McElroy, Tropospheric Chemistry: A Global Perspective, J. Geophys. Res. 86, 7210-7254, 1981.

Martin, A. and F. R. Barber, Sulfur Dioxide, Oxides of Nitrogen and Ozone Measured Continuously for Two Years at a Rural Site, Atmos. Environ., 15, 567-578, 1981.

Martinez, J. R. and H. B. Singh, Survey of the Role of NO_X in Non-Urban Ozone Formation, Final Report on SRI Project 6780-8, prepared for Monitoring and Data Analysis Division, Office of Air Quality Planning and Standards, Research Triangle Park, NC 27711, 1979.

McFarland, M., D. Kley, J. W. Drummond, A. L. Schmeltekopf and R. H. Winkler, Nitric Oxide Measurements in the Equatorial Pacific Region, Geophysical Res. Lett., 6, 605-608, 1979.

Messina, S. R., Analysis of the Relationship Between Meteorology and Air Pollution at Deuselbach, West Germany, Thesis, Graduate School, University of Maryland, College Park, Md., 20742, 1985.

Noxon, J. F., Tropospheric NO_2, J. Geophys. Res., 83, 3051-3057, 1978.

Parrish, D. D., E. Williams, R. B. Norton, and F. C. Fehsenfeld, Measurement of Odd-Nitrogen Species and O_3 at Point Arena, California, EOS, 66, 820, 1985.

Parrish, D. D., E. J. Williams, M. P. Buhr, R. B. Norton, F. C. Fehsenfeld, B. W. Gandrud, B. A. Ridley, J. D. Shetter, Partitioning of Odd-Nitrogen Species at a Rural, Eastern U.S. Site, EOS, 67, 891, 1986.

Parrish, D.D., M. Trainer, E.J. Williams, D.W. Fahey, G. Hübler, C.S. Eubank, S.C. Liu, P.C. Murphy, D.L. Albritton, and F.C. Fehsenfeld, Measurement of the NO_x-O_3 Photostationary State at Niwot Ridge, Colorado, J. Geophys. Res., 91, 5361-5370, 1986b.

Pickering, K. E., R. R. Dickerson, G. J. Huffman, J. F. Boatman and A. Schanot, Trace Gas Transport in the Vicinity of Frontal Convective Clouds, J. Geophys. Res., submitted, 1987.

Platt, U. and D. Perner, Direct Measurements of Atmospheric CH_2O, HNO_2, O_3, NO_2 and SO_2 by Differential Optical Absorption in the Near UV, J. Geophys. Res., 85, 7453-7458, 1980.

Pratt, G. C., R. C. Hendrickson, B. I. Chevone, D. A. Christopherson, M. V. O'Brien and S. V. Krupa, Ozone and Oxides of Nitrogen in the Rural Upper Midwestern U.S.A., Atmospheric Environment, 10, 2013-2023, 1983.

Research Triangle Institute, Investigation of Rural Oxidant Levels as Related to Urban Hydrocarbon Control Strategies, EPA-450/3-75-036, 1975.

Ridley, B. A., M. A. Carroll, and G. L. Gregory, Measurements of Nitric Oxide in the Boundary Layer and Free Troposphere over the Pacific Ocean, J. Geophys. Res., 92, 2025-2047, 1987.

Roe, J.M., A Climatology of a Newly-Defined Tropopause Using Simultaneous Ozone-Temperature Profiles, Air Force Geophysics Laboratory Report Number AFGL-TR-81-0190, 1981.

Russell, P.B., E.F. Danielsen, and R.A. Craig, The NASA spring 1984 Stratosphere-Troposphere Exchange Experiments: Science objectives and operations, EOS 66, 235, 1985.

Schiff, H.I., D. Pepper, and B.A. Ridley, Tropospheric NO Measurements up to 7 km, J. Geophys. Res., 84, 7895-7897, 1979.

Singh, H.B., L.J. Salas, B.A. Ridley, J.D. Shetter, N.M. Donahue, F.C. Fehsenfeld, D.W. Fahey, D.D. Parrish, E.J. Williams, S.C. Liu, G. Hübler, and P.C. Murphy, Relationship Between Peroxyacetyl Nitrate and Nitrogen Oxides in the Clean Troposphere, Nature, 318, 347-349, 1985.

Spicer, C. W., J. R. Koetz, C. W. Keigly, G. M. Sverdrup, and G.F. Ward Nitrogen Oxides Reactions Within Urban Plumes Transported Over the Ocean. EPA Draft Report on Contract No. 68-02-2957, 1982.

Spicer, C. W., Nitrogen Oxide Reactions in the Urban Plume of Boston, Science, 215, 1095-1097, 1982.

Stedman, D. H. and M. J. McEwan, Oxides of Nitrogen at Two Sites in New Zealand, Geophys. Res. Letters, 10, 168-171, 1983.

Tallamraju, R., National Oceanic and Atmospheric Administration, Boulder, CO, private communication, 1987.

Torres, A. L. and H. Buchan, Tropospheric Nitric Oxide Measurements Over the Amazon Basin, J. Geophys. Res., submitted 1987.

NO_x SOURCES AND THE TROPOSPHERIC DISTRIBUTION OF NO_x DURING STRATOZ III

D.H. Ehhalt and J.W. Drummond*
Institut für Atmosphärische Chemie,
Kernforschungsanlage Jülich
Postfach 1913
D-5170 Jülich, F.R.G.

ABSTRACT The global sources of atmospheric nitrogen oxide are briefly reviewed. About 50 x 10^{12} g Nitrogen are emitted annually mostly in the form of NO. Anthropogenic emissions account for about 30 x 10^{12} gN/yr. Most of the emissions take place over the continents and in the Northern Hemisphere. Meridional distributions of NO and NO_x measured during June 1984 show elevated concentrations in the upper troposphere over the continents between 60°N and 30°S latitude. Their patterns suggest lightning and upward transport from the atmospheric boundary layer as the major source processes for NO_x in the upper troposphere.

1. INTRODUCTION

The atmospheric distribution of a given trace gas depends on the distribution of its sources. The shorter its residence time in the atmosphere, the closer its concentration field will map the geographical distribution of its sources. Thus, the measurement of regional concentration fields should help to identify source areas and place constraints on their possible contribution to the atmospheric trace gas budget. In the following we will examine the meridional data on NO concentration measured by Drummond et al., 1987, during STRATOZ III, in June 1984 and investigate which source processes inject NO into the free troposphere. To set the stage we will first outline the known NO sources of global importance and present estimates of the respective source strengths and their spatial and temporal variations. In a second part we will introduce the tropospheric NO distributions and in a third we will discuss what constraints, if any, the measured NO fields place on the potential sources.

*Present address: Unisearch Associates Inc., Concord, Ontario, Canada, USA

I. S. A. Isaksen (ed.), Tropospheric Ozone, 217–237.

2. SOURCES OF NO_x

NO_x enters the atmosphere primarily in the form of NO. There is a large variety of sources which inject NO into all regions of the atmosphere making the estimate of the total source strength rather uncertain and the geographical pattern of the source distribution rather complex.

NO budgets have been published by several authors (Böttger et al. 1978; Ehhalt and Drummond, 1982; Logan, 1983). A summary of the major sources and their global emission rates based on these references is listed in Table I.

Not much new has been learned, since these budgets have been published. In fact, the biogenic release of NO from soils is the only source term which has received considerable attention in the meantime (see WMO, 1985, for a summary). Still the emission rates from various soils under various conditions range over a few orders of magnitude and the uncertainties remain large. But it has been found that the NO release from soils depends on temperature, to a lesser degree on moisture and nitrate in the soil. Thus the emission rates show diurnal and seasonal cycles (Anderson and Levine, 1987).

It is clear from table I that the surface sources totalling 38×10^{12} gN yr^{-1} dominate the global emission rate. Yet much of that NO_x is expected to remain within the atmospheric boundary layer, such that those sources which inject NO directly into the higher atmosphere are of great importance for the NO_x concentration in the free troposphere. Among these lightning seems to be the dominating term.

The global emission rate of NO is about 50×10^{12} gN yr^{-1} with an uncertainty range of ± 50 % as indicated by the upper and lower bounds of the source strength estimates in Table I. It is also noted that virtually all of the emissions take place over the continents. Surface emissions from the oceans appear negligible.

Even the major atmospheric sources are concentrated over the continents: Thunderstorms occur mainly over continents and the atmospheric NH_3 concentrations are significantly higher over continents.

Finally we note that a large fraction of the NO emissions are anthropogenic. In addition to the fossil fuel burning, 70 % of the burning of biomass are caused by man. Thus a subtotal of 60 % of atmospheric NO_x are derived from human activities.

Besides the global values the spatial distribution of the NO emissions is of interest. To obtain a survey of the latter we take a short cut. We argue that the global deposition pattern of nitrate should map, in reasonable

Table I: Global sources of NO_x and their strengths. The units are 1×10^{12} gN yr^{-1}. The emission rates for fossil fuel burning are from Logan (1983) and refer to the year 1979. The other data are from Ehhalt and Drummond (1982). For details see Böttger et al. (1978), Ehhalt and Drummond (1982) and Logan (1983).

		lower bound	mean	upper bound
I.	Surface Sources			
A.	fossil fuel burning			
	coal	3.4	6.4	9.4
	oil	2.4	3.1	3.8
	gas	0.7	2.3	3.9
	transportation	6	8.0	10
	industrial sources	0.6	1.2	1.8
	subtotal for A	13.1	21	28.9
B.	soil release	1	5.5	10
C.	biomass burning			
	savanna	1.8	3.1	4.3
	deforestation	.8	2.1	3.4
	fuel wood	1	2	3
	agricultural refuse	2	4	6
	subtotal for C	5.6	11.2	16.7
	subtotal for surface sources	19.7	37.7	55.6
II.	Atmospheric Sources			
	NH_3 oxidation	1.2	3.1	4.9
	lightning	2	5	8
	high flying aircraft	.2	.3	.4
	NO_y transported from stratosphere	.3	.6	.9
	subtotal for atmospheric sources	3.7	9.0	14.2
III.	Total Emission	23.4	46.7	69.8

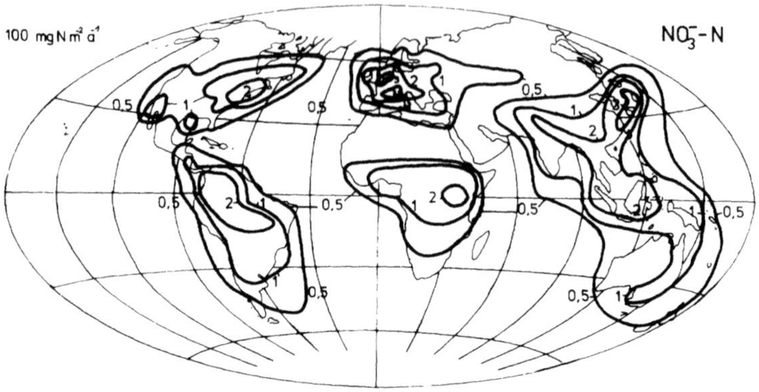

Figure 1: Isolines of nitrate deposition averaged over the years 1950 to 1977 (after Böttger et al., 1978). The units are in 100 milligrams of elemental nitrogen per square meter per year.

approximation that of the NO sources, because nitrate is the chemical form in which NO is eventually removed from the atmosphere and because the atmospheric lifetimes of NO_x and NO_y are so short that NO_x and NO_y are not dissipated widely after emission, certainly not on a global scale.

Figure 1 shows that distribution derived from precipitation data from 1950 to 1977 (see Böttger et al., 1987, Ehhalt and Drummond, 1982). The isolines of constant deposition are given in units of 100 mg $N/m^2/yr$ or 1 kg N/ha/yr. The contours for the northern hemisphere (N.H.) are reasonably well established, because it contains a reasonable number of stations. The southern hemisphere (S.H.) has far fewer stations, and the isolines there are highly uncertain (Böttger et al, 1978). Still they probably approximate the true distribution. The contours of highest deposition clearly outline the areas of high industrial activity in North America, in Europe and in Asia. But they also indicate quite high sources in the continental tropics. In fact, the deposition rates in the tropics are surprisingly high. They point to stronger emissions in that region than accounted for by Table I. Again we note that the deposition over the oceans is low and hence that the sources are mainly continental.

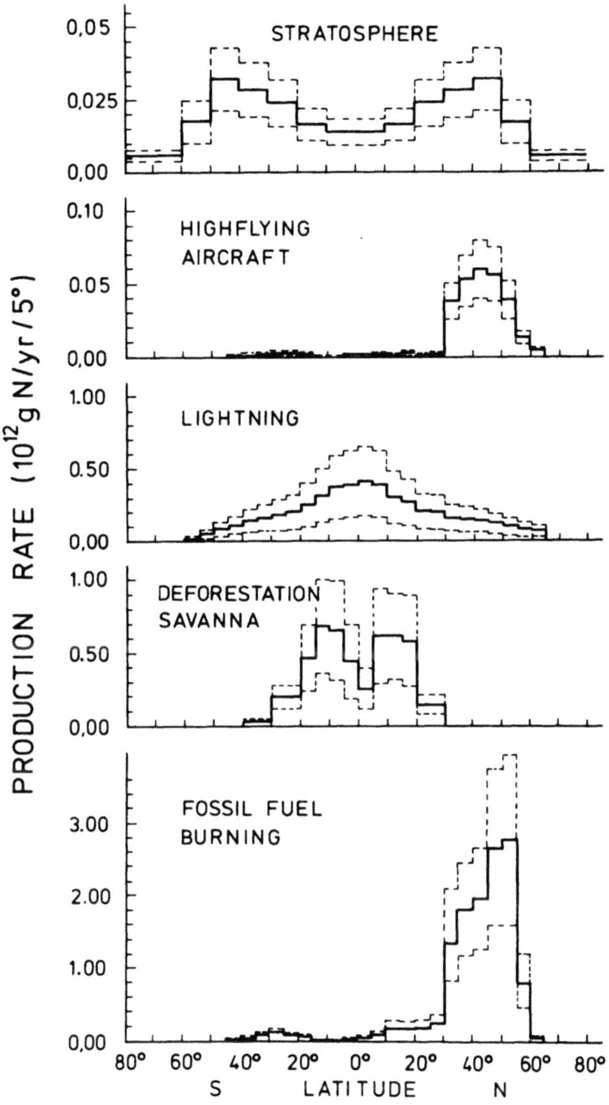

Figure 2: Estimated production rates of NO as a function of latitude for five sources (from Ehhalt and Drummond, 1982). The mean, upper and lower bounds were calculated using the detailed latitudinal distributions given by Böttger et al., 1978. Note that the scales for the production rates vary up to a factor of 40.

For a few select sources, mainly atmospheric ones, Figure 2 shows the estimated annually averaged latitudinal variation of the emission rates. It clearly indicates that fossil fuel burning is the major source. It is mainly located in the N.H. On the other hand deforestation and Savanna burning are tropical sources. The lightning source too dominates in the tropics with lower values in the midlatitudes. All these sources are mainly continental.

The stratospheric input of NO_y in both hemispheres was assumed to be distributed like the radioactive fallout in the N.H. resulting from stratospheric nuclear bomb explosions. It is quite likely that the stratospheric input into the S.H. is lower than that into the N.H. This source and that from highflying aircraft are much weaker than the others but they do inject some NO_y also over the ocean.

Besides the spatial, the temporal variations of the NO emission rates are of interest. Most of the NO sources show a seasonal variation. That of soil release has already been mentioned. Biomass burning which goes on during the dry season, lightning which occurs mainly during summer in midlatitudes and during the rainy season in the tropics, and stratospheric injection which takes place during late winter and early spring are highly seasonal. Even fossil

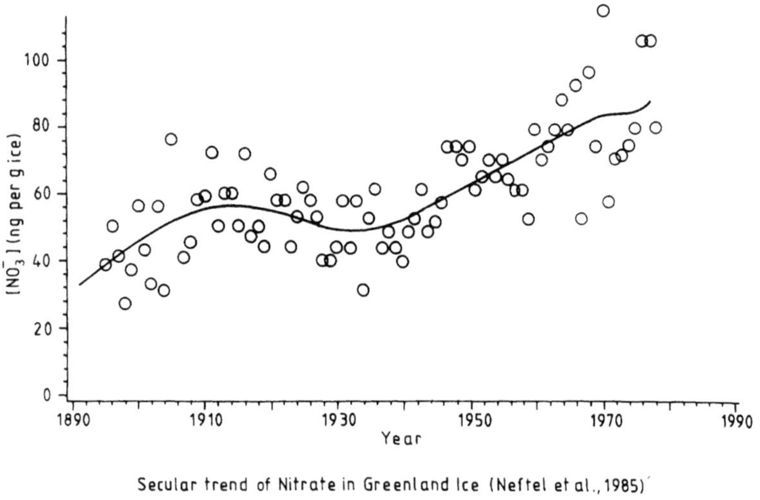

Figure 3: Secular trend of the nitrate concentration in Greenland snow (from Neftel et al., 1985).

fuel burning when averaged over most industrialized countries shows a seasonal pattern with about 20 % higher emissions in winter (Rotty, 1987).

In addition, the anthropogenic sources exhibit a secular increase. This is most easily demonstrated by the history of nitrate deposition in snow from Greenland, an area remote from all sources. That secular trend is shown in Figure 3 (Neftel et al., 1985).

The NO_3^- in the dated snow clearly shows an average increase: From about 40 ng NO_3^-/g ice in 1890 i.e. at the beginning of the industrial era to about 80 ng NO_3^-/g ice in 1970. NO_3^- in Greenland snow has roughly doubled its concentration which is consistent with our previous estimate that about 60 % of the NO amount are of anthropogenic origin. The current rate of increase is 1.6 %/yr. Consequently the anthropogenic fraction is increasing by about 3 % a year superimposed on a constant base of natural emissions.

3. THE MERIDIONAL DISTRIBUTION OF NO AND NO_x

The NO measurements were made in June 1984 during the STRATOZ III mission of the Caravelle 116 of the Centre d`Essais en Vol, Bretigny, France and have been published by Drummond et al. (1987).

1-D PROFILES

The mission`s flight track is shown in Figure 4. The plane landed about every 3000 km which allowed the sampling of vertical profiles during the approach to and take off from the airports. Their positions are indicated in Figure 4. Since many of the airports are located close to big cities or are themselves significant sources of NO, the air in the immediate vicinity of the airports was often polluted. These data were considered not representative of the regional atmospheric boundary layer (ABL) and were omitted. Thus in some cases the vertical profiles only extend between the upper boundary of the ABL and 12 km, the maximum flight altitude. Naturally, most of the airports are based on continents. Thus virtually all of the vertical profiles measured are under the influence of the continental NO sources. None of our individual vertical profiles of NO concentration can be considered maritime.

Examples of the resulting NO profiles are shown in Figures 5 and 6. They cover the latitudes from 70°N to 40°S.

All of these profiles show elevated NO concentrations in the upper troposphere. In the cases of profile 3A and profile 21A (Figure 5) these elevated concentrations can be assigned to definite sources. In profile 3A the high NO values are due to stratospheric air. In fact the aircraft had penetrated into the stratosphere at around 10 km altitude as indicated by the steepening gradient in potential temperature. This is corroborated by the high O_3 values observed simultaneously (see Marenco, this volume).

Figure 4: Flight track of the STRATOZ III mission.

Profile 21A contains the highest NO values found in the upper troposphere - about 3 ppb. In this case we strongly suspect that the aircraft intersected an airmass lofted from the atmospheric boundary layer. The high levels of NO are correlated with high levels of CO and low concentrations of O_3 characteristic of polluted boundary layer air (Marenco, this volume). In the other vertical NO profiles shown in Figure 6 as in most profiles measured by Drummond et al. (1987) the high NO values in the upper troposphere cannot be related that readily to a definite source.

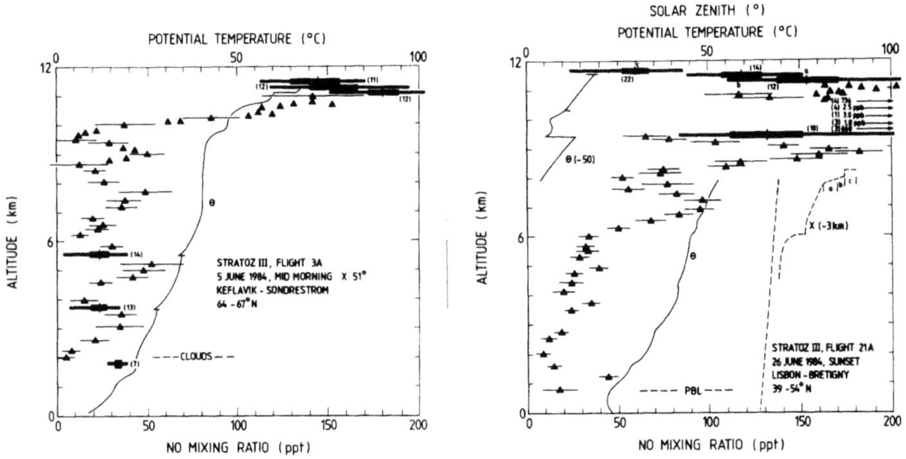

Figure 5: Vertical profiles of the NO mixing ratio during the flight sections 3A and 21A. The full triangles represent 3 minute running averages sampled during ascent (Δ) or descent (∇), and daytime (X ≤ 85°). The error bars represent the mean standard deviation. The bars indicate longer time averages obtained during periods of level flight: the central vertical line represents the mean value, the heavy bar the standard deviation of the mean, the thin bar the range of the 3 minute running averages during level flight. The number in parenthesis gives the time interval of level flight in minutes. NO data falling outside the plot are indicated by arrows - the associated numbers giving the actual mixing ratio. Also given are the potential temperature, Θ, and the solar zenith angle, X, the latter as curve when close to dawn or dusk.

The vertical NO profiles, however, show rather regular features which can be summarized as follows:

a) All of the (daytime) profiles in the northern hemisphere show an increase in the upper troposphere similar to the ones in Figures 5 to 6.

b) All of the (daytime) profiles in the tropical latitudes of the S.H. show such an increase as indicated by Figure 6.

c) This, together with usually elevated NO levels in the atmospheric boundary layer, results in C shaped profiles of NO for the atmosphere over the continents with low values between 2 and 8 km altitudes.

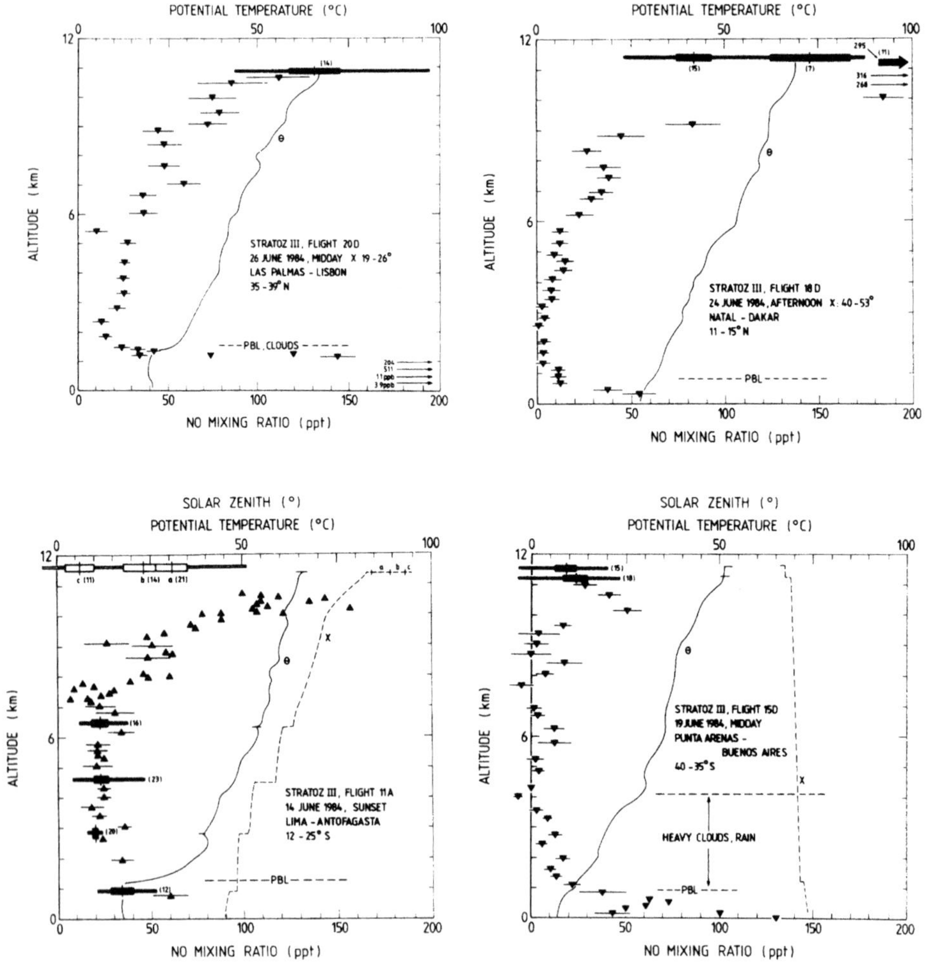

Figure 6: Vertical profiles of the NO mixing ratio during the flight sections 20D, 18D, 11A, 15D. See Figure 5 for explanation of the symbols.

2-D CROSS-SECTIONS

In order to give a better overview and to also present the essence of the data collected at cruising altitude, the NO measurements are summarized in the form of meridional cross sections (Figure 7). For clarity the figures do not contain the individual data points, but rather the isolines of constant mixing ratio interpolated between them. Such interpolation always implies a certain measure of subjectivity. It arises, in part, from the difficulty in assigning local mean values to data, which show large fluctuations. To minimize the fluctuations, the data are grouped into 4 sections: Southbound and northbound flights, with winds blowing either from the land or off the ocean. The first division is made because southbound and northbound flights were about two weeks apart and separated substantially in geographical location (Figure 4). The second distinction takes account of the fact that most of the NO emissions are located over the continents and that part of the cruising altitude flights were over the ocean. Thus the 2-D cross sections contain information on the NO concentration over the ocean. The southbound section is composed of the flights between Sondrestrom, Greenland, and Punta Arenas, Chile, the northbound section of the fligths between Punta Arenas and Paris. The latter section also includes the flights between Paris and Sondrestrom; but note that these took place at the beginning of the mission about three weeks prior to the other northbound flights (see Figure 4). The border between the two sections of northbound flights is marked by a vertical dashed line in Figure 7. The distinction between continental and oceanic air is somewhat preliminary; it is derived from a one day back trajectory based on the available wind data.

Figure 7 is essentially the projection of the NO concentrations measured along the flight track onto the meridional plane. Because the flight track is oriented essentially North-South (see Figure 4) this projection leads to little distortions at most latitudes - except for the most northern latitudes where the flight track is virtually East-West. Indeed the narrowly spaced vertical isolines of constant NO mixing ratio between 1 and 8 km altitude in panels a and c of Figure 7 are due to the projection of an essentially East-West gradient onto the meridional plane and should not be construed as a feature of the latitudinal gradient of NO.

Despite the fact that the individual panels in Figures 7 show a high degree of variability internally and with respect to each other, they share a few common large scale features which confirm and extend those noted in the 1-D profiles:

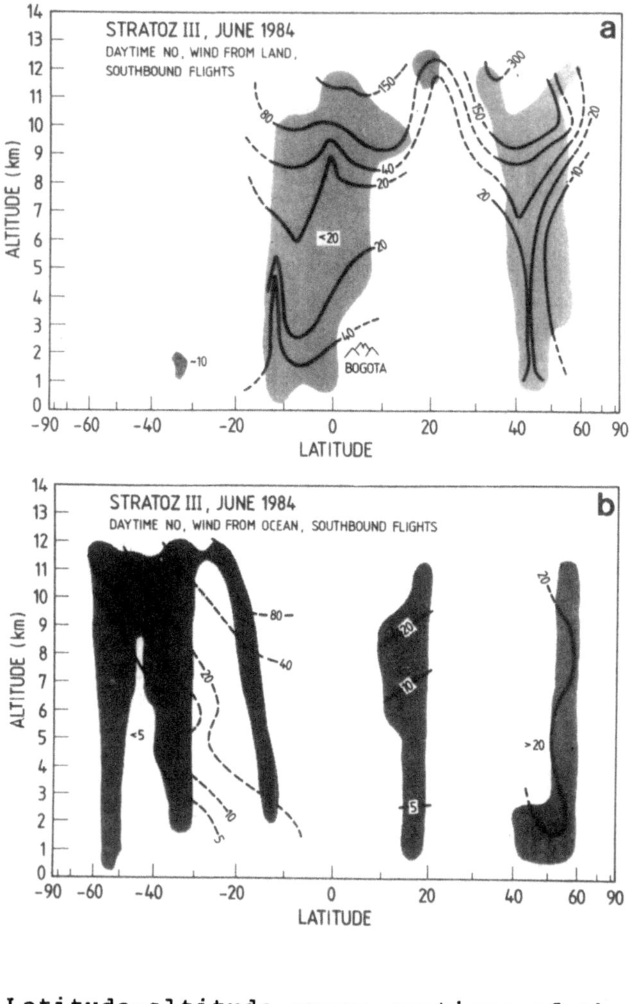

Figure 7: Latitude-altitude cross sections of the NO mixing ratio during STRATOZ III for the south- and northbound flights, and on and off shore winds. The contours of constant mixing ratio are given in ppt. Shaded areas indicate the regions over which data with the indicated general wind direction were collected.

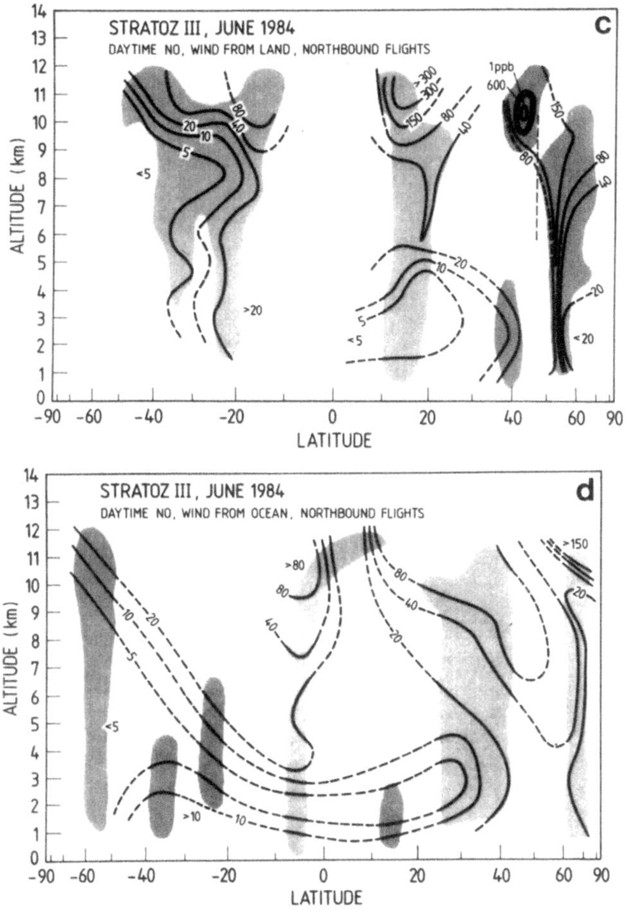

Figure 7: Latitude-altitude cross sections of the NO mixing ratio during STRATOZ III for the south- and northbound flights, and on and off shore winds. The contours of constant mixing ratio are given in ppt. Shaded areas indicate the regions over which data with the indicated general wind direction were collected.

1. The NO concentration in the southern hemisphere is generally lower than in the northern hemisphere. This implies weaker sources of NO_x in the S.H., at least during winter.
2. Air from over the ocean shows lower NO mixing ratios than air advected from the continents, especially at high altitudes in the N. H. In particular air sampled over the ocean, as over Bermuda or between Natal and Dakar shows low NO at high altitudes.
3. There is a trough of higher NO concentrations in the upper troposphere over the continents reaching from 60°N to 20°S latitude. It exceeds substantially the NO concentrations found in the middle troposphere and strongly suggests large scale input of NO_x into the upper troposphere.

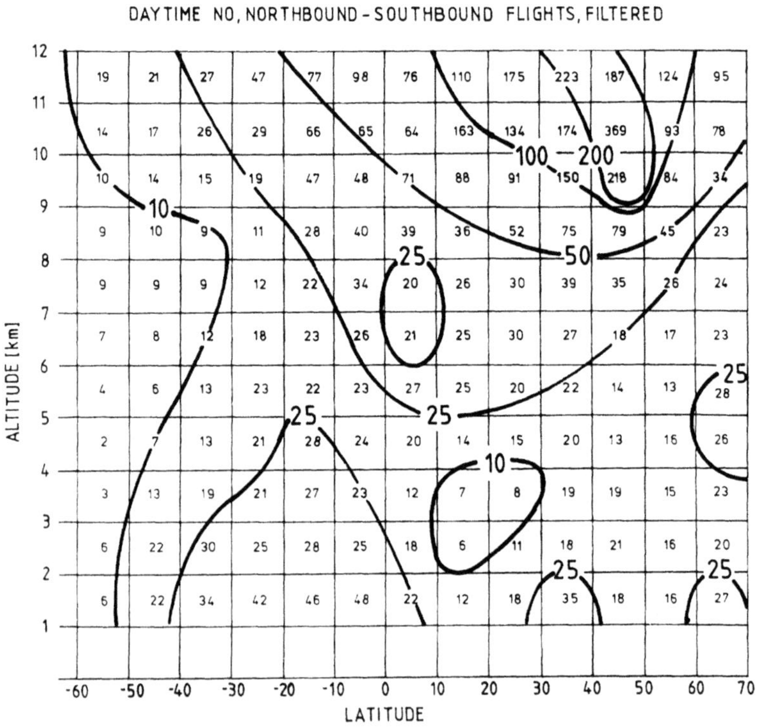

Figure 8: Average meridional distribution of the NO concentration during STRATOZ III. The NO concentrations are given in ppt. Each number represents the average of all concentration measurements falling into a 1 km altitude, 10° latitude grid box.

To emphasize the large scale features we have condensed the data even further by averaging all data into a grid of 1 km altitude and 10° latitude boxes (Figure 8). In addition the data were "filtered" to remove excursions and to fill in eventual gaps. This was done by an averaging procedure assigning 1/2 of the total weight to the central box and 1/16 to each of the adjacent boxes. The high values measured in the boundary layer were excluded to prevent propagation of the high boundary values into the atmospheric layer between 1 and 2 km altitude. This plot retains the essential features already mentioned: An area of high concentrations in the upper troposphere, lower values between 2 and 8 km altitudes, and low values in the southern hemisphere. The feature of high concentrations in the ABL, of course, has been removed.

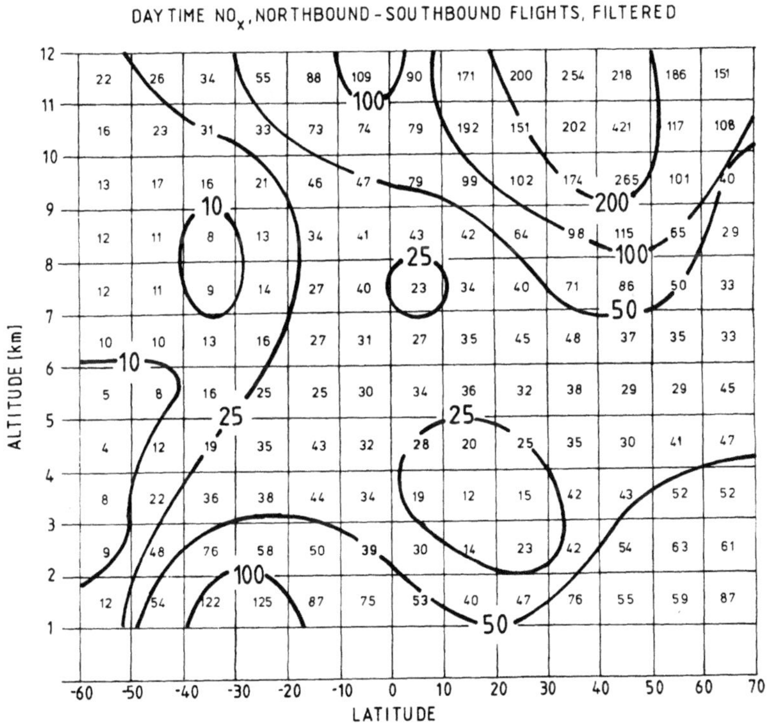

Figure 9: Average meridional distribution of the NO_x concentration derived from the NO measurements during STRATOZ III. The NO_x concentrations are given in ppt. Each number represents the average of all NO_x data points falling into a 1 km altitude, 10° latitude grid box.

Although suggestive, the NO distribution does not provide a quantitative mapping of the sources, because atmospheric NO is not a conserved quantity. It interconverts within minutes with NO_2. We have, therefore, calculated the corresponding concentrations of NO_x which is a much better conserved tracer, by using the simultaneously measured O_3 distribution (Marenco, this meeting) and calculated photolysis rates of NO_2. The latter were partly checked experimentally, because the photon flux from below had been monitored during the flights. In the upper troposphere about 80 % of the NO_x are present as NO, in the ABL only 30 %, such that the NO_x fields are somewhat distorted with respect the NO field (Figure 9). Nevertheless the overall pattern of the NO_x distribution closely resembles that already seen for NO (Figure 8).

4. THE TROPOSPHERIC BURDEN OF NO_x

To provide still another perspective the NO_x data are displayed as column densities, i.e. as number of molecules per m^2 between 1 and 12 km altitude, along the flight track (Figure 10). Obviously this is only possible at those locations where vertical profiles had been obtained, i.e. the burden shown in Figure 10 is more or less representative of continental profiles. We observe the highest values over Europe, reasonably high values over the Western Atlantic along east coast of North America, and similar high values along the west coast of South America. The latter comes somewhat as a surprise.

To compare the pattern of the tropospheric NO_x burden with the geographical pattern of the NO sources, the nitrate deposition rates from Figure 1 are also displayed. The correlation is surprisingly good with two exceptions, one along the west coast of South America and one along the west coast of North Africa where the column densities are relatively high compared to the deposition rates. However, both are areas of very low precipitation where one would expect such a deviation in the ratio of wet deposition of nitrate to the atmospheric NO_x burden.

5. CONSTRAINTS ON THE SOURCES OF NO_x

The various projections of the data presented in the previous section demonstrate systematic patterns in the vertical and horizontal distribution of NO_x. These patterns place certain constraints on the possible sources of atmospheric NO_x. In particular, the rather universal C-shape of the vertical NO profiles over the continents

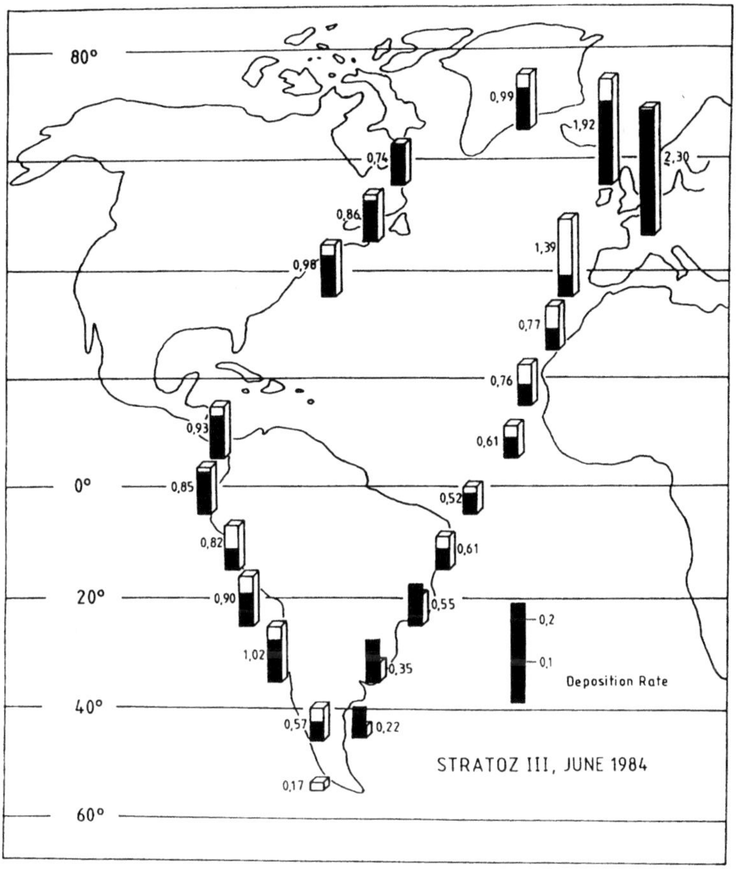

Column density NO_x (10^{19}n/m^2), 1-12km

Figure 10: Free tropospheric column density of NO_x along the flight track of STRATOZ III. The column density is open bars and numbers given in units of 10^{19} NO_x molecules per m². For comparison the nitrate deposition rate (in 0,1 g/m²/yr) at the same locations (from Figure 1) is shown as black bars.

requires widespread sources at the earth's surface, widespread sources in the upper troposphere and a widespread sink in the middle troposphere. (The sink is provided by conversion of NO_x to NO_y and subsequent rainout. It will be treated elsewhere.) Obviously our data, which cover essentially the altitudes above 1 km, lend themselves primarily to the study of those processes that inject NO_x into the free troposphere.

A priori there are four source processes which inject NO_x into the free troposphere and could contribute to the observed high NO levels in the upper troposphere. These are:

- NO emissions from high flying aircraft, which average globally about 0.3 x 10^6 tN/yr.
- Downward flux of NO_y from the stratosphere which in June should still be active in the northern hemisphere and just beginning in the southern hemisphere. This flux averages globally 0.6 x 10^6 tN/yr.
- Fast upward transport of NO_x from the planetary boundary layer by deep convection (cf. Chatfiled and Crutzen, 1984; Gidel, 1983). That transport has not yet been quantified, but the surface sources are the strongest totalling 40 x 10^6 tN/yr (see table 1).
- NO produced from lightning, much of which would be injected at the altitudes of the outflow layers of thunder clouds. The annual production of NO by lightning has been estimated to about 5 x 10^6 tN/yr.

We note that apart from the stratospheric flux, all these NO inputs are concentrated over the continents.

In the following the possible contribution of each of these sources to the high NO_x levels in the upper troposphere will be examined.

Except for part of the route over the North Atlantic the Caravelle did not follow the flight routes of commercial aircraft. Thus in most flights the danger of direct NO contamination by the exhaust from other aircraft was very low. Moreover, such contamination should have also shown up in the CO, CH_4, and non methane hydrocarbon (NMHC) measurements made during STRATOZ III and there was no indication of this. Finally, aircraft emissions are a minor source globally, and very low south of 30°N (cf. Figure 2). Thus, we would argue that apart from occasional and localized additions aircraft emissions contributed little to the high NO_x-levels observed in the upper troposphere over the continents. Their contribution should have been even less for the NO_x column densities in the free troposphere.

The contribution of the stratospheric source seems also to have been minor. Although those flights which penetrated into the stratosphere gave elevated NO levels, the stratospheric NO concentrations - averaging about 150 ppt - were not high enough to explain the often higher NO concentrations in the upper troposphere found during other flights. Moreover, it is difficult to explain the high NO levels at 8 to 12 km altitude over the continental tropics by such a source. The general upwelling in that area prevents effective downward mixing. In addition, since the tropopause is at a much higher altitude than in mid-latitudes, descending air from the stratosphere should be more strongly diluted by tropospheric air by the time it reaches 10 km altitude. Finally, in contrast to the flights over the continents all flights over the ocean showed low NO values, about 20 ppt, in the upper troposphere. This is not consistent with a more or less longitudinally homogeneous downward flux of NO_y as expected from the stratospheric source. Thus we are led to the conclusion that the uniformly high values of NO_x observed in the upper troposphere over the continents from at 60° N to 30° S latitudes cannot be explained by the stratospheric source nor by the exhaust from high flying aircraft, although both may contribute regionally in a minor way. (In fact, the values found at high altitudes over the ocean may serve as an upper limit for the contribution by the stratospheric source during June 1984.)

The fact, that the NO_x concentrations in the upper troposphere were high over the continents only, points to a major contribution from "continental" sources. This is corroborated by the column densities between 1 and 12 km altitude (Figure 10) which are also higher over the continental areas with high emissions. Thus it is the continental source processes, lightning and upward transport from the ABL, which remain as likely explanation for the high NO concentrations in the upper troposphere. There is direct evidence for both. Fast upward transport from the ABL has been observed for NO_x (cf. Figure 5) and for shortlived NMHC (Ehhalt et al. 1985). Lightning production of NO_x has been observed by Noxon, 1976. Thus both processes must contribute. In the tropics lightning could very well be the major source for high NO_x in the upper troposphere. But over the heavily populated and industrialized areas of Europe and North America we would argue that the upward transport of NO_x is the major source process. This seems to be also indicated by the high column densities observed in the free troposphere over Europe (Figure 10).

At present no quantitative statement can be given about the contribution of the various sources to the high NO levels in the upper troposphere. The situation will improve, however, when the other trace gas data from STRATOZ III become available. These data can be used to trace the origin of the NO. The NO_x input from high flying aircraft, for example, is accompanied by a mix of non methane hydrocarbons characteristic of jet exhaust, which can easily be distinguished from other NMHC sources (Ehhalt et al., 1985). By the same token, air transported upward from the planetary boundary layer should exhibit the NMHC pattern characteristic of that region as well as elevated

CO levels. Input of NO from the stratosphere is correlated with high O_3 and potential vorticity, but low CO. Lightning remains as the only source which produces NO but affects no other trace gas in a significant way. Thus each source appears to be sufficiently characterized by other tracers, to eventually allow a reasonable estimate of each contribution to upper tropospheric NO.

6. CONCLUSION

Elevated concentrations of NO were found in the upper troposphere over the continents between 60°N and 30°S latitude. The vertical and horizontal distribution of the NO_x concentration indicate that most of the NO_x concentration observed must be injected by fast vertical transport from the atmospheric boundary layer and by lightning. The NO concentrations found in the upper troposphere over the continents are high enough to allow the production of ozone. This is all the more important, since during STRATOZ III and the earlier STRATOZ II occasional high concentrations of light NMHC have also been observed in the upper troposphere (cf. Ehhalt et al., 1985). Thus there are large patches of relatively polluted and thus photochemically active air in the upper troposphere over the continents.

REFERENCES

Anderson, I.C. and Levine J.S., 1987, 'Simultaneous Field Measurements of Biogenic Emissions of Nitric Oxide and Nitrous Oxide', *J. Geophys. Res.* **92**, 965-976.

Böttger, A., Ehhalt, D.H. and Gravenhorst, G., 1978, 'Atmosphärische Kreisläufe von Stickoxiden und Ammoniak', Kernforschungsanlage Jülich, *Report-Jül-1558*.

Chatfield, R.B. and Crutzen, P.J., 1984, 'Sulfur Dioxide in Remote Oceanic Air: Cloud Transport of Reactive Precursors', *J. Geophys. Res.* **89**, 7111-7132.

Drummond, J.W., Ehhalt, D.H. and Volz A., 1987, 'Measurement of Nitric Oxide between 0-12 km Altitude and 67°N - 60°S Latitude Otained During STRATOZ III', submitted to *J. Geophys. Res.*

Ehhalt, D.H. and Drummond, J.W., 1982, 'The Tropospheric Cycle of NO_x', *Chemistry of the Unpolluted and Polluted Troposphere*, 219-251, Ed.: H.W. Georgii, W. Jaeschke, published by Reidel, Dordrecht, Holland.

Ehhalt, D.H., Rudolph, J. Meixner, F. and Schmidt, U., 1985, 'Measurements of Selected C_2-C_5 Hydrocarbons in the Background Troposphere: Vertical and Latitudinal Variations', *J. Atmos. Chem.* 3, 29-52.

Gidel, L.T., 1983, 'Cumulus Cloud Transport of Transient Tracers', *J. Geophys. Res.* **88**, 6587-6599.

Logan, J.A., 1983, 'Nitrogen Oxides in the Troposphere: Global and Regional Budgets', *J. Geophys. Res.* **88**, 10875-10807,.

Neftel, A., Beer, J. Oeschger, H., Zürcher, F. and Finkel, R.C., 1985, 'Sulphate and Nitrate Concentrations in Snow from South Greenland 1895-1978',*Nature* **314**, 611-613.

Noxon, J.F., 1976, 'Atmospheric Nitrogen Fixation by Lightning', *Geophys. Res. Lett.* 3, 463-465

Rotty, R.M., 1987, 'Estimates of the Seasonal Variation of Fossil Fuel CO_2 Emissions', *Tellus* **39 B**, 184-202.

WMO, 1985, 'Atmospheric Ozone, Assessment of our Understanding of the Processes Controlling its Present Distribution and Change', *World Meteorological Organization Global Ozone Research and Monitoring Project,* Report NO. **16**, **Vol I.**

CALCULATION OF THE DISTRIBUTION OF NOx COMPOUNDS IN EUROPE

Øystein Hov[1)], Anton Eliassen[2)] and David Simpson[3)]

[1)] NILU, P.O. Box 64, N-2001 Lillestrøm, Norway
[2)] The Norwegian Meteorological Institute, P.O. Box 320, Blindern, 0314 Oslo 3
[3)] Warren Spring Laboratory, Gunnels Wood Road, Stevenage, Hertfordshire SG1 2BX, UK

ABSTRACT. A lagrangian model to calculate the European NOx-budget, is described. Emissions of NOx, SO_2 and NH_3 are specified on the 39x37 EMEP-grid with a grid distance of 150 km at 60^0N latitude. The numerical weather forecast model at the Norwegian Meteorological Institute has provided data for wind, atmospheric stability, temperature, relative humidity and cloud cover. The chemical description includes 10 different species in the gas and aerosol phases. Calculations have been carried out of daily values of these species for the year 1985, and some comparison of calculations and measurements are shown in the paper.

1. INTRODUCTION

The formation of ozone (O_3) in the troposphere takes place through the photolysis of nitrogen dioxide (NO_2) at wavelengths below 400 nm whereby atomic oxygen in the ground state $O(^3P)$ is formed. $O(^3P)$ rapidly recombines with oxygen to form O_3. Most of the nitrogen oxides (NOx = NO + NO_2) are emitted as nitric oxide (NO), but NO is converted to NO_2 in the matter of a few minutes through the reaction with O_3. Net production of O_3 takes place when NO is converted to NO_2 through reaction with organic or inorganic peroxy radicals. Based on the reaction between NO_2 and the hydroxyl radical (OH), the lifetime of NOx can be estimated to be one day or less in the summer and perhaps up to one week in the winter. Nitric acid (HNO_3) formed through the reaction between NO_2 and OH is efficiently scavenged through dry and wet removal. A global mean value from the start of a trajectory until the first passage of a major cloud system, often connected with precipitation, is 9 days for December through February and 7 days for June through August (Hamrud and Rodhe, 1986). Over the Atlantic temperate region, the average time to encounter a major

I. S. A. Isaksen (ed.), Tropospheric Ozone, 239–261.

cloud system is 4 days in winter and 5 days in summer. This gives an indication about the lifetime of nitrate in the atmospheric boundary layer over Europe (less than one week), while nitric acid is efficiently dry deposited (lifetime of one day or less).

Peroxyacetylnitrate (PAN) and related compounds (peroxypropionylnitrate-PPN, peroxybenzoylnitrate - PbzN, etc.) are sinks for NOx, being formed through the reaction between NO_2 and peroxy radicals derived from acetaldehyde and higher aldehydes. PAN is not very water soluble and its deposition velocity is less than that for NO_2. The thermal decomposition of PAN is highly temperature dependent, but at moderate temperatures in the atmospheric boundary layer it is sufficiently stable to be transported over large distances, perhaps 1000 km or more (Hov, 1984). In the upper troposphere PAN can have a lifetime of several weeks, and dominate among the species derived from NOx (Singh and Salas, 1983).

NO, NO_2, PAN, HNO_3 and nitrate aerosol are the most important NOx-species in the troposphere. The sources for NOx are predominantly at or close to the ground. The chemical lifetimes of the NOx-species in the atmospheric boundary layer are shorter than, or comparable to, the typical exchange time of air between the ground and the free troposphere (2-4 d over Europe). This means that the bulk of the NOx-emissions appear in the atmospheric boundary layer only, and typically remain there for a few days before removal through dry or wet deposition, allowing for transport over a distance of the order of 1000 km. In Figure 1 is shown the volume weighted arithmetic mean annual concentration of nitrate in precipitation for 1985 over Europe, as measured daily at about 70 sites in EMEP (Co-operative programme for monitoring and evaluation of the long range transmission of air pollutants in Europe), taken from Schaug et al. (1987). From a maximum value of about 1 mg N/l in the countries with the highest emissions in continental Europe, the values fall to less than 0.3 mg N/l over Scotland and northern Scandinavia. This shows that the transport and removal of the most important species derived from NOx-emissions take place on a continental spatial scale.

NOx is essential for the chemistry of ozone and other trace gases in the free troposphere. NOx can be transported out of the atmospheric boundary layer in situations where the vertical transport is faster than the removal at the ground. In regions with strong convective mixing, e.g. in and around thunderstorm clouds, air can be brought from the ground and almost to the tropopause in a few hours (Dickerson et al., 1987). NOx is produced by lightning, and there is also a downward flux of NOx through the tropopause. NOx in the free troposphere plays a central role in many of the processes causing species like ozone, methane, carbon monoxide and hydroxyl in the atmosphere to change slowly.

The "global budget" of NOx is the sum of a number of regional or continental budgets for NOx. The deposition of NOx-species in Europe probably originates to a large extent from emissions within Europe itself. Recently results were published from the application of the Geophysical Fluid Dynamics Laboratory (GFDL) general circulation/transport model in Princeton to calculate the accumulated deposition of North American NOx-emissions (Levy and Moxim, 1987). It was shown that of 7.5 Mt(N) combustion nitrogen emitted annually in the United States and Canada, at most 0.2 Mt was predicted to reach Europe. This supports the view that North America and adjacent Atlantic waters have a NOx-budget quite separate from that of Europe, which in turn is separated from that of Asia and the far East.

In this paper, a model to calculate the European NOx-budget for time periods from one day and up to a year or more, is described. The model is developed within EMEP at the Norwegian Meteorological Institute.

2. DESCRIPTION OF EMEP NOx-MODEL

The EMEP grid area is shown in Figure 2. The grid distance is 150 km at 60^0N, and there are 39 grid elements in the x-direction and 37 in the y-direction.

The model is receptor-oriented and lagrangian, and the chemical development in an air parcel is calculated as it moves along the trajectory and picks up emissions. The emissions are assumed to be completely mixed in a layer whose thickness is equal to the mixing height. The mixing height field is obtained from radiosonde data from about 100 stations in Europe. For each radiosonde report, the mixing height was taken as the height up to the lowest stable layer with base above 200 m. A layer was defined as stable if the temperature decrease with height was less than about half of the dry-adiabatic value (Eliassen and Saltbones, 1983). If no stable layer was found below 2500 m, the radiosonde report was discarded. The mixing heights estimated in this way were then objectively analysed using a weighting function technique (Atkins, 1974) to give grid values.

Radiosondes are launched at 00 and 12 GMT. Only the 12 GMT (or 12-15 LT in Europe) reports were used in the calculations, representing an "enveloping" mixing height under which both old and recently emitted pollution is distributed.

Exchange of pollution between the boundary layer and the free troposphere takes place in the following way (Eliassen and Saltbones, 1983): At 12 GMT, the mixing height of an air parcel is defined by the objectively analysed value for that grid cell and date. As the air parcel moves along the trajectory, the mixing height is given by

$$h(t) = h_1 + \int_{t_1}^{t} w(t)dt \qquad (1)$$

where h_1 is the 12 GMT value, and w is the vertical velocity calculated every 6 h by the Norwegian Numerical Weather Prediction model for the 925 mb level. The trajectories are calculated using wind fields from the same model for the 925 mb

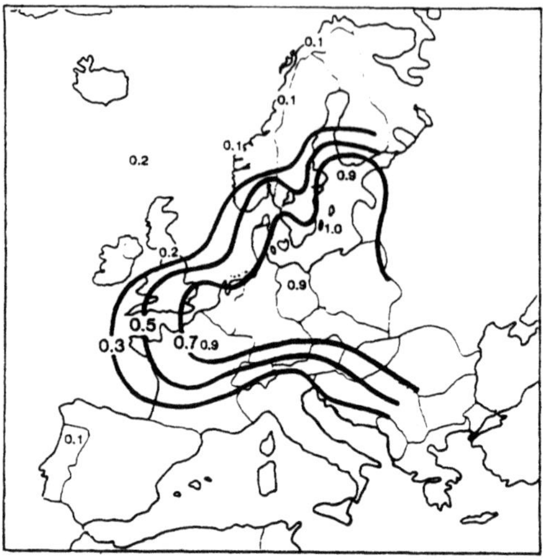

Figure 1: Volume weighted arithmetic mean annual concentrations of nitrate in precipitation 1985 (mg(N)/l), based on the daily observations at about 70 sites in the EMEP network (Schaug et al., 1987).

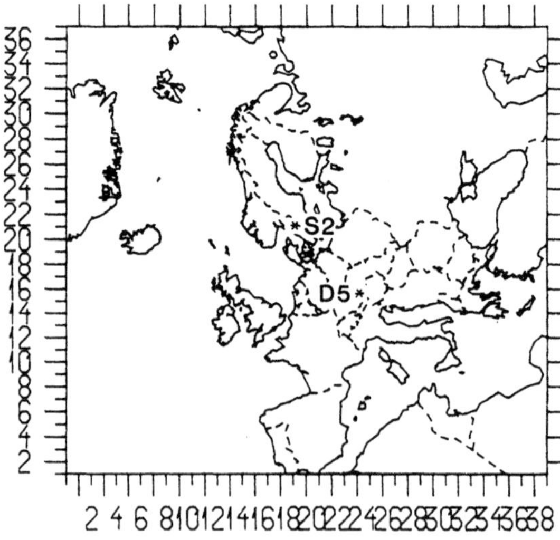

Figure 2: Map of the EMEP grid, 39x37 grid squares, 150 km grid size at 60^{0} N latitude.
S2 = Rörvik in Sweden, D5 = Brotjacklriegel in the Federal Republic of Germany.

pressure level. If eq. (1) gives a mixing height < 1000 m, the vertical velocity w(t) is reduced from the 925 mb-value by the expression

$$w(h < 1000\ \text{m}) = w(925\ \text{mb})\ \frac{h}{1000} \tag{2}$$

One day later (at $t_2 = t_1 + 24$ h) the calculated mixing height along the trajectory is $h(t_2)$, while the objectively analysed mixing height h_2 can either be smaller or larger than $h(t_2)$. If $h_2 > h(t_2)$ there is a dilution of boundary layer air; and the concentration c_{ABL} is modified to c'_{ABL}:

$$c'_{ABL} = c_{ABL}\ \frac{h(t_2)}{h_2} + c_a \left(1 - \frac{h(t_2)}{h_2}\right) \tag{3}$$

where c_a is the aloft (free tropospheric) concentration. If $h_2 < h(t_2)$, then the atmospheric boundary layer (ABL) concentration is not modified. This approach to model ABL-free troposphere exchange can also be used to estimate the fraction and the chemical form of the emissions which are transported out of the ABL. This can be done by integrating the emissions along the trajectories and sum up the losses $c_{ABL}\ (h(t_2) - h_2)$ every 12 GMT when $h(t_2) > h_2$.

In Table 1 is given a list of the chemical species which are calculated in the model from the continuity equation

$$\frac{Dc}{dt} = \frac{Q(1-\varepsilon)}{h} + P - \left(L + \frac{v_d}{h} + \Lambda\ \frac{p}{h}\right)c \tag{4}$$

where c is the concentration, Q emission term, h mixing height, P and Lc chemical production and loss, v_d dry deposition velocity, Λ scavenging ratio, and p precipitation intensity. ε is the fraction of Q deposited locally.

TABLE I. Chemical species calculated in the EMEP NOx-model.

NO	NH_4NO_3	sulphate
NO_2	nitrate	$(NH_4)_2SO_4$
HNO_3	NH_3	
PAN	SO_2	

Data for precipitation intensity are generated as a combination of observed 6-hour precipitation amounts for land areas, analysed objectively (Cressman, 1960) to give grid values for 6 hourly time intervals, and for ocean areas where observations are scarce, the calculated precipitation on the 150x150 km^2 grid from the numerical weather forecast model is used. On the average about two reports of precipitation amounts are received per 150 km grid square over land.

2.1. Chemistry

NOx is assumed to be emitted as 95% NO and as 5% NO_2 (by volume). NO and NO_2 are close to equilibrium through

$$NO + O_3 \rightarrow NO_2 + O_2 \qquad k_5 = 2.1 \times 10^{-12} \exp(-1450/T) \; (cm^3/\text{molecules x s}) \tag{5}$$

$$NO_2 + h\nu \rightarrow NO + O(^3P) \qquad J = A \exp(-\sec\theta \cdot B) \tag{6}$$

A and B are constants, θ solar zenith angle. At noon, mid summer at 60^0N with a clear sky, $J \approx 6 \times 10^{-3} s^{-1}$. The photolysis rate coefficient is reduced according to the extent of cloud cover according to the formula

$$J_{CLF} = J_{CLF=0} \left(1 - \frac{CLF}{2}\right) \tag{7}$$

where CLF is the fraction of the sky which is cloud covered ($0 \leq CLF \leq 1$). It is taken from the numerical weather forecast model every 6 h for each grid cell, where it is calculated as the maximum cloud cover from the heights 300, 500, 850 and 1000 mb.

PAN is formed through

$$CH_3COO_2 + NO_2 \longrightarrow PAN \qquad k_8 = 3.2 \times 10^{-12} \; cm^3/(\text{molecules x s}) \tag{8}$$

$$PAN \xrightarrow{\Delta} CH_3COO_2 + NO_2 \qquad \Delta = 7.94 \times 10^{14} \exp(-12350/T)(s^{-1}) \tag{9}$$

Nitric acid is formed through

$$NO_2 + OH \longrightarrow HNO_3 \qquad k_{10} = 1.1 \times 10^{-11} \; cm^3/(\text{molecules x s}) \tag{10}$$

O_3, OH and the acetylperoxy radical (CH_3COO_2) are not calculated in the model, but are prescribed from the monthly average values of a 2-dimensional, global tropospheric model (Isaksen and Hov, 1987), for every 10^0 latitude at the 750 m height level. In Figure 3 is shown ozone and hydroxyl as a function of time of year and latitude as taken from the global 2-d tropospheric model.

Nitrate is assumed to be formed in water droplets through the reaction sequence

$$O_3 + NO_2 \longrightarrow NO_3 + O_2 \qquad k_{11} = 1.2 \times 10^{-13} \exp(-2450/T) \tag{11}$$

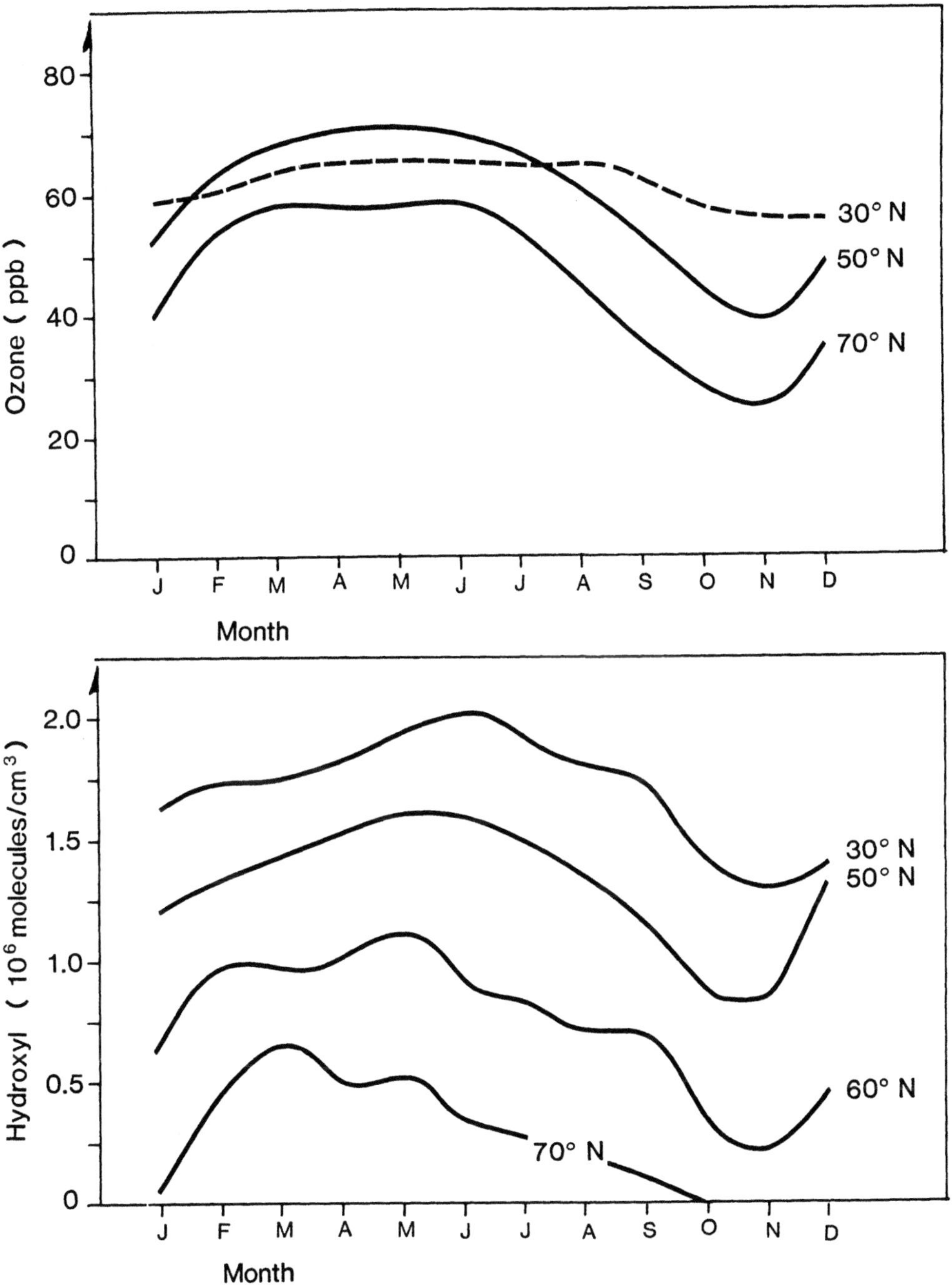

Figure 3: Ozone and hydroxyl as a function of month used in the model calculation, taken from 2-d global tropospheric model (Isaksen and Hov, 1987).

$$NO_3 + NO_2 \rightleftharpoons N_2O_5$$

$$N_2O_5 + H_2O_{(aq)} \longrightarrow 2H^+ + 2NO_3^-$$

Reaction 11 is only important at night, during daytime NO_3 and N_2O_5 are rapidly photolysed to NO and NO_2. At night N_2O_5 dominates in concentration over NO_3, and in the model it is assumed that at night, nitrate is formed at a rate determined by the rate of reaction 11. This somewhat overestimates the nitrate production. (aq) denotes aqueous phase.

There is formation of nitrate in the coarse particle mode, e.g. soil dust, at high relative humidity. This is modelled assuming a first order decay

$$HNO_3 \longrightarrow NO_3^- \qquad k_{12} = 1 \times 10^{-5} s^{-1} \qquad (12)$$

and a reverse reaction with half the rate coefficient of k_{12}. It is important to distinguish between HNO_3 and nitrate in the calculations since the dry removal of HNO_3 at the ground is much more efficient than of nitrate.

Ammonium nitrate is formed in the equilibrium

$$HNO_{3(g)} + NH_{3(g)} \overset{K}{\rightleftharpoons} NH_4NO_{3(s)} \qquad (13)$$

where (g) denotes gas and (s) solid phase.

K is determined by Stelson and Seinfeld (1982), $K = [NH_3][HNO_3]$ and

$$\ln K = \begin{cases} 70.78 - \frac{24220}{T} - 6.1 \ln \frac{T}{298} & rh < rh_d \ (i) \\ \ln K^* - \frac{20.75 + \ln K^*}{101 - rh} \left(\frac{rh - rh_d}{100 - rh_d}\right) & rh \geq rh_d \\ \text{where } K^* \text{ is calculated from (i);} & \\ \ln rh_d = \frac{856.23}{T} + 1.2306 \ \text{(delisquiscence)} & \end{cases} \qquad (14)$$

To account for the formation of ammonium nitrate, ammonia is required in the model calculation, and since sulphate competes with nitric acid for the ammonia available, sulphur dioxide and sulphate also enter the calculation.

SO_2 is transformed to sulphate, and the rate of transfer is determined by an empirical expression which is dependent on latitude and the time of the year (Eliassen and Saltbones, 1983):

$$k_t = 3 \times 10^{-6} + 2 \times 10^{-6} \sin\left(2\pi \frac{\tau}{T_o} + \theta\right), \ s^{-1} \qquad (15)$$

where T_o = 1 year, τ is time of year, θ chosen to minimize k_t at winter solstice.

In the presence of ammonia, below rh = 70% there is an aerosol phase NH_4HSO_4 and liquid phase with NH_4HSO_4 and H_2SO_4 dissolved, while above 70-80% relative humidity the dominant aerosol phase is ammonium sulphate $(NH_4)_2SO_4$.

In the calculation, the simplification is made that NH_4HSO_4 and $(NH_4)_2SO_4$ are produced with equal probability regardless of relative humidity, and the amount produced is determined by the concentration of NH_3 and sulphate.

The scavenging ratios applied are given in Table 2.

TABLE 2. Scavenging ratios used in the model calculation

Species	Λ (dimensionless)
NO	0
NO_2	0
PAN	0
HNO_3	1.4×10^6
NO_3^-, NH_4NO_3	1.4×10^6
NH_3	3×10^5
$SO_4^=$, NH_4HSO_4, $(NH_4)_2SO_4$	7×10^5
SO_2	2×10^5

The local dry deposition corrections (ε in eq. 4) are taken to be 4% for NO_2, 15% for SO_2 and 17% of the grid element emission for NH_3.

2.2. Dry deposition

The deposition of SO_2, NO_2, PAN and NH_3 is controlled by the aerodynamic resistance of the atmosphere, and also by the physical and chemical characteristics near and at the surface, notably stomatal openings and laminar layer resistance.

The deposition of HNO_3 is assumed to be entirely controlled by aerodynamic and laminar resistance. The deposition of aerosol species is assumed to take place at a velocity of 0.1 cm/s, a value which is appropriate for particles below 1 µm. This value is so small that the aerodynamic resistance and the effects of stability and surface characteristics can be ignored, and the value of 0.1 cm/s is applied everywhere and at all times.

The deposition velocity of SO_2 is taken to be 0.8 cm/s for mid-summer, daytime, neutral conditions at mid latitude when applied to the SO_2 concentration at 1 m above the ground. To reflect the influence of changes in biological activity with latitude and season, the following function is used to calculate the 1 m height dry deposition velocity for a given latitude and time of the year (Lehmhaus et al., 1986):

$$v_d = \begin{cases} v_d^o \; (\delta \sin^2\tau + \frac{r}{R} \cos^2\tau) \text{ over land} \\ v_d^o \; \text{over sea for } SO_2 \text{ and } NH_3\text{, zero over sea for } NO_2 \text{ and PAN} \end{cases} \qquad (16)$$

NO_2

where v_d^o is 0.8 cm/s for SO_2, 0.4 cm/s for NO_2, 0.2 cm/s for PAN and 1.0 cm/s for NH_3. $\tau = \pi(\frac{\text{day of the year} - 32}{366})$, so that sin τ has a maximum of 1 on 1 August, cos τ = 1 on 1 February.

$$\delta = \begin{cases} 1.0 \text{ during daytime (06-00h)} \\ 0.25 \text{ at night (00-06 h)} \end{cases} \qquad (17)$$

δ simulates stomatal closing during nighttime. r is the distance of the grid square from the north pole, R the distance from equator.

Businger´s constant flux relationships for the surface layer are used to calculate the dry deposition velocity applicable to the concentration at 50 m height, which is more representative of the boundary layer average concentration than the 1 m height concentration.

$$v_d(z_2) = v_d(z_1)(1 + \frac{v_d(z_1)}{k \cdot u_*} (\ln(\frac{z_2}{z_1}) - \Psi(\frac{z_2}{L}) + \Psi(\frac{z_1}{L})))^{-1} \qquad (18)$$

where z_2 = 50 m, z_1 = 1 m, k is von Karmans constant 0.37. u_* friction velocity, L Monin-Obukhov length and $\Psi(\frac{z}{L})$ Businger functions. The friction velocity u_* and the Monin-Obukhov length L are derived from four output parameters every 6 h from the Norwegian Meteorological Institute numerical weather forecast model: p_s (non-reduced, actual surface pressure), T_2 (temperature at 2 m height), τ (surface stress) and H (sensible heat flux density).

For HNO_3 the dry deposition is controlled only by aerodynamic and laminar resistance, and the 50 m deposition velocity is calculated from

$$v_d \; (50 \text{ m}) = \frac{1}{r_a(50 \text{ m}) + r_b} \qquad (19)$$

where r_a is the aerodynamic resistance from 50 m to the surface, and r_b is the laminar resistance of the surface, and

$$r_a(z) = \frac{\ln(\frac{z}{z_o}) - \Psi(\frac{z}{L})}{k \; u_*} \qquad (20)$$

where z_o is the surface roughness specified for each grid cell, and $r_b = 25\ sm^{-1}$ (4 cm/s deposition velocity at 0 aerodynamic resistance).

The dry deposition velocity fields for SO_2 and HNO_3 at 50 m height as an average for September 1985 are shown in Figure 4. For HNO_3 v_d is 0.8 cm/s or less over sea and 1-2 cm/s over land, for SO_2 mostly 0.4-0.6 cm/s. In winter v_d for HNO_3 is somewhat higher over the sea and lower over land than in summer, reflecting the shift in atmospheric stability from summer to winter. For SO_2 in winter there is a marked fall off with latitude in v_d, determined in parts by eq. (16), and there is a marked sea-land difference (also from eq. (16)).

Figure 4 shows that the deposition velocities calculated for 50 m height can be very different from measured 1 m height values. This is important for the calculation of species like SO_2, NH_3 and HNO_3 where dry removal is a major sink.

2.3. Emissions

The NOx- and SO_2-emission data employed are taken from the official submissions from the countries to the ECE Secretariat in Geneva or to the MSC-W at the Norwegian Meterological Institute when these are available. The NH_3-emissions are taken from Buijsman et al. (1985). In the future it is expected that official NH_3 emission data will gradually become available. The emission data employed in the calculations are given in Table 3. The emissions are distributed in the EMEP grid either as specified by the countries or when no information is available on the distribution of NOx, they are distributed the same way as the SO_2-emissions. NOx- and SO_2-emissions are split into low and high sources, the latter applicable to stacks taller than 100 m.

2.4. Initial and free troposphere concentrations

Initial concentrations of NO, NO_2, the sum of HNO_3 and nitrate, SO_2 and PAN were taken from the global 2-d model calculations by Isaksen and Hov (1987). Of the sum of HNO_3 and nitrate, 30% was taken to be HNO_3 and the rest nitrate. The initial concentration of NH_3 was given a low value. The free tropospheric concentrations of the species in the model were needed to calculate the dilution of ABL concentrations when the lagrangian mixing height was lower than the grid value at 12 GMT (eq. 3). The global model values were monthly averages for every 10^0 latitude, and were interpolated in space according to the location of the grid point. Initial and free troposphere concentrations used for January and July and for 60^0 and 40^0N, are shown in Table 4.

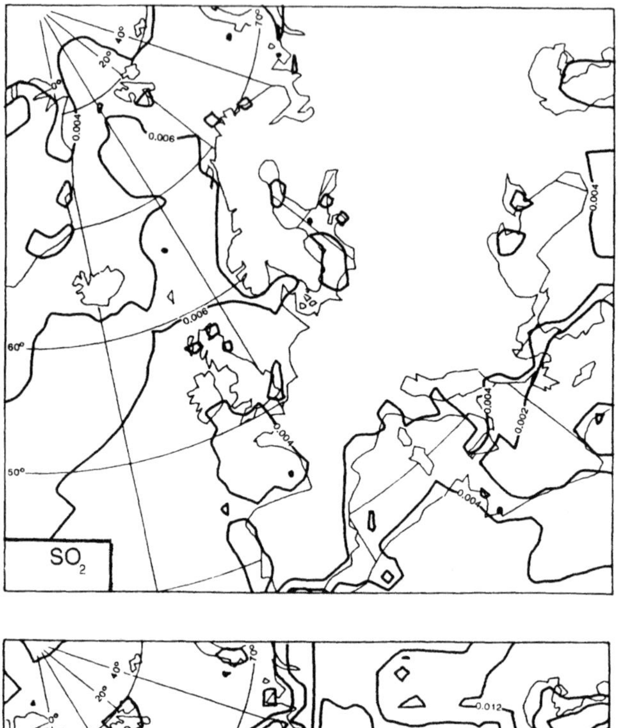

Figure 4: Mean dry deposition velocity fields for SO_2 and HNO_3 at 50 m height for September 1985 (average of fields calculated every 6 h).

TABLE 3. Emissions employed in the calculations

	NOx as NO_2	SO_2 as S	NH_3
Albania	[9]	[25]	[20]
Austria	216	85	[71]
Belgium	385	234	[79]
Bulgaria	150	[570]	[123]
Czechoslovakia	1127	1575	[167]
Denmark	238	163	[110]
Finland	250	185	[43]
France	1693	923	[701]
German Dem. Rep	[800]	[2000]	[202]
German Fed. Rep of	2900	1200	[365]
Greece	150	180	[94]
Hungary	300*	710	[126]
Iceland	12	3	[2]
Ireland	68	69	[116]
Italy	1462	1575	[355]
Luxembourg	22	7	[5]
Netherlands	522	158	[142]
Norway	215	50	[34]
Poland	840	2150	[399]
Portugal	192	153	[46]
Romania	[390]	100	[292]
Spain	950	1625	[227]
Sweden	305	136	[51]
Switzerland	214	48	[52]
Turkey	[175]	161	[582]
USSR	2930	5550	[2650]
United Kingdom	1690	1770	[398]
Yugoslavia	190	900	[196]
Sum	18395	21585	7648

Units: NOx emissions in thousand tonnes NO_2
SO_2 emissions in thousand tonnes S
NH_3 emissions in thousand tonnes NH_3

Emissions are for 1985 or for the closest available year.
Data in brackets are unofficial estimates.
*Preliminary estimate.

TABLE 4. Initial and free troposphere concentrations (ppb)

	January		July	
	40^0N	60^0N	40^0N	60^0N
NO	0.2	0.3	0.2	0.1
NO_2	0.8	1.1	0.4	0.2
HNO_3 +nitrate	1.4	0.8	1.2	0.6
SO_2	0.9	1.4	0.9	0.6
PAN	0.1	0.3	0.6	0.3
O_3	56	44	64	60
OH *)	$1.4x10^6$	$6.9x10^5$	$1.6x10^6$	$8.3x10^5$
CH_3COO_2 *)	$3.7x10^6$	$2.9x10^6$	$2.7x10^6$	$9.1x10^5$

*) in molecules/cm^3

3. CALCULATED AND MEASURED NOx-SPECIES AT SOME SITES IN EUROPE IN 1985

One objective for the model work is to calculate the wet and dry deposition of NOx-species in the EMEP grid for 1985, and to attribute the deposition in each grid cell to the emitting countries. This means that for each of the species in the model, track is kept of the relative contribution from each country which is traversed during the 96 h trajectory, as well as the influence of the initial concentrations and mixing in of free tropospheric air.

Here will be shown some examples from the model validation where observed and measured concentrations are compared. In the EMEP network about 100 stations in rural areas in Europe report daily measurements of SO_2 and sulphate in air and precipitation, many also report nitrate and ammonium in precipitation, while fewer measure NO_2, nitrate or ammonium in air. An example of a measured field was given in Figure 1 which showed annual averaged, precipitation weighted nitrate in rain for 1985, based on daily measurements when precipitation has fallen at about 70 sites.

Another set of data which are used for model validation, were obtained at 6 EMEP sites in the Nordic countries (Rörvik and Aspvreten in Sweden, Utö in Finland, Keldsnor in Denmark and Birkenes and Jergul in Norway) during a 3-month period from mid August 1985. Then an intensive measuring campaign took place, measuring NO_2, and sum nitrate and sum ammonium in air and precipitation on a daily basis.

Measurements and calculations for Rörvik in Sweden (S2, see Figure 2) and Brotjacklriegel in the Federal Republ of Germany (D5, see Figure 2) are shown. Rörvik is a coastal site, while Brotjacklriegel is located at 1016 m.a.s.l near the border to Czechoslovakia. In Figure 5 is shown measured and observed NO_2 on

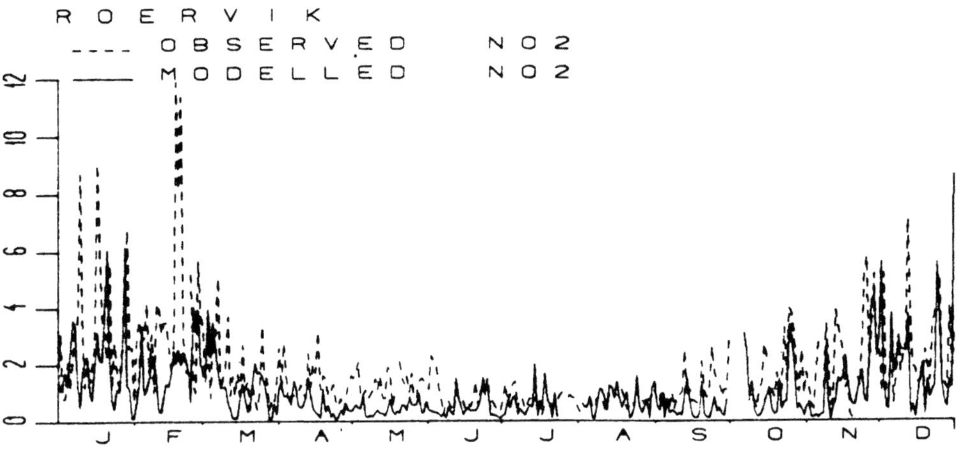

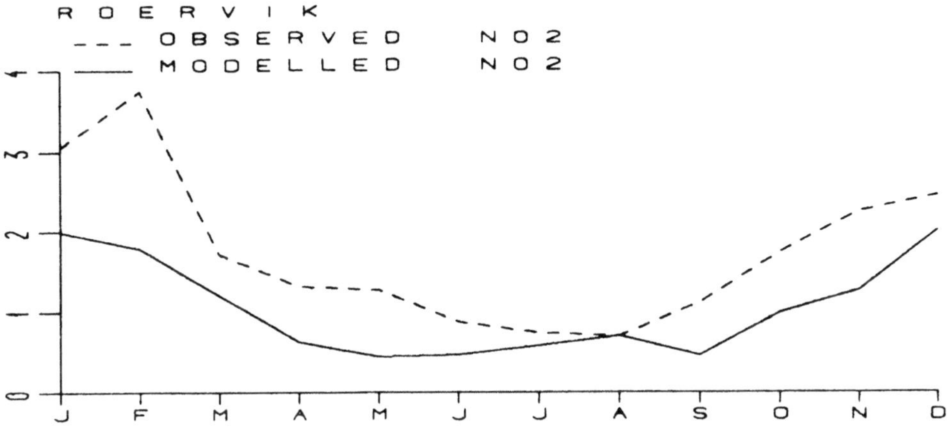

Figure 5: Measured and calculated concentrations of NO_2 on a daily and monthly basis at Rörvik in Sweden for 1985, in $\mu gN/m^3$ (see Figure 2)
(Full line: modelled, dashed line: observed).

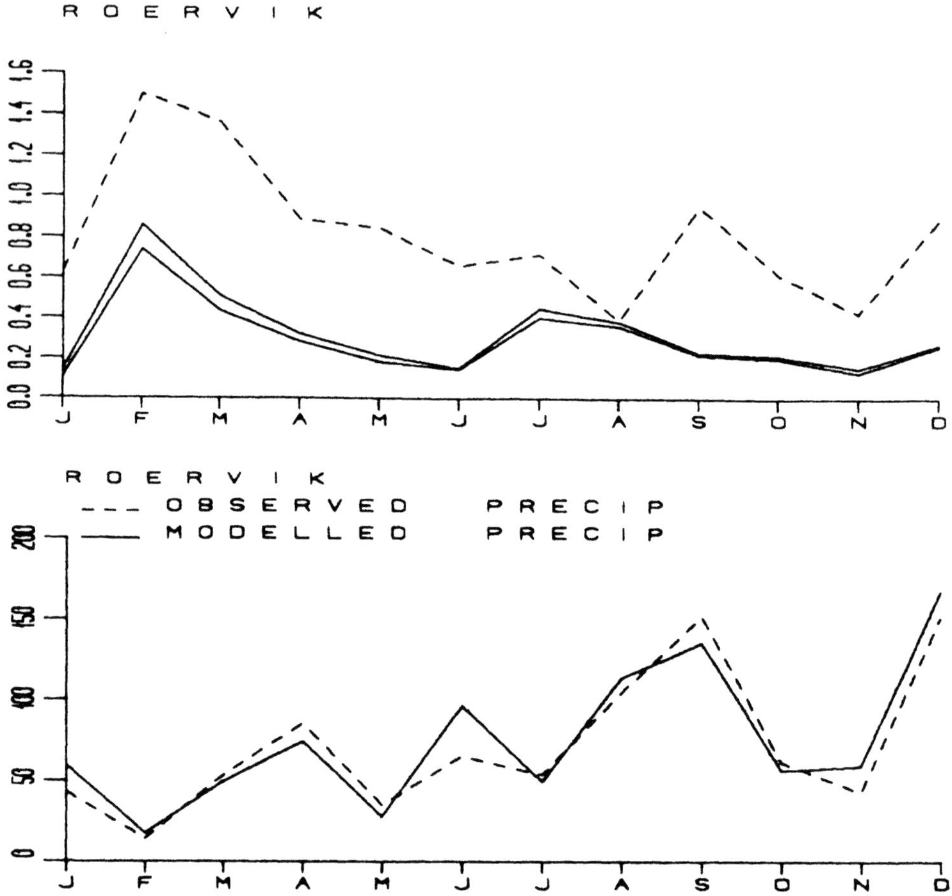

Figure 6: Measured nitrate in precipitation at Rörvik in Sweden on a monthly basis for 1985, in mg(N)/l, together with calculated nitrate in precipitation and sum of wet and dry nitrate deposition on wet days, also in mg(N)/l. Also shown is the monthly precipitation in mm as measured at the site (Rörvik) and the average value for the grid square around Rörvik. (Full line: modelled, dashed: observed).

a daily and monthly basis at Rörvik. The episodicity and the annual variation are reproduced quite well, but some episodes with high concentrations in January and February were not picked up by the calculations, explaining the differences in the monthly means. The mean level of NO_2 is about 1 $\mu gN/m^3$ (≈ 1.5 ppb) during the summer and 2-5 times as high during the winter months.

In Figure 6 is shown the concentration of nitrate in precipitation at Rörvik, and also shown is the result when both dry and wet deposition is added together on days with precipitation. This was done to estimate the maximum amount of nitrate in rain that the model could explain, assuming that the dry deposited nitrate in the sampler was washed down in the precipitation sample. It is seen that there is no clear seasonal variation in nitrate in rain, the concentration is high when the rainfall amount is low (February, see Figure 6), and vice versa. The observed precipitation at the site is seen to compare very well with the amount given as representative for the grid cell where Rörvik is located.

In Figure 7 the results for SO_2, sulphate, NO_2, sum nitrate and ammonium in air are given for the mid August-October period in 1985 with extended measurements. Calculated SO_2 and sulphate compare quite well with measurements, but the lowest calculated SO_2 concentrations are much lower than the measured values. Perhaps too little SO_2 was present initially in those trajectories, or perhaps more SO_2 was mixed in from aloft during ABL-free troposphere exchange, or perhaps the measurements are not too accurate at low concentrations?

In Figure 7c and 7d are shown airborne NO_2 and sum nitrate on a daily basis, and it is seen that sum nitrate calculated and measured compare well. The sum of nitrate exceeds that of NO_2 by a factor 2-3 during episodes (the October episode was not calculated due to lack of meteorological data), while on most days sum nitrate is comparable to or not more than a facator of 2 bigger than NO_2.

The close agreement between measured and observed total nitrate in air is interesting when compared with the factor of 5 and 3 difference for September and October, respectively, between observed and modelled nitrate in precipitation (Figure 6). The scavenging ratio adopted for nitrate is so high ($1.4x10^6$) that 75% of the nitrate in air is rained out if 1 mm of rain falls in 6 h and the mixing height is 1000 m, and all of the nitrate is rained out if just a few mm of rain falls in 6 h. This indicates that the amount of nitrate in precipitation does not reflect the nitrate concentration in air as measured at the site or calculated as an ABL-average. The calculated low nitrate values in precipitation are not easy to explain, but the nitrate concentrations in the air which is drawn into the cloud systems and take part in the processes where raindroplets are formed may be different from those at the ground. The results shown in Figures 6 and 7 may indicate that the nitrate concentrations are higher where the raindroplets are formed than closer to the ground.

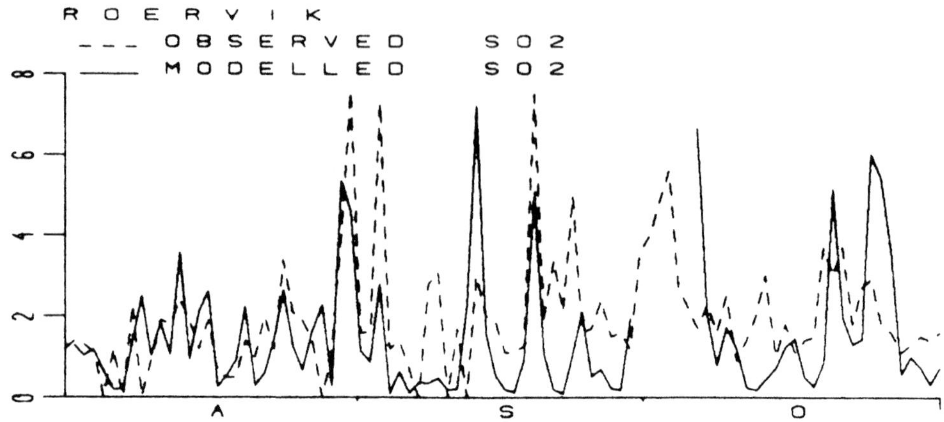
ROERVIK
OBSERVED SO2
MODELLED SO2
8
6
4
2
0
A
S
O

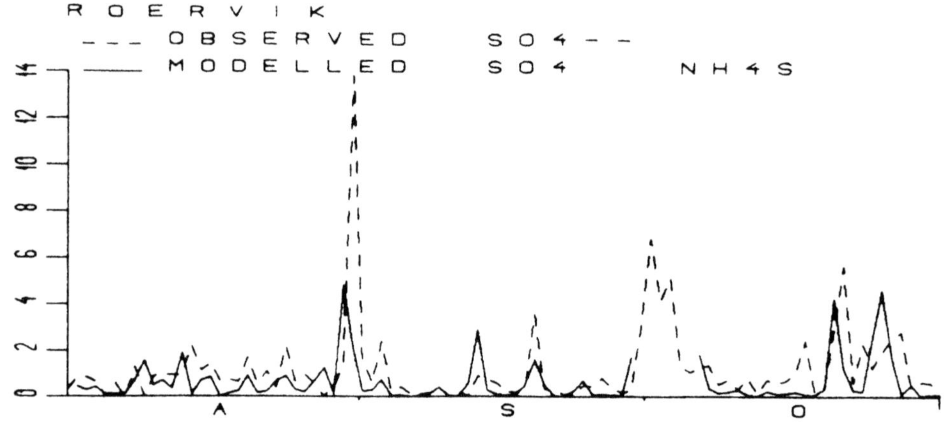
ROERVIK
OBSERVED SO4--
MODELLED SO4 NH4S
14
12
10
8
6
4
2
0
A
S
O

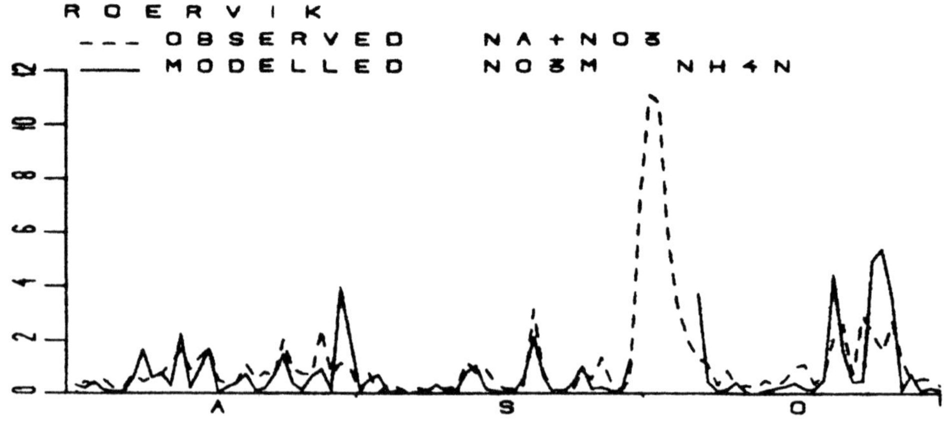
ROERVIK
OBSERVED NA+NO3
MODELLED NO3M NH4N
12
10
8
6
4
2
0
A
S
O

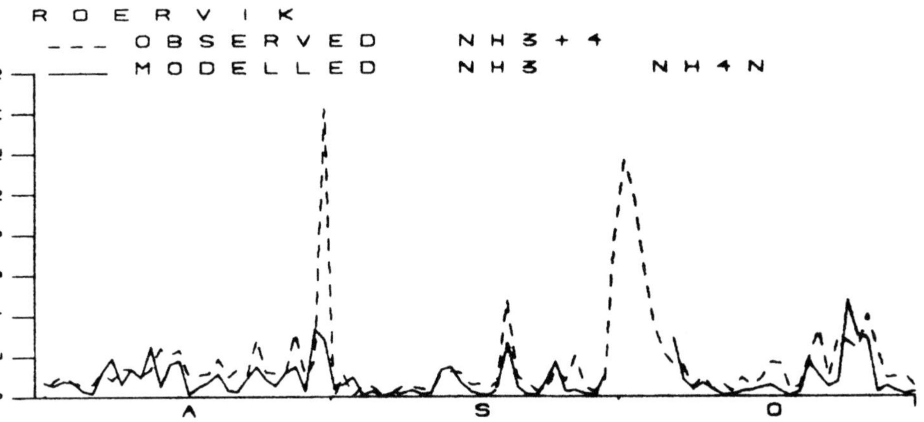

Figure 7: Measured and calculated SO_2, sum sulphate in air (in $\mu gS/m^3$), NO_2, sum nitrate and sum ammonium in air (in $\mu gN/m^3$) on a daily basis at Rörvik during the intensive measuring campaign in the fall of 1985, financed by the Nordic Council of Ministers.
(Full line: modelled, dashed: observed).

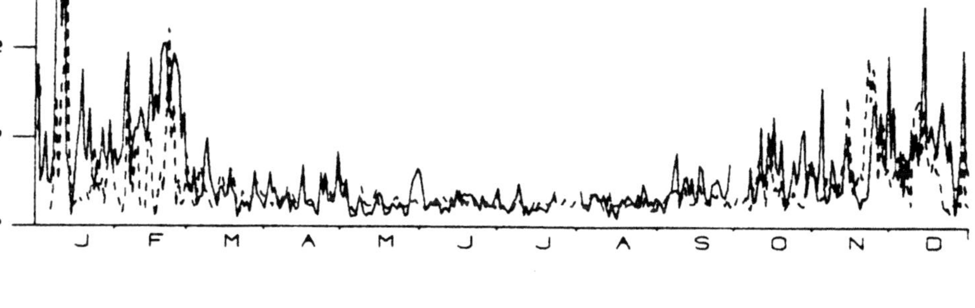

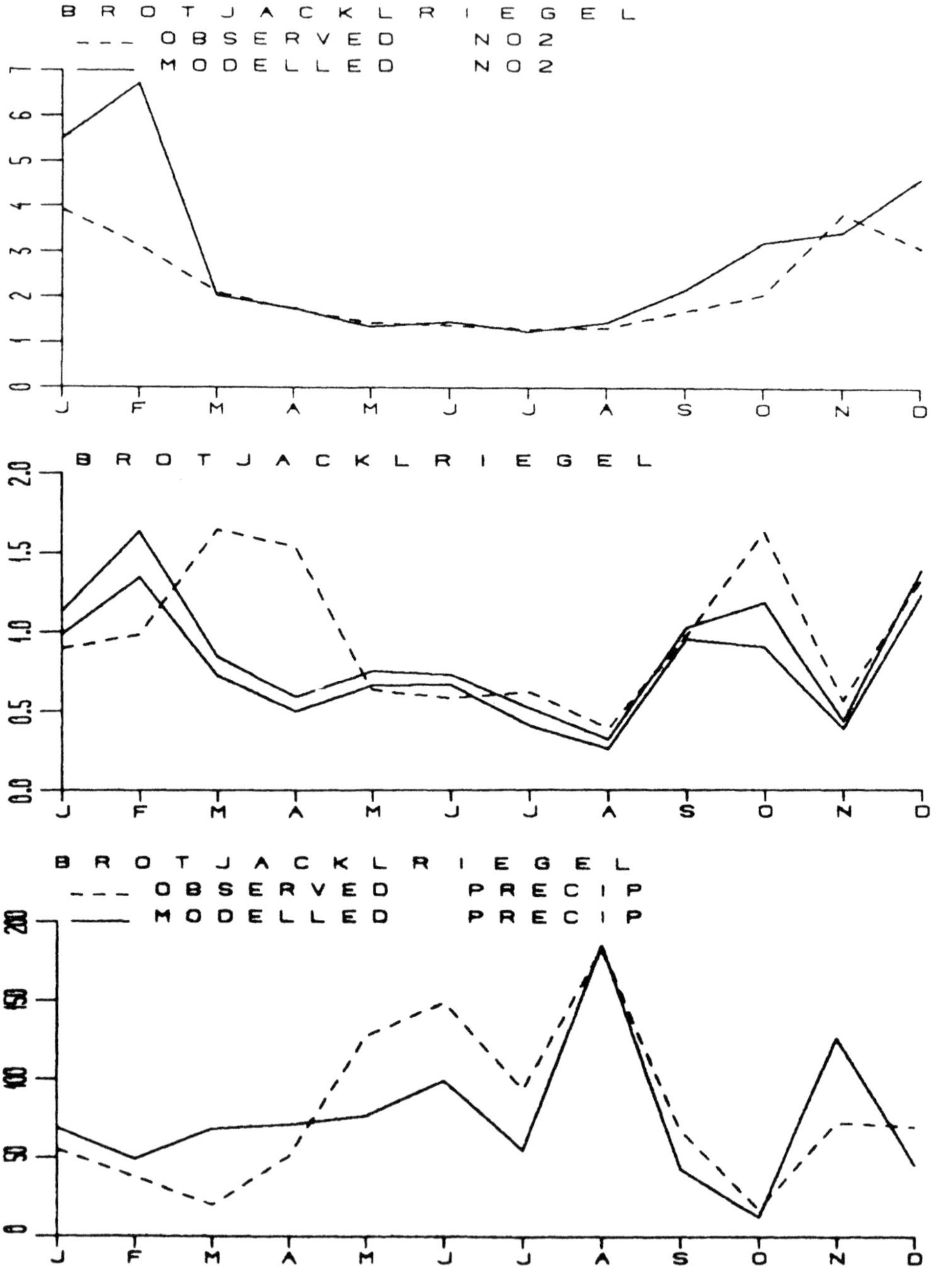

Figure 8: Same as Figures 5 and 6 for Brotjacklriegel (cpr. Figure 2).
(Full line: modelled, dashed: observed).

In Figure 7e is shown the measured and calculated sum of ammonium on a daily basis (sum of ammonium nitrate and ammonium-sulphate). The agreement is quite good.

In Figure 8 are shown the comparisons for Brotjacklriegel. There is a summer minimum and winter maximum in NO_2 in the same way as at Rörvik, but the concentrations are higher at Brotjacklriegel. The episodicity is quite similar. The monthly means are quite similar except in February where there is a factor 2 difference between the model and the observations. Brotjacklriegel is a high altitude site which may be above the top of the boundary layer in winter.

Monthly precipitation at the site as an average value for the grid cell is shown in Figure 8d, while in Figure 8c is shown measured nitrate in rain and calculated total nitrate in rain, and total nitrate in rain plus the dry deposition of nitrate on wet days. Measurements and calculations compare better than for Rörvik. Measurements of nitrate in air are not available, and it is therefore difficult to know if nitrate in air as measured on the site reflects the nitrate concentration in rain better than at Rörvik.

4. CONCLUSIONS

The model is defined in such a way that NOx-species are transported at least as far as SO_2 and sulphate and less NOx is deposited close to the source than what is found for SO_2. NO, which is the main component being emitted, is not dry deposited, and NO_2 into which it is rapidly converted, has a dry deposition velocity which is much smaller than that of SO_2. Nitric acid is efficiently dry deposited, but much of NO_2 is quickly converted to nitrate which is deposited at a rate comparable to sulphate.

The agreement found between measured and observed values of NO_2, and sum of nitrate in air and precipitation is perhaps sufficient to conclude that the model contain the most important features of the NOx-budget in the ABL.

The model is defined so that ammonia and ammonium have a shorter transport distance than SO_2 and sulphate, mainly because more NH_3 is deposited locally and the dry removal for NH_3 is higher than for SO_2. The model results may again justify the conclusion that the underlying hypothesis is correct.

ACKNOWLEDGEMENT

This work is sponsored by the Nordic Council of Ministers (AE&ØH), the Royal Norwegian Research Council for Science and Technology (ØH), and by the UK Department of Environment (DS). Discussions with R.G. Derwent, Harwell and Arne Semb, NILU, are acknowledged. Emission data were assembled by Jørgen Saltbones, and the meteorological data were calculated and processed by

Thor Erik Nordeng and Trond Iversen at the Norwegian Meteorological Institute. Calculated tropospheric values of species derived from NOx and SO_2 were obtained from Ivar S.A. Isaksen, University of Oslo.

REFERENCES

Atkins, M.J. (1974) The objective analysis of relative humidity. Tellus, 26, 663-671.

Buijsman, E., J.F.M. Maas and W.A.H. Asman (1985) Ammonia emission in Europe. Instituut voor Meteorologie en Oceanografie, Rijksuniversiteit Utrecht, Report IMOU-R-85-2.

Cressman, G.P. (1980) An operational objective analysis system. Mon. Weat. Rev., 87, 367-374.

Dickerson, R.R., G.J. Huffman, W.T. Luke, L.J. Nummermacker, K.E. Pickering, A.C.D. Leslie, C.G. Lindsey, W.G.N. Slinn, T.J. Kelly, P.H. Daum, A.C. Delany, J.P. Greenberg, P.R. Zimmerman, J.F. Boatman, J.D. Ray and D.H. Stedman (1987) Thunderstorms: An important mechanism in the transport of air pollutants. Science, 235, 460-465.

Eliassen, A. and J. Saltbones (1983) Modelling of long-range transport of sulphur over Europe: A two-year model run and some model experiments. Atmosphreic Environment, 17, 1457-1473.

Hamrud, M. and H. Rodhe (1986) Lagrangian time scales connected with clouds and precipitation. J. Geophys. Res., 91, 14377-14383.

Hov, Ø. (1984) Modelling of the long-range transport of peroxyacetylnitrate to Scandinavia. J. Atm. Chem., 1, 187-202.

Isaksen, I.S.A. and Ø. Hov (1987) Calculation of trends in the tropospheric concentration of O_3, OH, CO, CH_4 and NOx. Tellus, 39B, 271-285.

Lehmhaus, J., J. Saltbones and A. Eliassen (1986) A modified sulphur budget for Europe for 1980. EMEP/MSC-W report 1/86.

Levy, II, H. and W.J. Moxim (1987) Fate of US and Canadian combustion nitrogen emissions. Nature, 328, 414-416.

Schaug, J., J.E. Hanssen, K. Nodop, B. Ottar and J. Pacyna (1987) Summary report from the chemical co-ordinating centre for the third phase of EMEP, EMEP-CCC-Report 3/87. NILU, P.O. Box 64, N-2001 Lillestrøm.

Singh, H.B. and L.J. Salas (1983) Peroxyacetyl nitrate in the free troposphere. Nature, 302, 326-328.

Stelson, A.W. and J.H. Seinfeld (1982) Relative humidity and temperature dependence of the ammonium nitrate dissociation constant. Atmospheric Environment, 16, 983-992.

ATMOSPHERIC CHEMISTRY OF NO_x AND HYDROCARBONS INFLUENCING TROPOSPHERIC OZONE

Dr. R.A. Cox
Engineering Sciences Division
Harwell Laboratory
Didcot Oxfordshire U.K. OX11 ORA

ABSTRACT A review is given of the atmospheric chemistry of NO_x, hydrocarbons and other volatile organic compounds, which can influence the in-situ production and loss of ozone in the troposphere. The photochemistries of ozone and of nitrogen dioxide in the clean troposphere, and their roles in ozone and free radical production and loss are outlined. The interconversion reactions between the various organic and inorganic N-containing molecules, necessary to describe the atmospheric NO_x budget are discussed. The kinetics and mechanism for oxidation of organics following attack by OH are described and problems in the modelling of the complex mixture of hydrocarbons in the atmosphere are discussed. Finally, the factors controlling the balance between production and loss of free radicals and ozone are highlighted.

1. INTRODUCTION

Theories of tropospheric ozone up to the end of the 1960's were based on a large number of *ad hoc* observations of atmospheric ozone made at ground level and aloft, especially during the years following World War II. The accepted picture was of downward transport of ozone from the stratosphere to the upper troposphere, circulation within the tropospheric weather systems, with eventual removal by destruction at the earth's surface (Junge 1962)[1]. The possibility that tropospheric ozone could be modified by important in-situ chemical sources and sinks was first suggested by Crutzen (1973)[2] and by Chameides and Walker (1973)[3], although the formation of high concentrations of ozone from the NO_x and hydrocarbons present in polluted urban air under the influence of sunlight, "photochemical smog" had been recognised two decades earlier (Leighton (1961))[4]. Although much research had been carried out to understand the chemistry underlying the urban smog phenomenon, it was not until the early 1970's that progress was sufficient to point to its potential significance for the atmosphere as a whole. Since that time there has been a steady improvement in knowledge of the mechanisms and rate coefficients of many of the reactions involved in the atmospheric chemistry of ozone, NO_x and hydrocarbons, so that today we have a firm picture of the underlying chemistry of *in-situ* ozone formation and

I. S. A. Isaksen (ed.), Tropospheric Ozone, 263–292.

removal in the troposphere. The chemistry of OH and HO_2 radicals, of a variety of different molecules in NO_X family as well as representatives of several classes of volatile organic components, needs to be treated in detail. Some of the important chemical details will be highlighted in this paper.

2. OZONE PRODUCTION AND LOSS PROCESSES

2.1 Photochemistry of Ozone - Ozone Loss in the Clean Troposphere

In the troposphere ozone absorbs solar radiation in two regions of the visible and ultraviolet spectrum i.e. the weak Chappius bands (410-850 nm) and in the Huggins bands, which lie on the long wavelength side of the strong Hartley region (290-350 nm). The cut-off at $\lambda \simeq 290$ results from the complete absorption of light of shorter wavelengths by ozone absorption in the Hartley band at higher altitudes. The relationship between ozone absorption and solar radiation is illustrated in Figure 1. At $\lambda \geq 310$ nm, O_3 photolysis leads to production of oxygen atoms in the ground state

$$O_3 + h\nu\ (\lambda > 310\ \text{nm}) \rightarrow O\ (^3P) + O_2 \tag{1a}$$

which recombine with O_2 to form O_3 again

$$O\ (^3P) + O_2 + M \rightarrow O_3 + M \tag{2}$$

Thus no O_3 loss occurs following photolysis of these wavelengths. At $\lambda < 310$ nm, (oxygen atoms in the excited state) $O\ (^1D)$ are formed:

$$O_3 + h\nu\ (\lambda < 310\ \text{nm}) \rightarrow O\ (^1D) + O_2 \tag{1b}$$

and, although the excited atoms are rapidly quenched to the ground state by O_2 and N_2, a fraction react with water vapour to form OH:

$$O\ (^1D) + H_2O \rightarrow 2\ OH \tag{3}$$

This initiates a series of reactions in which the 'odd oxygen' (re $O_3 + O\ (^3P) + O\ (^1D)$ is removed e.g.

$$OH + CO \rightarrow H + CO_2 \tag{4}$$

$$H + O_2 + M \rightarrow HO_2 + M \tag{5}$$

$$OH + CH_4 \rightarrow CH_3 + H_2O \tag{6}$$

$$CH_3 + O_2 + M \rightarrow CH_3O_2 + M \tag{7}$$

$$CH_3O_2 + HO_2 \rightarrow CH_3OOH + O_2 \tag{8}$$

overall $\quad O_3 + CO + CH_4 \rightarrow CO_2 + CH_3OOH$

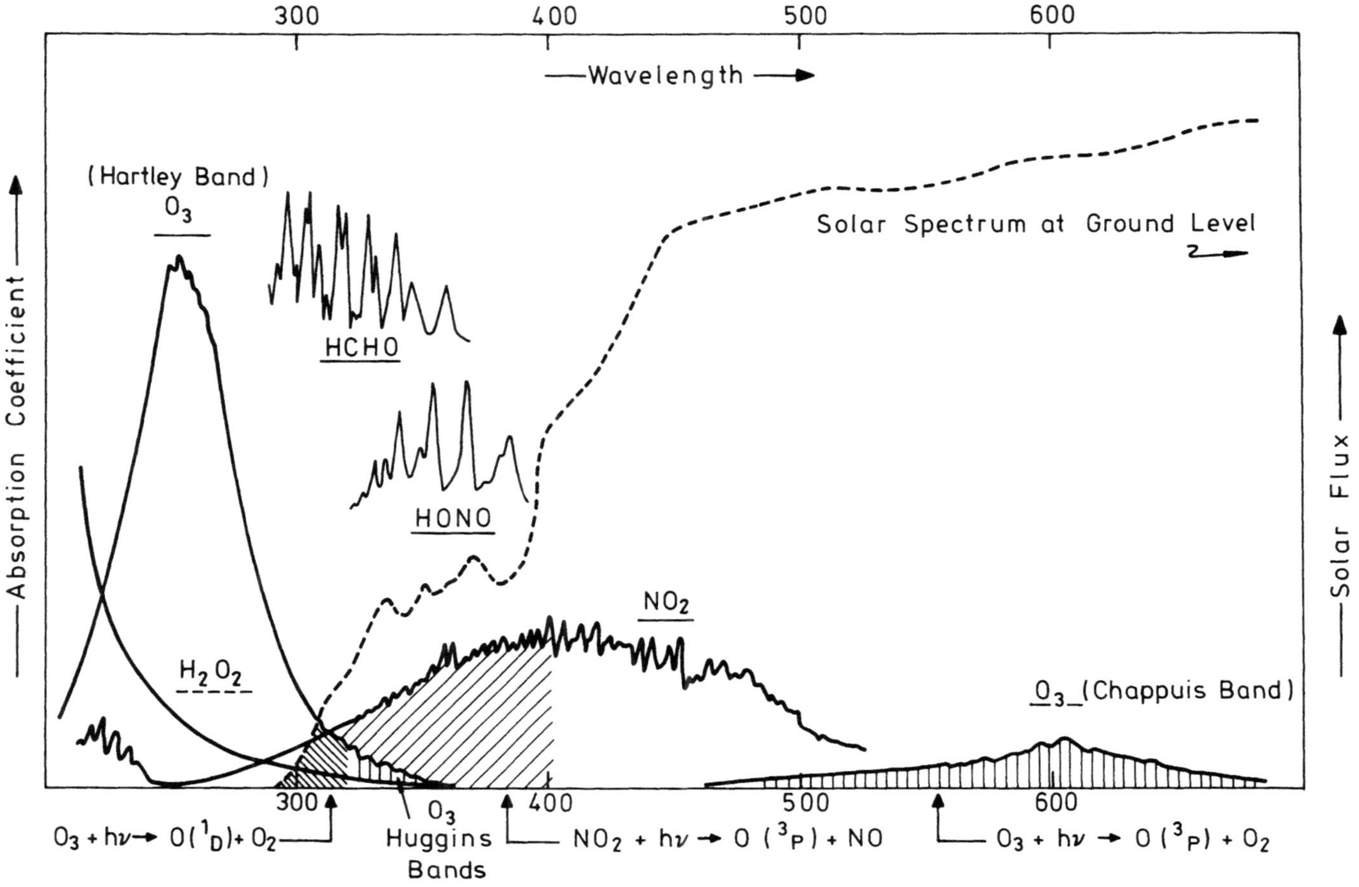

Fig. 1 PHOTOCHEMICAL PROCESSES IN THE LOWER ATMOSPHERE.

CO and CH_4 are the major reaction partners for OH in the background troposphere and this constitutes an important removal process for these molecules as well as a loss process for ozone.

Further loss of ozone results from the reaction of HO_2 with ozone:

$$HO_2 + O_3 \rightarrow OH + 2O_2 \tag{9}$$

which, coupled with reactions (4) and (5), constitutes a chain reaction with overall loss of O_3:

$$\text{overall} \quad CO + O_3 \rightarrow CO_2 + O_2$$

Because reaction (9) is rather slow, the catalytic cycle is inefficient and does not lead to a large 'amplification' in the ozone loss rate above that resulting from HO_x formation in reaction (3). Because of the dependence of the solar UV at $\lambda < 310$ nm on solar zenith angle, cloudiness, etc. and the variability in water vapour concentration, the ozone loss rate by photodissociation will depend very much on season, latitude and altitude. The average photochemical life-time of ozone in the troposphere is approximately 2 weeks. A possibly more important ramification of the photolysis of ozone is the daytime production of HO_x radicals [5] and molecules (OH, HO_2 and peroxides) and their interaction with the other trace gas cycles, which will be discussed below.

2.2 Photochemistry of NO_2 - Ozone Production

Nitrogen dioxide absorbs solar radiation throughout the visible and near UV region of the spectrum (see Fig. 1). Photodissociation occurs only at wavelengths $\leq$ 400 nm, when ground state oxygen atoms are produced:

$$NO_2 + h\nu\ (\lambda \lesssim 400\ \text{nm}) \rightarrow O\ (^3P) + NO \tag{10}$$

and ozone is formed in reaction (2). The other photofragment, NO reacts rapidly with O_3, reforming NO_2:

$$NO + O_3 \rightarrow NO_2 + O_2 \tag{11}$$

NO_2 photolysis is rapid (half life ~ 100 s in average clear sky conditions) and consequently a photostationary state can exist in daytime between NO_2, NO and O_3 through reactions (10), (11) and (2):

$$[O_3] = \frac{J_{10}}{K_{11}} \frac{[NO_2]}{[NO]} \tag{i}$$

Net ozone production results from NO_2 photolyis when NO is converted to NO_2 without loss of O or O_3. In the troposphere this role is mainly played by peroxy radicals, RO_2, which react rapidly with NO by an O atom transfer reaction:

$$RO_2 + NO \rightarrow RO + NO_2 \tag{12}$$

Here R = H or an organic radical, which is derived from the breakdown of hydrocarbons or related molecules. These reactions are the essential core of the chemical production of ozone in the troposphere. In the clean troposphere the oxidative breakdown of long lived volatile organic compounds, e.g. CO, CH_4, simple aliphatics, halogenocarbons, etc. provides the required peroxy radicals. In the polluted boundary layer emissions resulting from hydrocarbon usage as fuels, solvents, etc. gives rise at relatively high concentrations of a host of volatile organics, which provide the peroxy radicals.

No other significant chemical sources of tropospheric ozone have been discovered. Other gaseous pollutant molecules may produce $O(^3P)$ by photolysis in the near UV, e.g. OClO and SO_3, but these are not present in the atmosphere in significant amounts. Certain reactions of peroxy radicals produce ozone directly e.g.

$$CH_3CO_3 + HO_2 \rightarrow CH_3COOH + O_3 \qquad (13)$$

but these are unlikely to carry sufficient flux to influence the ozone budget.

The presence of nitrogen oxides also provides new routes for ozone loss. Reaction of O_3 with NO provides only a temporary loss since O_3 is regenerated from the product NO_2 by photolysis. Consequently NO_2 is often included in the total "odd oxygen" budget of the troposphere. Reaction of NO_2 with ozone can provide a loss process for ozone, depending on the fate of the NO_3 radical produced:

$$NO_2 + O_3 \rightarrow NO_3 + O_2 \qquad (14)$$

In sunlight, the major loss for NO_3 is by photolysis which can occur via two channels:

$$NO_3 + h\nu \rightarrow NO_2 + O\ (^3P) \qquad (15a)$$

$$\rightarrow NO + O_2 \qquad (15b)$$

Only the second of these, which is the minor channel, leads to overall loss of ozone. At night-time NO_3 is removed by reaction with NO_2 to give nitrogen pentoxide

$$NO_3 + NO_2 + M \rightarrow N_2O_5 + M \qquad (16)$$

Any loss of N_2O_5 without NO_2 regeneration will constitute a loss of odd oxygen. The details of the NO_x chemistry will be discussed later.

3. FACTORS AFFECTING QUANTITATIVE EVALUATION OF THE TROPOSPHERIC OZONE BUDGET

As will be evident from the previous discussion the factors influencing chemical production and loss of ozone in the troposphere are very complex. In the present assessment emphasis will be placed on the ozone production process and we will then go on to consider the competition between production and loss.

The presence of NO_2 is a requisite for ozone production and therefore the distribution of NO_2 in the troposphere is a major factor in determining the potential ozone production rates at different locations. Moreover in addition to NO and NO_2, which are closely coupled by the reactions already described, 'odd nitrogen' may be stored in a variety of reservoir molecules e.g. nitrogen pentoxide (N_2O_5), nitric acid (HNO_3), organic nitrates (e.g. peroxyacetyl nitrate), etc. Each of these reservoirs has its associated chemistry and interconversion between NO, NO_2 and the various reservoirs occurs on a range of time scales. These time scales determine the spacial distribution of total odd nitrogen (NO_y) and consequently that of NO and NO_2 (NO_x), following emission of nitrogen oxides from a particular source. A quantitative knowledge of the nature and rate of the interconversion reactions of NO_y is therefore a major part of any description of tropospheric chemistry relating to ozone production.

As already indicated, ozone generation results from the reaction with NO, of peroxy radicals derived from the oxidation of hydrocarbons and other volatile organics. The distribution of these molecules in the atmosphere is therefore a key factor in ozone formation. There are a wide variety of volatile organics in the atmosphere deriving from man made and natural sources. Although the chemistry of the oxidative breakdown of these organics follows a general mechanistic pattern, there are important differences in detail of the mechanisms and rates which have an important bearing on ozone formation. For example, the variation in reactivity of the organics leads to atmospheric residence times ranging from a few minutes to many years, with consequent differences in their atmospheric distribution. Another problem in dealing with the atmospheric chemistry of hydrocarbons is the representation of the large number of different organic species present in a realistic fashion. For ozone formation in the boundary layer, some sort of simplification of the chemical formulation is necessary but at the same time the essential features of the chemical kinetics must be maintained. In the perturbed background troposphere, simple models containing methane and CO chemistry only may not be sufficient to describe the ozone budget.

A common feature of the atmospheric chemistry of NO_x and hydrocarbons is the controlling functions of free radicals of the HO_x family (OH, HO_2). It follows that any adequate description of in-situ tropospheric ozone production and removal requires an accurate spacial and temporal distribution of the local steady state concentrations of OH and HO_2. The total HO_x radical concentration depends on the balance between the local production and loss rates, which are strong function of the chemical composition.

In the clean troposphere HO_x production is dominated by the ozone photolysis/water vapour mechanism already described, but in polluted air other radical sources such as photolysis of HONO, aldehydes and ketones need to be considered e.g.

$$HONO + h\nu \rightarrow OH + NO \quad (17)$$

$$HCHO + h\nu \rightarrow H + HCO \quad (18)$$

$$CH_3CHO + h\nu \rightarrow CH_3 + HCO \quad (19)$$

The subsequent reactions of the photo-fragments from these reactions give rise to HO_2 and OH.

The radical loss processes in unpolluted air are dominated by radical recombination reactions such as:

$$HO_2 + HO_2 \rightarrow H_2O_2 + O_2 \quad (20)$$

$$CH_3O_2 + HO_2 \rightarrow CH_3OOH + O_2 \quad (8)$$

but become dominated by reactions involving NO_x in polluted air e.g.

$$OH + NO_2 + M \rightarrow HNO_3 + M \quad (21)$$

Since available technology does not allow reliable measurement of tropospheric free radical concentrations to be made routinely, we have a heavy reliance on photochemical model calculations to provide quantitative estimates of free radical concentrations for assessment of the ozone budget. A correct chemical formulation of the production and loss processes is therefore critical.

4. TROPOSPHERIC NO_x CHEMISTRY

Almost all NO_x emitted to the atmosphere is in the form of nitric oxide, NO. However, NO reacts quickly in the atmosphere to form several oxidised species which can be interconverted on a range of time scales. The NO_x chemistry can be classified into three areas:

- gas-phase inorganic NO_x chemistry
- organic NO_x chemistry
- coupled gas-phase/condensed-phase NO_x chemistry.

The relationship between these three sets of processes is illustrated in Fig. 2. We now highlight some of the notable features of the NO_x interconversions.

4.1 The NO-NO_2 Photostationary State

Nitric oxide is converted to NO_2 by reaction with ozone, with a time constant of about 100 s, provided O_3 is in excess over the local NO concentration. In daylight a photostationary state is set up (equation (i)) which is influenced by light intensity (J_{10}) and temperature (k_{11} = 1.8 x 10^{-12} exp (- 1370/T) cm^3 $molecule^{-1}$ s^{-1}).

There have been a number of attempts to measure the stationary state <u>in-situ</u> from simultaneous measurements of O_3, NO, NO_2 and JNO_2 in the ground level atmosphere[6,7]. In air containing a few ppb of NO_x (NO + NO_2), the photostationary state relationship seems to hold quite

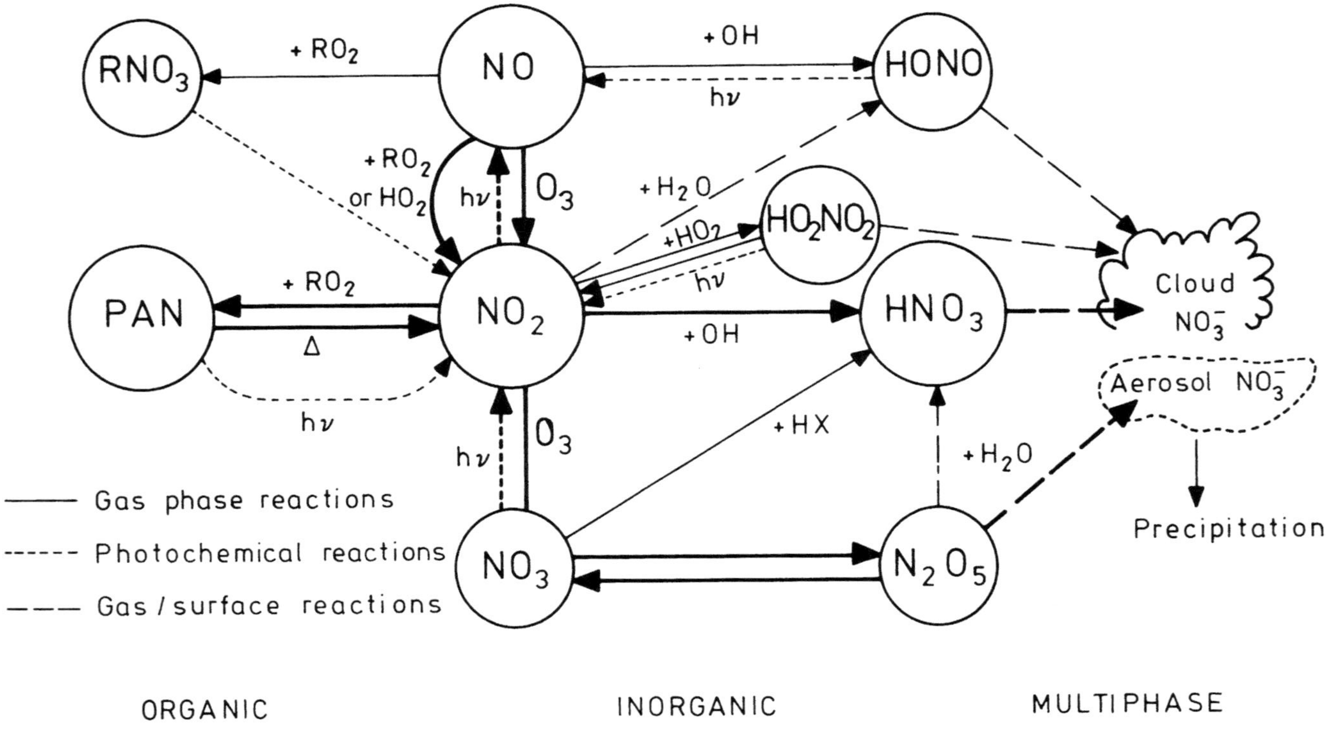

Fig. 2 SCHEMATIC DIAGRAM ILLUSTRATING TROPOSHPERIC NO_Y CHEMISTRY.

well. In very clean air containing < 1 ppb NO_x there appears to be a systematic departure from eqn. (i), indicating more rapid conversion of NO to NO_2 than expected from reaction (10).

Additional conversion of NO to NO_2 occurs through reaction of NO with HO_2 and RO_2 (mainly CH_3O_2 and other simple alkylperoxy radicals in clean air). The concentrations of these simple peroxy radicals in clean air are not sufficiently high to account for the departure from photostationary state, according to current knowledge. Possible chemical explanations include

- formation of complex and highly reactive peroxy radicals, e.g. from reactions involving natural hydrocarbons (terpenes)

- presence of halogen oxide radicals which could convert NO to NO_2 without ozone production via the sequence

$$XO + NO \rightarrow NO_2 + X \qquad X = Cl, Br \text{ or } I \qquad (22)$$

$$X + O_3 \rightarrow XO + O_2 \qquad (23)$$

Currently the terrestrial sources of terpenes and halogen oxide radicals are not well defined. Further work is necessary to understand the daytime NO, NO_2 and O_3 relationship in clean boundary layer air.

4.2 The Reactions of Peroxy Radicals with NO

The general reaction of RO_2 with NO involving O atom transfer

$$RO_2 + NO \rightarrow RO + NO \qquad (12)$$

appears to occur rapidly. Table I shows a summary of the measured rate data for these reactions. There are some detailed differences in the rate constant depending on R and the rate constants for a number of important peroxy radicals have not been measured.

At the present time it is assumed that the rate constants for the higher alkylperoxy radicals are the same as for HO_2. A complication in these cases is the occurrence of a second channel in the reaction, forming alkyl nitrates:

$$RO_2 + NO \rightarrow [ROONO] \rightarrow RONO_2 \qquad (12a)$$

This channel becomes increasingly important as the C-atom chain increases in length from C_2-C_6 (see Table II) and is more important for secondary alkylperoxy radicals. The branching ratio k_{12a}/k_{12} increases with increasing pressure in the range 50-760 Torr and decreases with increasing temperature, providing evidence for reaction through an intermediate vibrationally excited $ROONO^*$ molecule. It is not known

$$RO_2 \quad NO \rightarrow [ROONO]^* \xrightarrow{+M} RONO_2$$

$$\hookrightarrow RO + NO_2$$

whether the overall rate of the RO_2 + NO changes with pressure and temperature or whether the channel forming NO_2 is reduced at the expense of $RONO_2$ formation. Carter and Atkinson[8] have derived an expression for the yields of alkyl nitrates in the RO_2 + NO reactions for primary, secondary and tertiary, C_3-C_8 alkyperoxy radicals. It is probable that substituted longer chain peroxy radicals such as those produced in the breakdown of aromatics and oxygenated organics, also form nitrates in their reaction with NO, but there is little experimental data on this aspect.

Table I Rate Coefficients for the Reaction $RO_2 + NO \rightarrow RO + NO_2$

R	k_{300} cm^3 $molecule^{-1}$ $s^{-1} \times 10^{12}$	E/R K	Comment
H	8.3	240	well determined
CH_3	7.6	180	well determined
C_2H_5	8.9	-	more data required[a]
CH_3CO	140	-	relative rate study only[b]
CCl_3	17.0	330	well determined
CCl_2F	15.0	430	well determined
$CClF_2$	16.0	500	well determined

[a]Data from Plumb et al[9] [b]Cox and Roffey[10]; remainder from NASA[11]

Table II Branching Ratio for $RONO_2$ Formation in RO_2 + NO Reaction

R	k_{12a}/k_{12}
C_2H_5	<.014
2-C_3H_7	.042
2-C_4H_9	.090
2-C_5H_{11}	.130
2-C_6H_{13}	.209
2-C_7H_{15}	.300
2-C_8H_{17}	.323

Data from Carter and Atkinson[8]

Reaction 12a provides a reservoir for NO_x since the alkyl nitrates, although relatively stable towards photolysis and oxidation in the troposphere, are ultimately removed by OH attack or photolysis with release of NO_2 again:

$$OH + RONO_2 \rightarrow H_2O + R^1CHO + NO_2 \qquad (24)$$

$$RONO_2 + h\nu \rightarrow RO + NO_2 \qquad J \simeq 2 \times 10^{-6}\ s^{-1} \qquad (25)$$

4.3 Formation of the Oxyacids of Nitrogen

Formation of oxyacids is potentially an important sink mechanism for NO_x since the acids are water soluble and can be removed by precipitation elements. By far the most important process in this respect is the oxidation of NO_2 to HNO_3. This occurs by two routes i.e. directly via reaction with OH and by reaction with O_3, via NO_3:

$$OH + NO_2 + M \rightarrow HNO_3 + M \qquad (21)$$

$$O_3 + NO_2 \rightarrow O_2 + NO_3 \qquad (14)$$

followed by

$$NO_3 + NO_2 + M \rightleftarrows N_2O_5 + M \qquad (16)$$

$$N_2O_5 + H_2O \rightarrow 2HNO_3 \qquad (26)$$

or

$$NO_3 + HX \longrightarrow HNO_3 + X \quad (X = \text{organic radical, etc}) \qquad (27)$$

The direct route is well established and the life-time of NO_2 can be determined provided the OH concentration is known. The process is limited to daytime. The second route via NO_3 is less important in daytime because of the rapid photolysis of NO_3:

$$NO_3 + h\nu \rightarrow NO_2 + O(^3P) \qquad J_{15a} = 0.18\ s^{-1}\ *$$

$$\rightarrow NO + O_2 \qquad J_{15b} = 0.022\ s^{-1}\ *$$

(* values for overhead run at the surface of the earth, 470-700nm). At night-time NO_3 loss by reaction with HX to give HNO_3 and with NO_2 to give N_2O_5, becomes more important. The N_2O_5 route is potentially the most important for HNO_3 formation but is complicated by the rapid equilibration between NO_3 and N_2O_5 at temperatures near the earths surface, and also the reliance on heterogeneous reaction for hydration of N_2O_5. The evidence seems to point to the reaction being very important in fog and cloud situations but not elsewhere.

A possible significant minor loss process for NO_3 is reaction with HO_2, which could occur by either two exothermic channels

$$NO_3 + HO_2 \rightarrow HNO_3 + O_2 \qquad (28a)$$

$$\rightarrow OH + NO_2 + O_2 \qquad (28b)$$

Preliminary measurements[12] indicate a rate constant of approximately 4.0×10^{-12} at 298 K which would imply that it occurs 100-1000 times slower than photolysis of NO_3. Two routes would have very different effects; the first channel leads to removal of both NO_x and odd hydrogen radicals, whilst the second route regenerates OH and NO_2 and could therefore participate in cyclical reactions leading to ozone loss.

Nitrous acid, HONO, is formed mainly in daytime by reaction of OH with NO:

$$OH + NO + M \rightarrow HONO + M \qquad (29)$$

It only provides a temporary reservoir for NO_x since HONO undergoes photolysis with a half life of approximately 10 min:

$$HONO + h\nu \rightarrow OH + NO \qquad (17)$$

Formation of HONO by the reaction of NO_2 with water:

$$2NO_2 + H_2O \rightarrow HONO + HNO_3 \qquad (30)$$

is a heterogeneous reaction and can occur at the earths surface or on aerosol particles. Its rate has been measured in the laboratory[13] and found to be too slow to have significance as an NO_x loss mechanism. However, since HONO is photolysed rapidly, this reaction can provide a significant net source of OH radicals in polluted air near the ground.

Table III Arrhenuis Parameters and Lifetimes of the Peroxynitrates

pressure = 1 atm, 760 Torr; temperature = 300K

R	A (s^{-1})	E (kJ mol^{-1})	Lifetime (s)
H	$10^{15.3}$	90.3	2.7
CH_3	$10^{17.3}$	93.0	0.08
iC_3H_7	$10^{17.0}$	96.0	0.53
CH_3CO	$10^{16.1}$	111	2380.
C_2H_5CO	$10^{17.2}$	117	1520.
ClCO	$10^{16.8}$	115.9	2457.

Peroxynitrous acid, HO_2NO_2, is formed by reaction of HO_2 radical with NO_2. The rate is rapid but at temperatures prevalent near the earths surface HO_2NO_2 is unstable, decompositing thermally back to NO_2 and HO_2 with a lifetime of a few seconds. In the colder upper troposphere the thermal lifetime of HO_2NO_2 may be months and it therefore can provide a reservoir for NO_x. NO_2 is released from the reservoir by photolysis or by reaction with OH reducing the overall lifetime in the reservoir to a few days.

$$OH + HO_2NO_2 \rightarrow H_2O + NO_2 + O_2 \qquad (31)$$

$$HO_2NO_2 + h\nu \rightarrow \text{products} \qquad (32)$$

4.4 Organic NO_x Chemistry - Formation of PAN'S

The reaction of NO_2 with peroxy radicals leads to formation of peroxynitrates:

$$RO_2 + NO_2\ (+M) \rightarrow RO_2NO_2\ (+M) \qquad (33)$$

The stability of the peroxynitrates depends on the nature of the organic radical R, which is reflected in the Arrenhius activation energies for the thermal decomposition of peroxynitrates shown in Table III. It will be seen that the acetyl (PAN), propionyl (PPN) and chlorofomyl peroxynitrates are much more stable than the simple alkylperoxy derivatives. Consequently, the reservoir function of the alkylperoxynitrates is confined to cold regions of the upper troposphere as for HO_2NO_2. In surface air their rapid decomposition ($\tau \leq 1$ s at 300 K) leads to an equilibrium between RO_2 and NO_2 with only a small amount of NO_x sequestered as RO_2NO_2. On the other hand PAN and other hand PAN and other peroxyacyl nitrates have lifetimes with respect to decomposition of approximately 1 hr at 300 K, 2 days at 273 K and 148 days at 250 K. Thus at temperatures at or below 283 K PAN becomes a stable product of photochemical oxidation of simple organics like acetaldehyde, which yield acetyl radicals in their oxidative breakdown:

$$OH + CH_3CHO \rightarrow CH_3CO + H_2O \qquad (34)$$

$$CH_3CO + O_2\ (+M) \rightarrow CH_3CO_3 \qquad (35)$$

$$CH_3CO_3 + NO_2\ (+M) \rightleftarrows CH_3CO_3NO_2\ (PAN) \qquad (36)(-36)$$

Even at surface temperatures such as those occurring in photochemical pollution in the boundary layer, PAN's may build up to relatively high concentrations, through equilibration of the formation reaction.

Loss of PAN then occurs through alternate removal processes for the precursor radicals, the most important being the reaction of CH_3CO_3 with NO:

$$CH_3CO_3 + NO \rightarrow CH_3 + CO_2 + NO_2 \qquad (37)$$

The overall decomposition rate of PAN is therefore given by a complex expression, i.e.

$$-\frac{d[PAN]}{dt} = k_{-36}[PAN]\left\{1 - \frac{1}{1 + (k_{37}[NO]/k_{36}[NO_2])}\right\}$$

This equation implies that in the absence of NO, e.g. at night-time and in the absence of NO sources, PAN has essentially no homogeneous loss mechanism. Two additional homogeneous loss processes must be considered for PAN and for other peroxynitrates. Firstly, these molecules can be photolysed in the "tail" of their UV absorption bands which extend into the near UV region (λ > 300 nm). These spectra are not well known, particularly for the less stable peroxynitrates.

$$RO_2NO_2 + h\nu \rightarrow \text{products} \qquad (38)$$

Secondly the peroxynitrates react with hydroxyl radicals:

$$OH + RO_2NO_2 \rightarrow \text{products} \qquad (39)$$

The rate constants for these reactions are not known, except for PAN which has been measured recently[14], and shows rather low reactivity with OH. Although the products of photolysis and OH reaction with peroxynitrates have not been established, it can be safely assumed that these reactions lead to release of NO_x (NO, NO_2 or NO_3) and are consequently important aspects of the chemistry of peroxynitrates in the background troposphere where thermal decomposition is very slow. For PAN these reactions place an upper bound of approximately 3 months on its mean lifetime.

The other forms of organic NO_x are the organic nitrates, which although more stable, are much less abundant than the peroxynitrates. This arises because the major volatile organics do not readily form nitrates in their atmospheric degradation in the presence of NO_x. However, there are several minor processes which form nitrates i.e.

- reaction of larger alkylperoxy radicals and possibly substituted alkyl and aromatic peroxy radicals, with NO (referred to above)

- reaction of alkoxy radicals with NO_2 (a minor pathway in competition with oxidation of RO)

- addition of NO_3 to alkenes (may be important in polluted air at night-time)

- minor nitration pathways in oxidation of certain aromatics.

The main sinks for nitrates are photolysis in the near UV tail of the UV absorption band and reaction with OH. Data for these reactions are sparse.

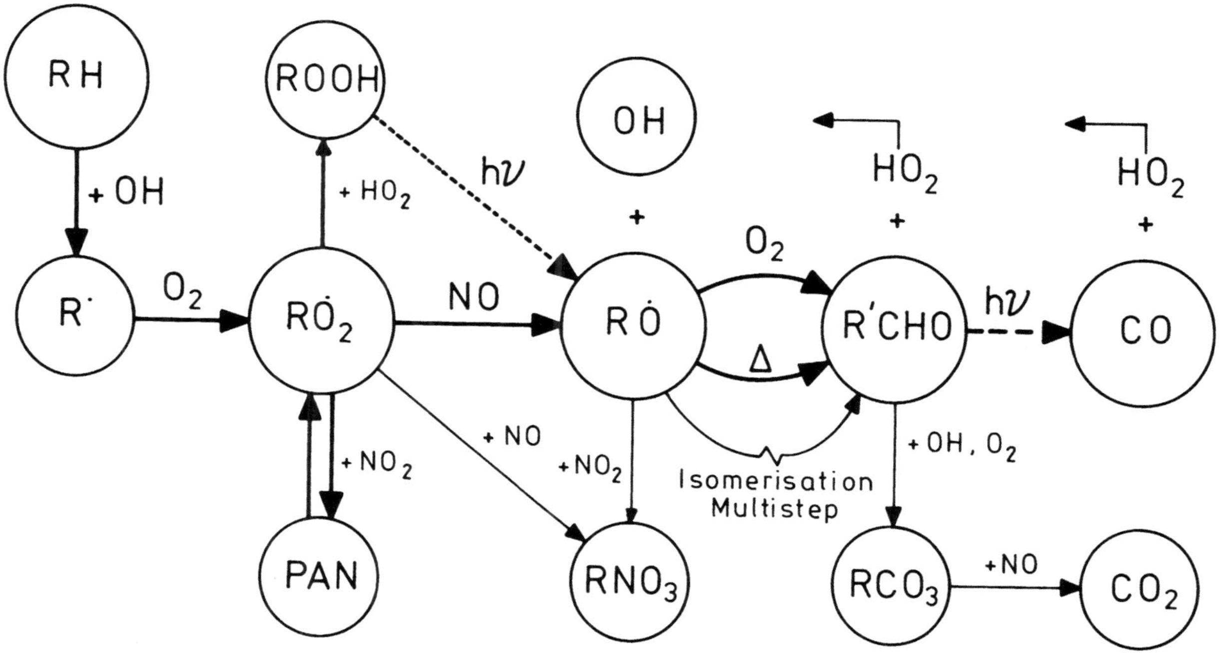

Fig. 3 SCHEMATIC DIAGRAM ILLUSTRATING TROPOSPHERIC ORGANIC CHEMISTRY.

5. OXIDATION OF VOLATILE ORGANICS

Hydrocarbons and related volatile organic compounds undergo gas phase oxidation in the atmosphere to form CO_2 and H_2O. The reactions are mainly initiated by the attack of OH radicals on the compound, followed by a series of elementary reaction pathways which are illustrated in Fig. 3. In this generalised reaction mechanism, peroxy radicals are produced at different stages of the oxidative breakdown, and so lead potentially to ozone production through reaction (12) + (10). There are several competing reactions in the scheme which influence the efficiency of ozone production. However, the overall impact of the oxidation of organics on the tropospheric ozone budget is largely determined by the reactivity of the molecules towards OH radicals, which determines the atmospheric lifetime of the organics.

5.1 Reaction of OH radicals with Volatile Organics

The kinetics and mechanisms of the gas phase reactions of the hydroxyl radical with volatile organic compounds under atmospheric conditions is a subject that has been extensively investigated experimentally and reviewed widely over the past 10-15 years. Rate constants for the elementary OH + organic reactions and their temperature dependencies have been established using a variety of well developed laboratory techniques, so that now there is an adequate data base for a large number of hydrocarbons, oxygenates, sulphur-, nitrogen- and halogen-containing organics, which are of significance for ozone formation and atmospheric chemistry generally. Furthermore, techniques have been developed with some success for estimating rate constants for OH radical reactions with organic compounds for which no experimental data are available, using structure/reactivity considerations. The available kinetic data have been comprehensively reviewed and evaluated by Atkinson[15]. Atkinson and Lloyd[16] have evaluated data for the subsequent reactions following the reaction of OH with selected organics.

The attack of OH on organics may be by abstraction of H or, in the case of an unsaturated substrate, by addition:

$$OH + RH \rightarrow H_2O + R \tag{40}$$

$$OH + \gt C{=}C\lt \rightarrow \gt CH(OH){-}\dot{C}\lt \tag{41}$$

The subsequent reactions of the initially formed radical follow the general pattern shown in Fig. 3, although when addition occurs, particularly in the case of aromatics, the mechanism is more complex due to competitive pathways for the initial radical R, produced following OH attack. An important feature of the scheme in Fig. 3 is that OH radicals are regenerated through the production of HO_2 at the end of the chain, followed by its reaction either with NO or O_3 to give OH.

Other notable features of the generalised oxidation schemes are summarised as follows:

- in air the initially formed radical R is rapidly transformed to RO_2. For aromatic radicals the RO_2 radical is unstable and R and RO_2 become equilibrated, opening up possible competitive pathways for R.

- RO_2 is converted efficiently to RO by reaction with NO, with a minor channel to form nitrates becoming important for larger R (C_n, $n \geq 4$).

- RO_2 also reacts reversibly with NO_2 to form PAN's of varying stability.

- Under conditions where NO_x concentrations are low, RO_2 reacts with HO_2 or another RO_2 leading to the formation of hydroperoxides (in the case of HO_2) or aldehydes, alcohols or RO radicals in the case of RO_2 + RO_2 reaction.

- Alkoxy radicals, RO, can undergo several competing reactions; a) reaction with O_2 gives HO_2 together with a carbonyl compound: this only occurs when the RO radical has an H-atom attached to the C-O group i.e. primary and secondary alkoxy radicals. b) thermal decomposition to form an aldehyde or ketone with fewer carbon atoms, together with a smaller radical: this predominates for tertiary alkoxy radicals. c) isomerisation by internal H abstraction from the 4- or 5- carbon position relative to the C-O position. This process seems to be dominant for alkoxyradicals derived from n-C_5 and n-C_6 alkanes.

- The carbonyl compounds produced from alkoxy radicals undergo photochemical dissociation, to produce HCO as well as other R type radicals; HCO reacts rapidly with O_2 to give CO and HO_2:

$$HCO + O_2 \rightarrow HO_2 + CO \qquad (42)$$

Carbonyl compounds are also removed by reaction with OH, producing acyl peroxyradicals, which on reaction with NO, produce CO_2 via the sequence:

$$RCO_3 + NO \rightarrow RCO_2 + NO_2 \qquad (43)$$

$$RCO_2 \rightarrow R^{\cdot} + CO_2 \qquad (44)$$

In this way the carbonyl C atom is oxidised directly to CO_2 without intermediate formation of CO.

5.2 Hydrocarbons in Chemical Models

An exact description of the detailed chemical degradation of all volatile organics in the atmosphere is impractical in atmospheric models, because of the complexity of the chemical mechanisms and the large number of different organic species present in the atmosphere. Some sort of simplification of the chemical formulation is necessary but at the same time the essential kinetic features of the free radical chain mechanism and the interaction with NO_x producing ozone, needs to be maintained.

For ozone production in the background troposphere simplification can be based on the exclusion of those volatile organics which are relatively short lived and are therefore oxidised in the boundary layer. This would include many of the 'reactive' hydrocarbons which are considered important for photochemical smog modelling, as well as natural hydrocarbons from vegetation such as terpenes and isoprene. A typical investory for background tropospheric ozone modelling might include:

CO CH_4	Carbon containing molecules in unperturbed troposphere
Alkanes	C_2-C_5
Alkenes	C_2-C_3, (plus isoprene if natural continental BL included)
Aromatics	benzene, toluene

(Together with main oxygenated degradation products from these molecules)

For modelling of photochemical ozone production in urban and regional boundary layer scales, several methods have been used to simplify the model representation of the chemistry of the large number of hydrocarbons which are emitted to the atmosphere. Leone and Seinfield (1985) [17] have classified the models as being of 'lumped structure', 'lumped molecule', or 'surrogate species', depending on whether structural components of the molecules (e.g. double bonds), representative molecules of each classification (e.g. alkenes) or hydrocarbons not necessarily related to the emissions, were used to represent the hydrocarbons released and degraded in the atmosphere.

A number of comparisons of the performance of these different methods of treatment of hydrocarbons in models have been made, although different criteria have been used for comparison. These studies have shown that different mechanisms predict different relationships between emission scenario and photo-oxidant production. Leone and Seinfield's analysis investigated in detail the origins of some of the differences. Some of these resulted merely from use of differing rate constants whilst others were due to use of completely different reaction sequences. Hough[18] has shown that mechanisms published since 1984 all produce similar results for ozone and nitric acid formation, for the same emission data base with a 1 day photochemistry simulation. Agreement for formation of PAN and hydrogen peroxide was less satisfactory. Hov[19] extended model comparison to 5 days photochemistry, for 7 different mechanisms with only chemical source and sink terms. Ozone production was comparable in all models on day 1 but diverged increasingly on subsequent days. A substantial part of the differences could be attributed to differences in the OH concentrations calculated in the models. The differences in OH were due to the effect of the model formulation on interconversion and sink processes of the free radicals, which control the overall reaction rate.

The development of ways in which large numbers of real components are represented by a limited number of model variables clearly requires further research. The application of lumped parameter models to the perturbed background troposphere, e.g. polar winter hydrocarbon chemistry, has not yet been attempted. A suitably modified lumped structure model might be usefully applied to this problem.

5.3 Peroxyradical Reactions in the Absence of NO

In the absence of NO, the normal propagation of the degradation of volatile organics through reaction (12) is stopped. Under these conditions the peroxyradicals, which are unreactive toward closed-shell molecules, react with other radicals. Although HO_2 reacts with O_3, the corresponding reaction of organic peroxyradicals with O_3 appears to be too slow to be significant:

$$RO_2 + O_3 \rightarrow RO + 2O_2 \qquad (45)$$

The most abundant peroxy radicals in the clean troposphere are HO_2 and CH_3O_2 and their major interactions are through the reactions:

$$HO_2 + HO_2 \rightarrow H_2O_2 + O_2 \qquad (20)$$

$$CH_3O_2 + HO_2 \rightarrow CH_3OOH + O_2 \qquad (8)$$

CH_3OOH is normally assumed to be the product of the CH_3O_2 + HO_2 reaction, by analogy with reaction (20). However, formation of HCHO + H_2O or CH_3OH + O_3 are both exothermic, and are therefore potential pathways for this reaction. Inconsistencies in the rate constant determinations for reaction (8) based on radical decay kinetics and CH_3OOH product formation suggests that more than route is operative There is no spectroscopic evidence for ozone formation in this reaction, unlike the CH_3COO_2 + HO_2 reaction.

The self reaction of CH_3O_2 which proceeds by all three of the following channels:

$$CH_3O_2 + CH_3O_2 \rightarrow HCHO + CH_3OH + O_2 \qquad (46a)$$

$$\rightarrow 2CH_3O^{\cdot} + O_2 \qquad (46b)$$

$$\rightarrow CH_3OOCH_3 + O_2 \qquad (46c)$$

is considerably slower than the CH_3O_2 + HO_2 and is therefore not of importance in the atmosphere.

Peroxy radicals derived from other volatile organic molecules in the troposphere will generally react with either HO_2 or CH_3O_2, which are the most abundant peroxyradicals. The rate constants and product pathways of these reactions have not been established in many cases, although a pattern of behaviour in emerging from the kinetic and mechanistic data that are becoming available. A summary of current kinetic information on peroxy radical kinetics is given in Table IV. It will be seen that the rate constants show a large variation with the structure and substitution of the organic fragment.

The hydroperoxides formed in reaction (8) and analogous reactions of other RO_2 radicals undergo further oxidation by photodissociation to form oxy-radicals + OH:

$$ROOH + h\nu \rightarrow RO^{\cdot} + OH \qquad (47)$$

The absorption spectra for ROOH are similar to H_2O_2 (see Fig. 1), exhibiting an extended 'tail' into the near UV region. The hydroperoxides are also oxidised by OH attack either on the organic radical or the peroxy H-atom, e.g.

$$OH + CH_3OOH \rightarrow H_2O + HCHO + OH \qquad (48a)$$

$$\rightarrow H_2O + CH_3OO \qquad (48b)$$

Since the overall effect of reaction (8) + (48) is the conversion

net $\quad OH + HO_2 \rightarrow H_2O + O_2$

it leads to loss of odd hydrogen radicals. The photolysis route regenerates the radicals removed in forming ROOH.

Table IV Kinetics of Peroxyradical Reactions in the NO-free Troposphere

for RO_2 + RO_2 → products
HO_2 → products } at 298 K
CH_3O_2 → products

Radical, R	Rate coefficient x 10^{12} cm^3 $molecule^{-1}$ s^{-1}		
	RO_2	HO_2	CH_3O_2
HO_2	-	2.7	6.4
CH_3O_2	-	6.4	0.40
$C_2H_5O_2$[a]	8.4 x 10^{-2}	6.3	*
$iC_3H_7O_2$[b]	6.9 x 10^{-4}	*	*
$tC_4H_9O_2$[c]	3.6 x 10^{-6}	> 1	0.010
CH_3COO_2[d]	18	11	16
$CH_2(OH)O_2$[e]	6	15	*
$CH_3CH(OH)O_2$[f]	*	*	0.9
$C_2H_3O_2$[g]	< 1	*	*

[a]Jenkin et al[20] [b]Kirsch et al[21] [c]Kirsch and Parks[22] [d]Veyret and Moortgat, unpublished [e]Cox and Veyret, unpublished [f]Moortgat et al[23] [g]Pagsberg et al, unpublished.

5.4 Reaction of other Oxidising Species with Volatile Organics

Although OH is the major attacking radical in the atmospheric oxidation of organics, minor but significant oxidation pathways involving attack by ozone, NO_3 and HO_2 need to be considered, since these can trigger the radical chain reactions leading to ozone production. Table V shows a summary of rate constants measured for these reactions with a selection of volatile organic compounds.

Reaction with ozone is only of importance for alkenes, since reaction with alkanes, aromatics and aldehydes is negligably slow under atmospheric conditions. The kinetics and mechanisms of ozone-alkene reactions have been reviewed extensively by Atkinson and Carter[24]. The rate constants and their temperature dependencies have been measured for most of the simple alkenes and for some dialkenes and terpenes. For typical background O_3 concentrations (~ 10^{12} molecule cm^{-3}) the life times of the alkenes range from 9 days for C_2H_4 to less than 15 mins for 2-3 dimethyl-2-butene and some terpenes. The mechanisms of the gas phase O_3- alkene reactions are complex with a variety of products including aldehydes, ketones, acids as well as inorganics such as CO, CO_2 and H_2O. The reaction also generates free radicals, as discussed below.

The ozone-alkene reaction only has a very minor influence on the ozone budget in the troposphere. Its importance lies in the modification in the nature and concentration fields of the more reactive volatile organic molecules in the boundary layer.

The reactions of NO_3 with organics are much faster than the ozone reactions and for some molecules such as substituted alkenes, terpenes and dimethyl sulphide, these provide significant loss mechanisms for these molecules in the night-time boundary layer (Winer et al)[27]. The reactions also provide a loss mechanism for NO_3 and hence for NO_x as discussed above. The mechanism of the reaction of NO_3 with alkenes has been shown to be an addition to the olefinic double bond e.g.

$$NO_3 + CH_3CH = CH_2 \rightarrow CH_3\dot{C}HCH_2ONO_2 \qquad (49)$$

Under atmospheric conditions, the radical formed is expected to add O_2 rapidly to form a peroxyradical which can subsequently participate in the normal oxidation chain involving NO and NO_2. With aldehydes and aromatics the NO_3 reaction involves H-abstraction and HNO_3 formation, providing a sink for NO_x e.g.

$$NO_3 + HCHO \rightarrow HNO_3 + HCO \qquad (50)$$

The only reaction of HO_2 which is potentially of significance in initiating the oxidation of organic molecules, is the reaction with carbonyl compounds, in particular with aldehydes and with dicarbonyls. The reaction takes place by addition and is facilitated by a rapid H-atom transfer in the initial adduct, to form a hydroxy substituted peroxy radical e.g.

$$HO_2 + HCHO \rightleftarrows [\dot{O}CH_2OOH] \rightleftarrows HOCH_2OO^{\bullet} \qquad (51)$$

The addition appears to be reversible and the rate of reaction depends on the competition between the reverse reaction and other reactions of the hydroxy-peroxy radical. Formic acid production via the reactions:

$$HOCH_2OO^{\bullet} + NO \rightarrow HOCH_2O^{\bullet} + NO_2 \qquad (52)$$

$$HOCH_2O^{\bullet} + O_2 \rightarrow HCOOH + HO_2 \qquad (53)$$

is a potentially significant pathway for HCHO oxidation. There is considerable controversy at the present time concerning the magnitude of the equilibrium constant, K^*. Other additional reactions of HO_2, e.g. with alkenes, have a large activation energy and are consequently too slow to be significant in the atmosphere.

Table V Rate Constants at 298 K for Reactions of Selected Volatile Organic compounds with O_3, NO_3 and HO_2

	k_{O_3} [d]	k_{NO_3} [e]	k_{HO_2} [f]
		molecule cm^{-3} x 10^{15}	
C_2H_4	0.00175	0.061	-
C_3H_6	0.0113	4.2	-
C_5H_{12}	-	0.024	-
C_5H_8 [a]	0.0143	323	*
C_6H_{12} [b]	1.16	31.000	6 x 10^{-4}
C_6H_6	-	<0.023	-
HCHO	-	0.323	70
CH_3CHO	-	1.34	0.01
CH_3SCH_3 [c]	-	500	*

- \- very slow
- * not measured, possibly significant
- a isoprene
- b 2,3-dimethyl-2-butene
- c dimethyl Sulphide
- d data take from Atkinson and Carter[24]
- e data mainly from Atkinson et al[25]
- f data from Veyret et al[26] for HCHO. Other sources unpublished work of Walker, U. of Hull, and Barnes, U. of Wuppertal

6. PRODUCTION AND LOSS OF FREE RADICALS

The central driving force of most atmospheric oxidation mechanisms is the odd hydrogen radicals, OH and HO_2. Since ozone production in the troposphere depends on the rate of production of RO_2 radicals, which in turn depends on the rate of production of R$^\bullet$ by attack of OH, the atmospheric concentration of OH and related radicals is a critical factor in ozone production. These radicals exist in steady states which are a balance between total production and loss of free radicals and the interconversion between them. A schematic illustration of these processes controlling OH and HO_2 in the clean troposphere is presented in Fig. 4.

The interconversion reactions will be already familiar: OH is converted to HO_2 via reaction with CO or via the CH_4 oxidation chain involving CH_3O_2 and CH_3O; HO_2 is converted back to OH via reaction with NO or with O_3. If the interconversion is rapid compared to production and loss, the ratio $[HO_2]/[OH]$ is given by the expression:

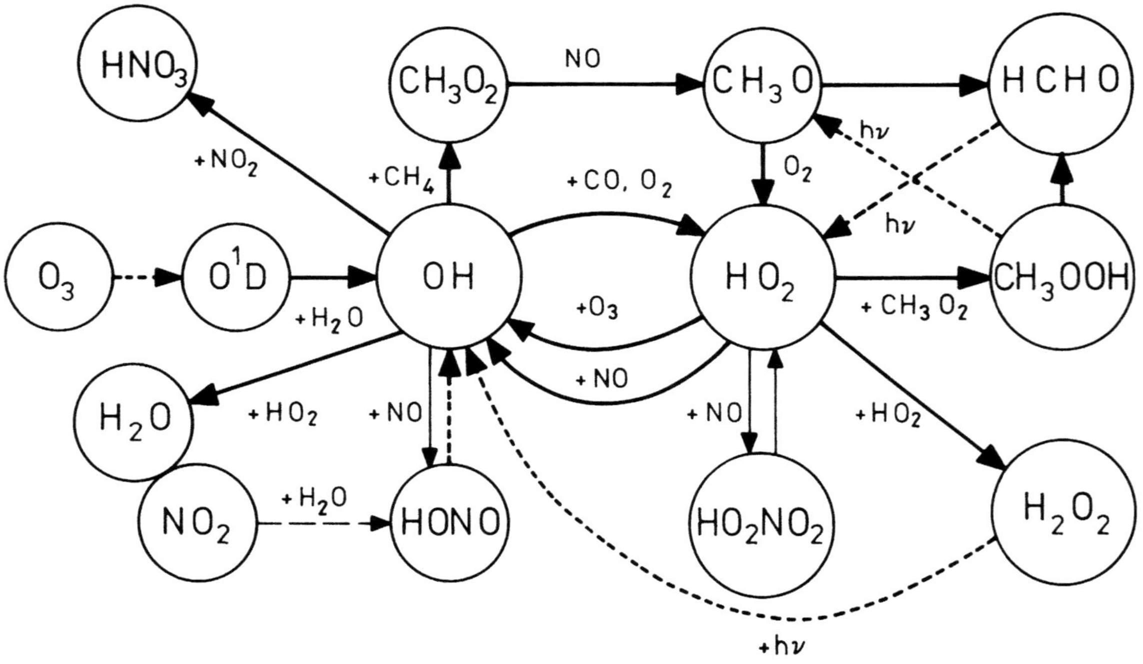

Fig. 4 REACTIONS GOVERNING CONCENTRATIONS OF OH AND HO_2.

$$\frac{[HO_2]}{[OH]} = \frac{k_4\,[CO] + k_6\,[CH_4]}{k_5\,[O_3] + k_{12}\,[NO]} \qquad (k_{12} \text{ for } R{=}H)$$

The presence of additional volatile organic molecules which react with OH, as in the perturbed atmosphere, will tend to increase $[HO_2]$ relative to $[OH]$. This effect will be counterbalanced by any increase in $[NO]$ in the perturbing air mass.

An important effect of changing the ratio $[HO_2]/[OH]$ is the resulting change in the partitioning between the different loss processes. The major loss process involving OH is formation of HNO_3, which is long lived and is removed by precipitation or deposition. In the NO_x free atmosphere there is a minor loss of OH through the reaction

$$OH + HO_2 \rightarrow H_2O + O_2 \qquad (54)$$

Loss mechanisms involving HO_2 (and the analogous organic peroxy-radicals) lead to the formation of hydro-peroxides and peroxynitrates. Whilst these molecules also may undergo physical removal, they can also undergo photodissociation or dissociate thermally, and thus only provide temporary loss of free radical entities. The potential oxidising capacity of the atmosphere is not lost when peroxy radicals are stored as temporary reservoirs in this way.

The sources of atmospheric free radicals are predominantly photochemical and consequently, radical driven chemistry and ozone production only occurs during daytime. In the clean troposphere, ozone photochemistry and the $O(^1D) + H_2O$ reaction provides a major source of odd-hydrogen radicals. The second most important source is photodissociation of formaldehyde and here resides a potential perturbing influence on radical concentrations since formaldehyde is intermediate in the oxidation of almost all volatile organic compounds. Note that HCHO is a true free radical source since its in-situ production does not involve loss of radicals, as for example H_2O_2 and the organic peroxides. The photolysis of HCHO is the origin of the amplifying effect of CH_4 oxidation on the tropospheric free radical concentrations. There is however a second channel in HCHO photolysis producing H_2 and CO directly, which does not constitute a free radical source:

$$HCHO + h\nu \rightarrow H + HCO \qquad (18a)$$

$$\rightarrow H_2 + CO \qquad (18b)$$

This latter channel, together with other loss processes for HCHO (i.e. reaction with OH or removal by precipitation) will tend to reduce the amplifying effect. Other carbonyl compounds e.g. CH_3CHO, acetone, provide minor photochemical radical sources in the background troposphere.

In the boundary layer, free radical production results from a variety of photolabile substances which are emitted to the atmosphere, mainly from man made sources. These include various carbonyl compounds, nitrous acid, nitrites, nitrates, and halogen compounds such as Cl_2 and CH_3I.

In addition to photochemical sources, there are several thermal sources of free radicals, which have significance for night-time atmospheric chemistry. This subject has been considered in more detail in recent years. These sources involve ozone either directly or indirectly through NO_3 formation in the $NO_2 + O_3$ reaction.

Ozone-alkene reactions are thought to produce free radicals by decomposition of the Criegee intermediate, CH_2OO formed following attack of ozone on the double and e.g.

$$O_3 + C_2H_4 \rightarrow HCHO + CH_2OO^* \qquad (55)$$

A fraction of the CH_2OO radicals are formed vibrationally excited and can decompose to produce H atoms e.g.

$$CH_2OO^* \rightarrow [HCOOH^*] \rightarrow CO_2 + 2H \qquad (56)$$

The overall yield of atoms or radicals depends on the nature of the parent alkene; for C_2H_4 the H atom yield is of the order of 4% of the overall O_3 + alkene reaction but for higher alkenes it may be larger.

Formation of radicals from NO_3 results from its attack on organic compounds e.g.

$$NO_3 + CH_3CHO \rightarrow HNO_3 + CH_3CO \qquad (57)$$

The subsequent reactions of the radicals produced from thermal sources at night-time follow the same pattern as in daytime except that the chain reactions will be terminated more rapidly due to generally lower abundance of NO, especially where local NO sources are absent.

7. CHEMICAL PRODUCTION AND LOSS OF OZONE

The dominance of chemical production or loss processes for ozone in the background troposphere is highly sensitive to the local NO_x and hydrocarbon composition. This is illustrated by considering the following reaction sequences for oxidation of CH_4 to formaldehyde, firstly when NO is present:

$$OH + CH_4 \rightarrow CH_3 + H_2O \qquad (6)$$

$$CH_3 + O_2 + M \rightarrow CH_3O_2 + M \qquad (7)$$

$$CH_3O_2 + NO \rightarrow CH_3O + NO_2 \qquad (12)$$

$$CH_3O + O_2 \rightarrow HCHO + HO_2 \qquad (58)$$

$$HO_2 + NO \rightarrow NO_2 + OH \qquad (12)$$

$$2(NO_2 + h\nu \rightarrow NO + O) \quad (10)$$

$$2(O + O_2 + M \rightarrow O_3 + M) \quad (2)$$

NET $CH_4 + 4O_2 \xrightarrow{h\nu} HCHO + H_2O + 2O_3$

when NO is about

$$OH + CH_4 \rightarrow CH_3 + H_2O \quad (6)$$

$$CH_3 + O_2 + M \rightarrow CH_3O_2 + M \quad (7)$$

$$CH_3O_2 + HO_2 \rightarrow CH_3OOH + O_2 \quad (8)$$

$$CH_3OOH + h\nu \rightarrow CH_3O + OH \quad (59)$$

$$CH_3O + O_2 \rightarrow HCHO + HO_2 \quad (58)$$

NET $CH_4 + O_2 \xrightarrow{h\nu} HCHO + H_2O$

Thus the competition between the two reactions of CH_3O_2 with NO and HO_2 determines whether the CH_4 oxidation leads to O_3 production for a null cycle. The sensitivity to the chemistry increases further when the subsequent oxidation of HCHO is considered in the absence of NO:

$$HCHO + h\nu \rightarrow H + HCO \quad (18)$$

$$2(H + O_2 + M \rightarrow HO_2 + M) \quad (5)$$

$$HCO + O_2 \rightarrow HO_2 + CO \quad (42)$$

$$2(HO_2 + O_3 \rightarrow OH + 2O_2) \quad (9)$$

$$OH + CO \rightarrow H + CO_2 \quad (4)$$

$$OH + HO_2 \rightarrow H_2O + O_2 \quad (54)$$

NET $HCHO + 2O_3 \xrightarrow{h\nu} H_2O + CO_2 + 2O_2$

Here 2 molecules of ozone are removed, whereas if HO_2 reacts with NO: net ozone production occurs:

$$2(HO_2 + NO \rightarrow NO_2 + OH) \quad (12)$$

$$OH + CO \rightarrow H + CO_2 \quad (4)$$

$$OH + HO_2 \rightarrow H_2O + O_2 \quad (54)$$

$$2(NO_2 + h\nu \rightarrow NO + O) \quad (10)$$

$$2(O + O_2 + M \rightarrow O + M) \qquad (2)$$

NET $$HCHO + 4O_2 \xrightarrow{h\nu} H_2O + CO_2 + 2O_3$$

In these cycles it is the competition between HO_2 reaction with O_3 and NO which determines the balance between production and loss of ozone. Thus the kinetics and mechanisms of the peroxy radicals reactions under atmospheric conditions are a key factor in the ozone budget in the background troposphere, where NO concentrations are low enough for the alternate reactions of the RO_2 radicals to compete.

When sufficient NO is present to propagate the oxidation cycle, the rate of ozone formation is determined by the rate of formation of the peroxy radicals which is in turn determined by the rate of OH attack or photochemical dissociation of the volatile organics present. The factors affecting this process can be summarised on the basis of the kinetic equation governing the peroxy radical production

$$\frac{d[RO_2]}{dt} = \sum_{voc} k_{OH}[OH][VOC] + \sum_{voc} J[VOC]$$

here k_{OH} is the rate constant for the OH + organic reaction and is a measure of the 'reactivity' factor for the particular molecule. Temp- erature and pressure dependencies of k_{OH} need also to be considered. J is the photodissociation rate, which is given by the product of the wavelength dependent solar flux, absorption cross section and quantum yield:

$$J = \int_{\lambda_1}^{\lambda_2} I_\lambda \, \sigma_\lambda \, \varphi_\lambda \, d\lambda$$

where the wavelength range $\lambda_2 - \lambda_1$ covers the envelope of the solar spectrum in the troposphere. The local [OH] is controlled by the chemical factors discussed in this paper and is coupled to hydrocarbon, NO_x and O_3 concentrations. Finally, the concentration field of the volatile organics is determined by their source pattern, transport and transformation within the atmosphere. It is clear from these considerations that it will require an intense research effort to draw together all the factors in order to describe quantitatively and in detail, the budget of tropospheric ozone.

ACKNOWLEDGEMENT

This review was written as part of a research programme in Atmospheric Chemistry supported by the U.K. Department of the Environment.

May 31st 1987

REFERENCES

1 Junge, C.E., Air Chemistry and Radioactivity. Academic Press, New York, 1963.

2 Crutzen, P.J., Pure Appl. Geophys **106-108** 1385, 1973.

3 Chameides, W.L., and Walker, J.C.G., J. Geophys Res **78** 8751, 1973.

4 Leighton, P.A., Photochemistry of Air Pollution. Academic Press, New York, 1961.

5 Levy, H., Science **173** 141 1971; Planet Space Sci. **20** 919, 1972.

6 Stedman, D., and Jackson, J.O., Int. J. Chem. Kinet. **Symp I** 492, 1975.

7 Calvert, J.G., Environ. Sci. Technol. **10** 248, 1976.

8 Carter, W.P.L., and Atkinson, R., J. Atm Chem. **3** 377, 1985.

9 Plumb, I.C., Ryan, K.R., Steven, J.R., and Mulcahy, M.F.R., Int. J. Chem. Kinet. **14** 183, 1982.

10 Cox, R.A., and Roffey, M.J., Environ. Sci. Technol, **11** 900, 1977.

11 NASA Panel for Data Evaluation 'Chemical Kinetics and Photochemical Data for Use in Stratospheric Modelling', Evaluation No.7, JPL Publication 85-38, July 1985.

12 Wayne, R.P., Smith, S.J., Cox, R.A., and Hall, I.W., "Laboratory Studies of the Nitrate Radical" in 'Physical Chemical Behaviour of Atmospheric Pollutants', CEC Air Pollution Research Report 2, D. Reidel, Dordrecht, 1987.

13 Jenkin, M.E., and Cox, R.A., Atm Environment (submitted).

14 Wallington, T.J., Atkinson, R., and Winer, A.M., Geophys Res. Lett, **11** 861, 1984.

15 Atkinson, R., Chem. Rev. **86** 69, 1986.

16 Atkinson, R., and Lloyd, A.C., J. Phys. Chem. Ref. Data **13** 315. 1984.

17 Leone, J.A., and Seinfield, J.H., Atmos. Environment **19** 437, 1985.

18 Hough, A.M., "An Intercomparison of Mechanisms for the Production of Photochemical Oxidants", to be published.

19 O. Hov 'Evaluation of the Photo-Oxidants-Precursor Relationship in Europe', CEC Air Pollution Research Report 1, Report AP/60/87. Commission of the European Communities, 1987.

20 Cattell, F.C., Cavanagh, J., Cox, R.A., and Jenkin, M.E., J. Chem. Soc. Far. 2 **82** 1999, 1986.

21 Kirsch, L.J., Parkes, D.A., Waddington, D.J., and Wooley, A., J. Chem. Soc. Faraday Trans. I **75** 2678, 1979.

22 Kirsch, L.J., and Parkes, D.A., J. Chem. Soc. Faraday Trans I **77** 293, 1981.

23 Moortgat, G.K., Burrows, J.P., Schneider, W.S., Tyndall, G.S., and Cox, R.A., 'A Study of the HO_2 + CH_3CHO Reaction in the Photolysis of CH_3CHO, and its Consequences for Atmospheric Chemistry', CEC Air Pollution Research Report 2 Physical-Chemical Behaviour of Atmospheric Pollutants, D. Reidel Dordrecht, 1987.

24 Atkinson, R., and Carter, W.P.L., Chem. Rev. **84** 437, 1984.

25 Atkinson, R., Plum, C.N., Carter, W.P.L., Winer, A.M., and Pitts, J.M. Jr., J. Phys. Chem. **88** 1210, 1984.

26 Veyret, B., Rayez, J-C, and Lesclaux, R., J. Phys. Chem. **86** 342, 1982.

27 Winer, A.M., Atkinson, R., and Pitts, J.N. Jr., Science **224** 156, 1984.

OZONE PRODUCTION IN THE BLACK FOREST: DIRECT MEASUREMENTS OF RO_2, NO_x AND OTHER RELEVANT PARAMETERS

A. Volz, D. Mihelcic, P. Müsgen, H.W. Pätz, G. Pilwat,
H. Geiss, and D. Kley
Institut für Chemie 2: Chemie der Belasteten Atmosphäre
Kernforschungsanlage Jülich, GmbH, Postfach 1913, FRG

ABSTRACT. The in-situ production of ozone was investigated from simultaneous measurements of RO_2-radicals, O_3, NO_x, and the NO_2-photolysis frequency during a field campaign in the Black Forest, FRG, in July 1986. During one day, the measured RO_2 concentrations agree well with those calculated from the photostationary state of NOx. On the second day, extremely high RO_2 concentrations were observed which are not reflected in a corresponding O_3 production. It is concluded that some RO_2 radicals have a significantly reduced reactivity with respect to NO than is commonly believed. Also, the measured ratio of HO_2 to organic peroxy radicals was much lower than that derived from accepted chemical decomposition schemes of hydrocarbons.

1. INTRODUCTION

The primary photochemical source of O_3 in the troposphere is the photolysis of NO_2 (I) at wavelengths between 300 and 420 nm, followed by the rapid recombination of the ground-state O-atoms with molecular oxygen (II). During daytime, process (I) has a characteristic time con stant of several minutes. Usually, reaction (I) does not lead to a net gain of O_3, since NO reacts rapidly with O_3 (III) returning the NO_2.

(I) $NO_2 + h\nu \rightarrow NO + O$; $J(NO_2) \approx 0.01 s^{-1}$ at $\chi_{sun} < 30^o$

(II) $O + O_2 + M \rightarrow O_3 + M$; $k_2 = 1.4\ (E\text{-}14) cm^3 s^{-1}$ at 1 atm

(III) $NO + O_3 \rightarrow NO_2 + O_2$; $k_3 = 1.8\ (E\text{-}14) cm^3 s^{-1}$ at 298 K

(IV) $RO_2 + NO \rightarrow RO + NO_2$; k_4 (see text)

However, in the presence of peroxy radicals (RO_2), NO is partially converted to NO_2 by reaction (IV) thus leading to a net production of O_3 through (I) and (II). The peroxy radicals are formed during the oxidation of CO and hydrocarbons by hydroxyl radicals (V, VI).

(V) $CO + OH + O_2 \rightarrow HO_2 + CO_2$

(VI) $R\text{-}H + OH + O_2 \rightarrow RO_2 + H_2O$

I. S. A. Isaksen (ed.), Tropospheric Ozone, 293–302.

Other potential sources for RO_2 are reactions of unsaturated hydrocarbons with O_3 or NO_3. A detailed discussion of the NO_x and hydrocarbon chemistry relevant to ozone is given by Cox in this volume.

The photostationary state between NO and NO_2 is established through (I) to (IV) and can be described by equation (1), where $J(NO_2)$ is the photolysis frequency for (I) and k_3 and $(k_4)_i$ are the rate coefficients for (III) and (IV), respectively.

$$(1)\quad [NO_2]\cdot J(NO_2) = [NO]\cdot[O_3]\cdot k_3 + [NO]\cdot\Sigma[RO_2]_i\cdot(k_4)_i$$

The second term on the right hand side of (1) directly represents the O_3 production, provided that the photolysis of NO_2, (I), is fast compared to other reactions of NO_2. This assumption is usually fullfilled during day-time. The notation $\Sigma[RO_2]_i\cdot(k_4)_i$ is used, because different RO_2 radicals may react with NO at different rates. Rate constants have been determined for the overall reaction of NO with HO_2 and a few organic peroxy radicals. They are all in the range of 5 to 9 (E-12)cm^3s^{-1}. It was, therefore, suggested to use an average rate constant k_4 of 7.6(E-12)cm^3s^{-1} for all organic peroxy radicals in photochemical calculations (Atkinson and Lloyd, 1984). In addition, it was found that formation of NO_2 is not the only possible path for the reaction of NO and RO_2. The larger organic peroxy radicals seem to form signif-icant amounts of organic nitrates (VII) (Atkinson, et al., 1987). Since this path does not lead to O_3 formation, the branching ratio between III and VII must be known.

$$(VII)\quad RO_2 + NO \rightarrow RONO_2$$

There are two ways to experimentally determine the in-situ production of O_3 from equation (1). They are implied by equations (2) and (3).

$$(2)\quad P(O_3) \equiv [NO_2]\cdot J(NO_2) - [NO]\cdot[O_3]\cdot k_3$$
$$(3)\quad P(O_3) \equiv [NO]\cdot\Sigma[RO_2]_i\cdot(k_4)_i \approx [NO]\cdot k_4\Sigma[RO_2]_i$$

Experiments have been performed to measure the O_3 production (Kelly, 1980) by accurately determining all the parameters in equation (2). In addition, the combination of field experiments with photochemical models was used (Parrish et al., 1986a; Trainer et al., 1987). The peroxy radicals were then calculated in the model from measurements of their precursors, that is, the different hydrocarbons and CO. These experiments place high demands on the accuracy of the respective measurements, since the O_3 production is usually only a small fraction of the total interconversion rate of NO_x. The second approach requires the knowledge of the concentrations of all hydrocarbons involved and their photochemical decomposition mechanisms. Both are only known in a crude manner.

We have seeked here a different approach by measuring the concentrations of RO_2 radicals directly together with that of NO. Simultaneously we have measured all the parameters in equation (2). The measurements, in principal, represent two independent determinations of the O_3 production rate. A comparison of the two rates then allows us to investigate experimentally the average reactivity of RO_2 radicals including the formation of nitrates.

2. EXPERIMENTS

The field experiments were performed during June and July 1986 on the Schauinsland , a mountain located about 10 km SSE of the city of Freiburg (48°N, 8°E; 1150 m ASL). They comprised simultaneous measurements of O_3, NO, NO_2, RO_2, hydrocarbons (C_2-C_{10}), H_2O_2, and J(NO_2). All the measurements were made at 10 m above ground to avoid a direct influence of the surface, such as emissions or deposition. The different techniques used are shown in Table 1.

Table 1: Measurements and Instrumentation

Species	Technique	Detect. Limit	Time Resol.	Manufact.
O_3	UV-Absorption	2 ppb	1 min	Dasibi
NO	Chemiluminescence	20 ppt	2 min	KFA
NO_2	Photolyt. Converter Chemiluminescence	40 ppt	2 min	KFA
NO_2	Chemiluminescence		1 min	Scintrex
NO_2 RO_2,HO_2	Matrix Isolation ESR-Spectroscopy	10 ppt	30 min	KFA
H_2O_2 org. Peroxides	Fluorescence	50 ppt	1 min	G.Kok
Hydrocarbons	Gaschromatography	10-50 ppt	120 min	KFA Siemens
J (NO_2)	Photodiode	-	1 min	KFA

The chemiluminescence monitor used for the measurements of NO (and NO_2 after conversion by photolysis; Kley et al., 1980) was of similar design as that described by Drummond et al., 1985, although with somewhat lower sensitivity. Calibration was performed with mixtures of NO in N_2 (Airco) which were in turn calibrated by gas-phase titration with O_3. NO_2 was additionally measured by Matrix Isolation/Electron Spin Resonance (MIESR, cf. Mihelcic et al., 1985). The same method was used for the determination of the peroxy radicals. It is the only available direct technique for RO_2 and has the advantage of an absolute, spectroscopic detection of the total amount of NO_2 and RO_2. Speciation of the different RO_2 radicals is only possible under certain conditions. It is, in principle, feasible to distinguish the HO_2 radical from other peroxy radicals because its ESR spectrum shows a different hyperfine structure. In practice, this distinction should be possible whenever the HO_2 concentration exceeds 30 % of the total RO_2.

The photolysis frequency of NO_2 was determined indirectly from measurements of the photon flux in the wavelength interval 300-400 nm. The instrument is similar to an Epply photometer but has a uniform response over 2π (Junkermann et al., 1987). The photometer was calibrated against a direct measurement of the NO_2 photolysis by using ESR-spectroscopy (Mihelcic et al., 1987).

3. RESULTS

In the following, results from two sunny days in July, 1986 are presented. The diurnal variations of the different trace gases obtained on the 2nd and 5th of July 1986 are shown in Figure 1. Both days were equally sunny with only light clouds, as is evident from the NO_2-photolysis frequency, which, on both days, reached maximum values around 0.009 s^{-1}. The meteorological situation was quite different. On July 2, south-easterly winds prevailed in Freiburg and on the Schauinsland during the night. Between 7:00 and 9:00 CET, the wind direction shifted from south-east to north-west, there remaining until 20:00 CET. On July 5, in Freiburg the same wind shift was observed as on July 2. On the Schauinsland, however, the wind direction was west to south-west throughout that day. There are no major nearby sources of pollution in this wind sector.

The different meteorological situations are reflected in the NO_x concentrations: On July 5, low and almost constant NO_x concentrations

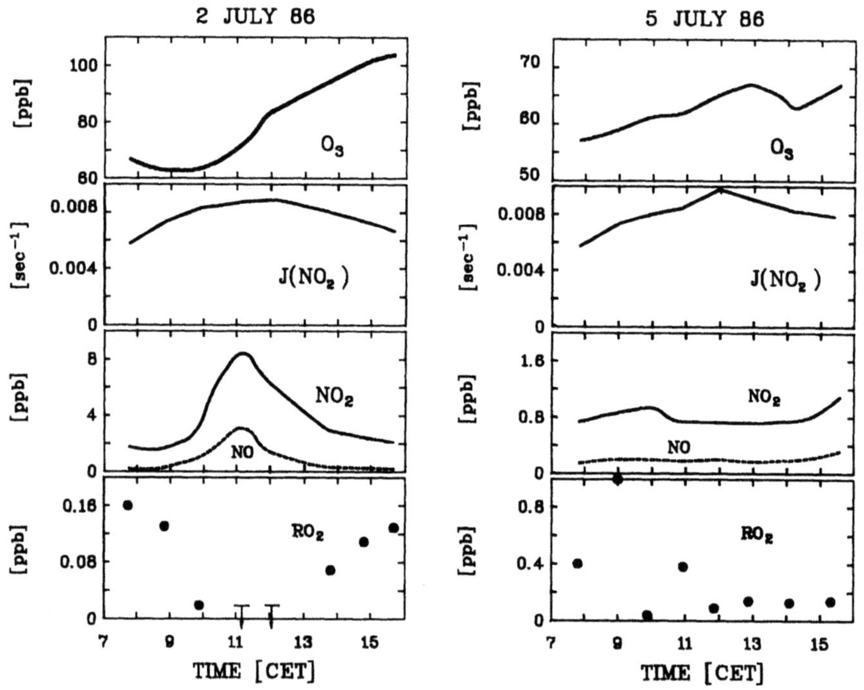

Figure 1. Diurnal variation of NO, NO_2, O_3, $J(NO_2)$, and RO_2 radicals on July 2 and 5. Location: Schauinsland, Black Forest, 48°N, 8°E, 1150 m ASL.

around 1 ppb were encountered throughout the day. About the same NO_x levels were found in the morning of July 2. However, around 10:00 CET, a break-through of polluted air from the area of Freiburg occured with NO_x levels increasing sharply to a maximum of 10 ppb around 11:00. During the afternoon, NO_x then decreased again to about the same level as had been observed in the morning.

Ozone also showed a different behaviour on the two days. While on July 5 only a weak increase from 60 to 65 ppb was observed during the day, the ozone concentration almost doubled on July 2 reaching values above 100 ppb in the afternoon. The fairly high ozone concentrations already present in the morning are typical for a mountain station, which resides above the inversion layer at night.

The concentrations of the RO_2 radicals also showed a different pattern on both days. On July 2, RO_2 and NO_x were anticorrelated. In the morning and afternoon, RO_2 concentrations around 150 ppt were found, while RO_2 was below our detection limit of 10-20 ppt during the high NO_x concentrations. This behaviour is in agreement with the fact that the reaction with NO is the major sink for RO_2.

On the afternoon of July 5 the RO_2 concentration was around 150 ppt as on July 2. Most surprising, however, were the extremely high levels found during the morning with a maximum value of 1 ppb around 9:00.

Another important result of the RO_2 measurements was the absence of a detectable structure in the spectra attributable to HO_2 radicals. As pointed out before, HO_2 should be detected in the total spectrum if its concentration amounts to 30 percent or more of the total RO_2.

4. DISCUSSION

From the data presented in the experimental section the in-situ production of O_3 was calculated according to equations (2) and (3). For NO_2, the MIESR-measurements were used, since that technique is free of interferences due to its spectroscopic nature. The MIESR-measurements represent averages over the sampling time of about 30 min. The other data were also averaged over the time-intervals during which the samples had been collected.

Figure 2 shows the calculated O_3 production rates in units of ppb/h for July 2 and 5. The open symbols give the results obtained from equation (2). The error bars represent 1 σ uncertainty limits They were calculated from the uncertainties of the individual measurements, and the uncertainty of k_3 (c.f. NASA, 1982). The latter dominates the total uncertainty of the O_3 production.

The full symbols give the O_3 production rates derived from the right hand side of equation (3). As stated above, the ESR spectra of most peroxy radicals lack a marked hyper-fine structure. The principal measurement obtained is the sum of all RO_2 radicals present in the sampled air. We have, therefore, adopted the average rate coefficient for process (IV) suggested by Atkinson (1984), $k_4 = 7.6\ (E\text{-}12) cm^3 s^{-1}$ and neglected the formation of nitrates (VII). The error bars were calculated from the uncertainties in the RO_2 and NO concentrations.

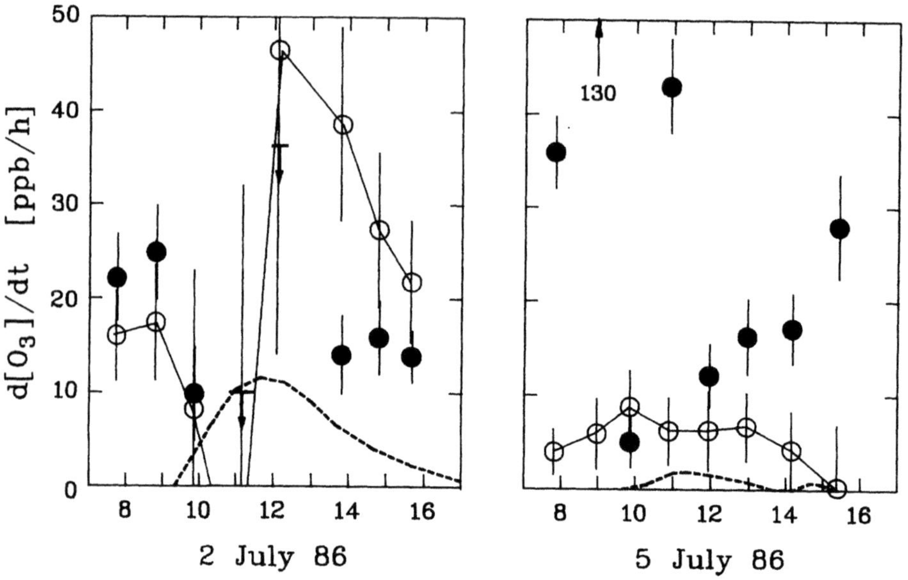

Figure 2. Observed increase of O_3 (dotted curve) and the O_3 production rate calculated from the measured RO_2 radicals (● symbols) and from the photostationary state of NO/NO_2 (○ symbols).

For the morning of July 2, the O_3 production rate calculated from equation (2) was 10-20 ppb/h. At 10:45 CET, that is when NO_x started to rise, it first decreased to -30 ± 60 ppb/h and immediately thereafter increased to 50 ± 40 ppb/h. During the afternoon, the O_3 production gradually decreased to 20 ppb/h.

The O_3 production calculated from equation (3), that is from the RO_2 measurements, showed a similar variation. As can be seen from Fig. 2, both production rates agree within the error bars, except for one time interval in the afternoon where the difference is alsmost 2 σ. The large uncertainties during the time, when NO_x was high, are due to the fact that the O_3 production in equation (2) is the small difference of two large terms. At the same time, only upper limits could be calculated from equation (3), because the RO_2 concentration was below the detection limit (arrows in Fig. 2).

The observed O_3 increase also showed a maximum around 12:00 CET. The absolute values, however, were about a factor of 5 lower than the calculated production rates. This difference cannot be explained by losses of O_3, namely its reaction with olefins, photolysis followed by formation of OH radicals, and dry deposition which together were estimated to have been less than 5 ppb/h for that particular day. Therefore, transport must be invoked in order to explain the difference between the calculated net production of O_3 and its observed increase.

The role of transport is indicated from the sudden onset of a decrease of NO_x after 11:15 CET (c.f. Fig. 1). A similar pattern was observed for the concentrations of 222-Rn which is continuously monitored by the Bundesamt für Zivilschutz on the Schauinsland (Weiss, priv. comm.). Radon also showed an increase at 10:00 CET and a gradual decrease in the course of the afternoon. This is consistent with transport of air from the Rhine valley to the Schauinsland followed by vertical or horizontal dilution.

On July 5, the situation was quite different. The O_3 production calculated from equation (2) was much lower than on July 2, around 6 ppb/h during most of the day. These fairly constant values went along with almost constant NO_x concentrations. The observed O_3 increase was again much smaller than the production, around 0-2 ppb/h. On that day, the difference of $\approx$ 5 ppb/h was consistent with the chemical losses of O_3.

Most striking was the large discrepancy between the O_3 production calculated from NO_x and $J(NO_2)$ and that from the RO_2 radical measurements. The latter yield much higher production rates, particularly in the morning with a value of 130 ppb/h at 9:00 (CET). It should be noted, that this represents an average over the sampling period of 30 minutes. It is clear that such a production rate should have been reflected in the observed O_3 concentration, which was not observed (c.f. Figure 1). The only possible conclusion is, therefore, that the assumption made for the calculation, namely that all RO_2 radicals have the same reactivity with respect to NO, was not valid under the conditions on July 5.

This is emphazised in Figure 3, where the measured RO_2 concentrations are compared to those calculated from equation (4), derived from equations (2) and (3).

$$(4) \qquad \Sigma[RO_2]_i = \frac{[NO_2]\cdot J(NO_2) - [NO]\cdot[O_3]\cdot k_3}{[NO]\cdot k_4}$$

On July 2, the calculated RO_2 agreed well with that measured by ESR spectroscopy, while on July 5, the measurements gave much larger RO_2 concentrations than calculated. This implies that a large fraction of the RO_2 radicals present on July 5 reacted with NO at a much lower rate than generally assumed based on the existing kinetic information. One could argue, that the discrepancy is due to the formation of nitrates (VII). However, this would constitute a significant loss of NO_x, especially for the extreme case on the morning of July 5 when the RO_2 and NO_x concentrations both were around 1 ppb. There, a fast reaction (VII) would have caused NO_x to disappear, while in reality the NO_x concentration remained unchanged.

We therefore conclude from these measurements, that under certain circumstances RO_2 radicals exist which react much more slowly with NO than is assumed in current chemical schemes of NO_x and hydrocarbon chemistry.

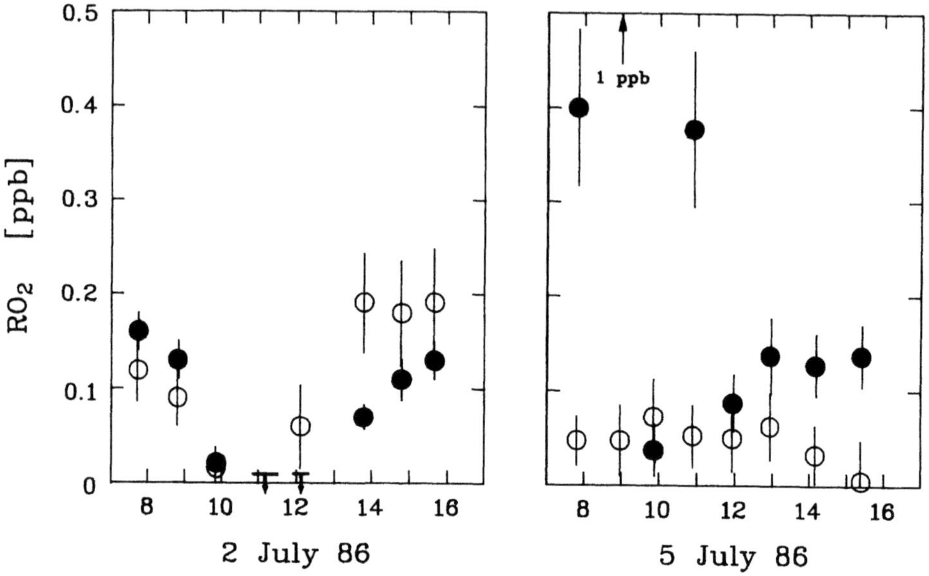

Figure 3. Comparison of the measured RO_2 concentration (● symbols) with that calculated from the photostationary state of NO/NO_2 (equation 4) (○ symbols)

Another important result of the RO_2 measurements was the absence of an HO_2 signal in the spectra. As pointed out above, the fraction of HO_2 was always below 30 % of the total RO_2, that is the ratio of HO_2 to organic peroxy radicals was always less than 0.5. This is a factor of 2 lower than the ratio of 1 predicted by photochemical models (Trainer et al., 1987).

For the high RO_2 concentrations on July 5 this could be do to the lower reactivity of the organic peroxy radicals with NO (see above). On July 2, however, all the RO_2 radicals reacted with NO at about the same rate as HO_2, since (i) the measured RO_2 concentrations were in good agreement to those calculated from equation (4), as is evident from Fig. 3 and (ii) the rate constant k_4 used here for the calculation of RO_2 is only 15 % lower than that for the reaction of HO_2 with NO (NASA, 1982).

There are two possible explainations for the lower HO_2 to RO_2 ratios found on July 2:

(i) HO_2 is produced less efficiently than assumed in the current decomposition schemes of hydrocarbons, and (ii) HO_2 is removed from the atmosphere more efficiently than RO_2. This would involve a process not included in the models. Additional losses of HO_2 were recently suggested by Mozurkewich et al., 1987 who demonstrated that HO_2 is removed efficiently by reaction with aerosols if certain metal ions such as Cu(II) are present. Clearly, our measurements present direct evidence for such a process.

A lower HO_2 to RO_2 ratio would also be in line with the results of Platt et al., 1987 and Perner et al., 1987 who always found their measured concentrations of hydroxyl radicals to be lower than those predicted by model simulations. Since OH and HO_2 are rapidly interconverted, additional losses of HO_2 not accounted for by the model could explain this discrepancy.

5. CONCLUSIONS

The in-situ formation of ozon was investigated from measurements of peroxy radicals and from measurements of the photostationary state between NO and NO_2. During one day, when relatively fresh pollution from the city of Freiburg was encountered, the two different determinations of $P(O_3)$ were in good agreement. This implies that the reactivity of the RO_2 radicals present in that air-mass was similar to the reactivity derived from laboratory studies of the decomposition of hydrocarbons in the presence of NO_x. However, the ratio of HO_2 to RO_2 was much lower than expected. Additional losses of HO_2 must be invoked in order to explain this discrepeancy.

During the other day, when air-masses with little input of fresh pollution were encountered, the measured concentration of RO_2 was found to be much higher than that calculated from the photostationary state of NO_x. Therefore, these RO_2 radicals must exhibit a substantially reduced reactivity towards NO than presently assumed.

A more detailed analysis of the data, including the measurements of hydrocarbons and peroxides, will be performed in the future with the help of numerical model simulations.

ACKNOWLEDGEMENT

The authors wish to thank R. Graul and the group of the UBA monitoring station on the Schauinsland for their hospitality and cooperation during the field campaign and W. Weiss of the Bundesamt für Zivilschutz for the information on Radon.

REFERENCES

Atkinson, R., and Lloyd, A.C., 1984, 'Evaluation of Kinetic and Mechanistic Data for Modeling of Photochemical Smog', *J. Phys. Chem. Ref. Data* **13**, 315-444

Atkinson, R., Aschmann, S.M., and Winer, A.M., 1987, 'Alkyl Nitrate Formation from the Reaction of a Series of Branched RO_2 Radicals with NO as a Function of Temperature and Pressure', *J. Atmos. Chem.* 5, 103-124

Cox, R.A., 'Atmospheric Chemnistry of NO_x and Hydrocarbons Influencing Tropospheric Ozone', *this volume*

Junkermann, W., Platt, U., and Volz, A., 1987, 'A new detector for the solar radiation flux leading to photolysis of ozone $J(O^1D)$ and related species', *to be published*

Kelly, T.J., Stedman, D.H., Ritter, J.A., and Harvey, R.B., 1980, 'Measurements of oxides of nitrogen and nitric acid in clean air', *J. Geophys. Res.* **85**, 7417-7425

Kley, D. and McFarland, M., 1980, 'Chemiluminescence detector for NO and NO_2', *Atmos. Tech.* **12**, 62-69

Mihelcic, D., Müsgen, P., and Ehhalt, D.H., 1985, 'An Improved Method of Measuring Tropospheric NO_2 and RO_2 by Matrix Isolation and Electron Spin Resonance', *J. Atmos. Chem.* **3**, 341-361

Mihelcic, D., Müsgen, P., and Volz, A., 1987, 'Absolute Calibration of a Photoelectric $J(NO_2)$ Sensor by ESR-Spectroscopy' (*to be published*)

Mozurkewich, M., Murry, P.H., Gupta, A., and Calvert, J.G., 1987, 'Mass Accommodation Coefficient for HO_2 Radicals on Aqueous Particles', *J. Geophys. Res.* **92**, 4163-4170

NASA, 1982, 'Chemical Kinetic and Photochemical Data for Use in Stratospheric Modeling', *J.P.L. Publication*, 82-57

Parrish, D.D., Trainer, M., Williams, E.J., Fahey, D.W., Hübler, G., Eubank, C.S., Liu, S.C., Murphy, P.C., Albritton, D.L., and Fehsenfeld, F.C., 1986, 'Measurements of the NO_x - O_3 photostationary state at Niwot Ridge, Colorado', *J. Geophys. Res.* **91**, 5361-5370

Perner, D., Platt, U., Trainer, M., Hübler, G., Drummond, J., Junkermann, W., Rudolph, J., Schubert, B., Volz, A., Ehhalt, D.H., Rumpel, K.J., and Helas, G., 1987, 'Measurements of Tropospheric OH Concentrations: A Comparison of Field Data with Model Predictions', *J. Atmos. Chem.* **5**, 185-216

Platt, U., Rateike, M., Junkermann, W., Hofzumahaus, A., and Ehhalt, D.H., 1987, 'Detection of Atmospheric OH Radicals', *Free Rad. Res. Comms.* **3**, 165-172

Trainer, M., Williams, E.J., Parrish, D.D., Buhr, M.P., Allwine, E.J., Westberg, H.H., Fehsenfeld, F.C., and Liu, S.C., 1987, 'Impact of Natural Hydrocarbons on Rural Ozone: Modeling and Observations', *submitted to Nature*

Trainer, M., Hsie, E.Y., McKeen, S.A., Tallamraju, R., Parrish, D.D., Fehsenfeld, F.S., and Liu, S.C., 1987, 'Impact of Natural Hydrocarbons on Hydroxyl and Peroxy Radicals at a Remote Site', *submitted to J. Geophys. Res.*

MODEL STUDIES OF BACKGROUND OZONE FORMATION

Shaw C. Liu
Aeronomy Laboratory/NOAA
325 Broadway
Boulder, CO 80303

ABSTRACT. Because of inhomogeneity in the NO_X distribution, the troposphere can be divided into several net production and net loss regions for O_3. The net production regions include the industrial area, the free troposphere, and the biomass burning area. The net loss regions are the oceanic boundary layer and clean continental boundary layer. In summer, the net production and the net loss are about the same magnitude as the O_3 flux from the stratosphere. In winter, the production of O_3 depends critically on nighttime reactions involving NO_3 and N_2O_5. Without these reactions, the O_3 production in winter can be as large as in summer.

Introduction

The formation of ozone in the background air depends strongly on the distributions of precursors of ozone, namely NO_X ($NO + NO_2$), CO, CH_4 and non-methane hydrocarbons (NMHC). Because CH_4 and CO are abundant throughout the troposphere NO_X usually becomes the rate-limiting precursor. In addition, NO_X plays a key role in the photochemistry of OH and peroxy radicals that are essential in the production of ozone (Crutzen, 1979). Therefore, determining the atmospheric distribution of NO_X is probably the most important task in evaluating the ozone formation.

Observations of NO_X show that the distribution of NO_X is highly inhomogeneous and has large temporal variation (Kley et al. 1981; Bollinger, 1982; Ridley et al. 1987). This is expected because of the extremely short chemical lifetime of NO_X and the widely scattered NO_X sources (Logan, 1983). In many parts of the world such as the boundary layer in the Pacific, the NO_X is so low that there is net photochemical sink for ozone (McFarland et al. 1979; Liu et al. 1983; Chameides, et al., 1987). On the other hand, in the industrialized countries the ozone budget is dominated by the photochemical formation. It is obvious that the inhomogeneity has to be accounted for in modelling the background ozone formation. Three-dimensional models such as the general circulation model (GCM) with adequate resolution and chemistry are

I. S. A. Isaksen (ed.), Tropospheric Ozone, 303–318.

needed for this purpose. This type of models are still under development. Alternatively, in the following sections we will try to evaluate the background ozone formation by dividing the troposphere into various production and sink regions according to observed NO_X distribution.

The definition of O_3 production and loss terms is not unique. It depends on how the so-called "odd oxygen" (O_X) is defined, especially in areas where oxidation of NMHC is important (Liu, 1977; Fishman et al. 1979; Logan et al. 1981; Levy et al. 1985).

In order to avoid confusion, in the following only the net O_3 production or loss (i.e., production minus loss) will be given for each of the regions. The loss includes photochemical loss and surface deposition. Thus, the net production or loss of each region represents the export or import for that region. These values will then be compared to stratospheric injection flux which has been estimated to be about 5 x 10^{10} $cm^{-2}s^{-1}$ (Danielson and Mohnon, 1977; Mahlman et al. 1980). The asymmetry is estimated to be about 50% in favor of the northern hemisphere (NH) (Mahlman et al. 1980). We will assume the stratospheric flux to be 7 x 10^{10} $cm^{-2}s^{-1}$ in the NH and 4 x $10^{10}cm^{-2}s^{-1}$ in the SH. In addition, for simplicity, the flux is assumed to be constant with season.

The seasonal variability of O_3 formation is an important problem. It has been used as an indicator of the origin of tropospheric O_3 (Fabian et al. 1968; Chatfield and Harrison, 1977; Liu et al. 1980; Logan, 1985). In addition, the observed spring O_3 maximum may be linked to the Arctic haze problem (Rahn and McCarffrey, 1979; Barrie et al. 1981; Isaksen et al. 1985; Penkett and Brice, 1986; Liu et al. 1987). Formation of O_3 in winter in the mid-latitude may also contribute to the spring maximum observed over many background or even clean stations (Liu, et al., 1987). However, nighttime reactions involving NO_X and NMHC may have a major impact on the chemistry of O_3 and other pollutants. It will be shown that uncertainty in the nighttime reactions can lead to a large error in the evaluation of O_3 production, especially for winter conditions.

Because of its strong influence on the distributions of O_3 and its precursors, atmospheric transport plays a major role in the O_3 formation. The photochemical lifetime of O_3 is relatively long, ranging from about a week in the boundary layer in the tropics to a few months for mid-latitude under winter conditions. (Liu, et al., 1987). Depending on the season, O_3 produced regionally may contribute significantly to zonal or even hemispheric O_3 distributions, making the distinction between polluted and clean regions difficult. The transport of NO_X and O_3 will be discussed by two separate papers in this workshop. We will limit our discussion on the transport processes that have direct impact on the background O_3 formation.

Current evaluations of O_3 budget rely almost entirely on model studies that need to either calculate or prescribe according to observation key species such as NO_X, NMHC, CO, H_2O_2, and HO_X ($OH+OH_2+H$). While reliable observational instruments are now available to test model results, the test for model calculated OH or HO_X is severely limited because of lack of sensitive instruments (Beck et al. 1987; Davis et al. 1987; Perner et al. 1987). Since the calculated O_3 production depends linearly on the concentrations of HO_X, it is possible that large errors may exist due to uncertainty in the HO_X abundance. On the other hand, the photochemical loss of O_3 which consists primarily of the reaction between $O(1_D)$ and H_2O can be calculated reliably. This provides a valuable constraint because the sources must balance the sink, at least on the global scale.

Ozone Formation in Summer

As discussed earlier, available measurements of NO_X indicate that the troposphere can be divided into several net O_3 production and net O_3 loss regions. This is particularly true in the planetary boundary layer (PBL) in summer where the lifetime of NO_X is less than a day (Liu et al. 1987). There is little doubt that most of the boundary layer in the developed countries is a major net O_3 production region (e.g., Hov, 1984; Fishman et al. 1986; Liu et al. 1987). This is probably also true for large areas of the developing countries. In these regions NO_X is mostly emitted from anthrogenic sources (Logan, 1983; Bollinger, 1982). Another important net O_3 production region is the free troposphere which is loosely defined here to be the region between 2 km and the tropopause. Based on an NO_X profile compiled by Kley et al. (1981), Liu et al. (1980) proposed the upper free troposphere to be a net O_3 production region. This was substantiated in a recent study by Chameides et al. (1987) based on simultaneous measurements of NO, CO, O_3, and H_2O in central and eastern Pacific in the fall of 1983. A third likely net O_3 production region is the tropical boundary layer during biomass burnings (Delany et al. 1985). The boundary layers of areas not mentioned above are probably net O_3 sinks. We will divide the O_3 sinks into clean oceanic region and clean continental region. In the following each of these regions will be discussed in more detail and an attempt will be made to derive the net O_3 production or loss for each region. The net production or loss will be converted to hemispheric average value to facilitate comparison with the stratospheric flux and with each other.

Elevated O_3 levels have been observed in summer over large rural areas in western Europe (Cox et al. 1975; Guicherit and Van Dop, 1977, Greenfelt and Schjoldager, 1984), in central and eastern U.S. (Research Triangle Institute, 1975; Vukovich et al. 1977) and in other industrialized countries. Observational and modelling studies show that high level rural O_3 is the product of long range transport of O_3 precursors and multi-day photochemical production and accumulation of O_3 (Guicherit and Van Dop, 1977; Vukovich et al. 1977; Isaksen et al. 1978; Fehsenfeld et al. 1983). Among the O_3 precursors, most of the

NO_X is emitted from anthropogenic sources (Logan, 1983). With regard to NMHC, the picture is less clear. Near the urban centers anthropogenic NMHC dominate (Lamb et al. 1987). In the rural areas covered by isoprene emitting deciduous forest natural NMHC contribute to most of the O_3 production (Trainer et al. 1987). On the other hand, with terpenes emitting coniferous forest, the impact of natural NMHC may be much smaller (Hov, et al., 1983). However, substantial uncertainties exist in the oxidation reaction schemes of natural NMHC, especially in the oxidation of terpenes.

The amount of O_3 produced in the rural areas of industrialized countries is large (Isaksen et al. 1978; White et al. 1983; Fishman et al. 1986; Liu et al. 1987). The size of the areas is difficult to estimate because of the transport of O_3 and its precursors. Fishman et al. (1986) gave an estimate of 5% of the NH. This estimate will be adopted here. The exact size is not critical to the outcomes of the following discussions. By extrapolation of observed O_3 in an urban area, White et al. (1983) obtain a net O_3 production from industrialized areas of about 1 to 2 x 10^{10} $cm^{-2}S^{-1}$. Fishman et al. (1986) estimated from a 1-dimensional model that the net O_3 production averages over the NH was 5 x $10^{10}cm^{-2}s^{-1}$. Using a Lagrangian airmass model approach Liu et al. (1987) obtained a similar value. However, the latter two estimates do not account for the surface deposition. If one assumes a deposition velocity of 0.5 $cm^{-2}s^{-1}$ (Aldaz, 1969; Galbally and Roy, 1980; Wesely et al. 1981; Lenschow et al. 1982) the surface destruction would reduce the net production by half. In addition, nighttime sink for NO_X via N_2O_5 or NO_3 (Ehhalt and Drummond, 1982; Platt et al. 1984; Noxon, 1983) could further reduce the net O_3 production of the two later estimates by reducing ambient NO_X concentrations (Liu et al. 1987). This will bring the latter two estimates down to about the same value as that of White et al. (1983). It should be noted that considerable uncertainty still exists in the nighttime reactions of NO_3 and N_2O_5. This will be discussed in more detail later.

In evaluating the O_3 production in the rural areas of industrialized countries it was realized that the relationship between O_3 and its precursors is highly nonlinear such that significant error may exist in simple models such as 1 and 2- dimensional models that do not account for the nonlinearity effect (Liu et al. 1987). Lesser degree of nonlinearity exists at urban levels of O_3 precursors (U.S. EPA, 1977; Hov and Derwent, 1981; Sakamaki et al. 1982). Since transport processes play a key role in determining the distributions of the precursors, it is essential that models with realistic transport processes are used to evaluate the results obtained by simpler models.

Measurements of NO and NO_2 in the free troposphere (Noxon, 1976; Schiff et al. 1979; Kley et al. 1981; Dickerson, 1984; Ridley et al. 1987; Chameides et al. 1987; Drummond et al. 1987; Torres et al. 1987) indicate that the NO_X distribution is rather uniform except for a few air masses clearly identified as the continental boundary layer air.

A general picture emerging from the measurements shows that the average level of NO in the clean background oceanic air in the NH is about 3 pptv below 2 km about 10 pptv in the lower free troposphere (2 to 4 km), about 15 pptv in the middle free troposphere (4 to 8 km), and about 20 pptv in the upper free troposphere (above 8 km). Drummond, et al (1987) found smaller NO values in the SH than the NH during their observation mission in June, 1984. However, this asymmetry could be due to seasonal variation in the NO concentration. Chameides et al. (1987) showed that at the levels of NO in the NH free troposphere cited above and the concentrations of CO, O_3, and H_2O observed simultaneously over central and eastern Pacific there was significant net O_3 production. The diurnally average value of the column net production between 2 and 10 km was estimated to be about 5 x 10^{10} $cm^{-2}s^{-1}$ during the fall, 1983 field operation. Chameides et al. also estimated a net O_3 loss of about 3 x $10^{10}cm^{-2-1}$ below 2 km. The values are in general consistent with but on the lower end of the estimates of an earlier model study by Liu et al. (1980).

The sources of upper tropospheric NO_x is an interesting problem. Stratospheric NO_x intrusion, high-flying aircraft emission, and lightning were first proposed to be the source (Liu et al. 1980; Liu et al. 1983; Ko et al. 1986). On the other hand, cloud transport of near surface anthropogenic NO_x to the upper troposphere was also thought to be a likely source (Gidel, 1983; Chatfield and Crutzen, 1984). Alternatively, peroxyacetyl nitrate (PAN) formation from anthropogenic NO_x and NMHC may be transported over long distance and to upper troposphere (Crutzen, 1979). Chameides et al. (1987) found there was no correlation between NO and CO observed simultaneously, suggesting a non-anthropogenic source for NO_x. However, their measurements were carried out in the fall when activities of cumulus clouds were infrequent. More measurements, especially during summer season, are needed to identify the sources of free tropospheric NO_x.

As mentioned in the introduction, a considerable error potentially exists in model calculated O_3 production due to the uncertainty in the HO_x photochemistry. There are some observational evidences suggesting that calculated OH concentrations are higher than values observed or derived from observations (Perner et al. 1987; Roberts et al. 1985; Parrish et al. 1986). Recent measurements of the sticking coefficient of HO_2 with aqueous particles and water droplets show values greater than 0.2 (Mozurkewich et al. 1987; Worsnop et al. 1987). If these values are applicable to tropospheric aerosols the concentrations of OH and HO_2 could be reduced by a factor of 2 or more depending on the ambient conditions. A reduction of OH and HO_2 of this magnitude would change the net O_3 production substantially, especially in the free troposphere where the photochemical production is comparable to photochemical loss. Of course, at present such a generalization of the aerosol removal process of HO_2 is not warranted because the sticking coefficient should depend on the chemical and physical characteristics of aerosols. For this reason the heterogeneous removal of HO_2 is not included in this study.

There is little information on the seasonal variation of NO_x in the free troposphere. If the NO_x distribution is assumed to be constant with season, model calculations show that the net O_3 production in the free troposphere is essentially constant with season while the net O_3 loss below 2 km in summer is about a factor of 2 higher than spring and fall (Liu and Trainer, 1987). Thus in summer the clean oceanic atmosphere below 2 km is a significant sink for O_3 with a net column loss rate of about 6×10^{10} $cm^{-2}s^{-1}$. On top of this sink we can add a loss of O_3 due to dry deposition of about 2.5×10^{10} $cm^{-2}s^{-1}$ that is estimated from an average O_3 level of 20 ppbv and a deposition velocity of 0.05 cm s^{-1} (Lenschow et al. 1982). In the NH, about half of the surface area is covered by oceans. Therefore the oceanic lower atmospheric O_3 sink averaged over the NH is about 4.3×10^{10} cm^{-2} s^{-1}.

Another O_3 sink region is the PBL in the clean continental region. Observations made at three rural stations in Canada indicate that the daytime surface O_3 mixing ratios in summer range from 25 to 35 ppbv (Logan, 1985). In the tropics, the surface O_3 values observed at Panama (Chatfield and Harrison, 1977), Natal and near Manaus (Kirchhoff, 1987; Logan, 1985) are about 15 ppbv in the days when there is no biomass burning. Ozonesonde measurements over the stations in Canada and the tropics show there is a persistent and significant increase of O_3 mixing ratio from the PBL to the free troposphere (Logan, 1985; Kirchhoff, 1987), suggesting downward flow of O_3 from the free troposphere or a net O_3 sink in the PBL. Assuming the average mixing ratio of O_3 in the region is 15 ppbv, the O_3 sink due to surface deposition alone would be as much as 2×10^{11} cm^{-2} s^{-1}. However, model calculations show that even at the relative small rate of biogenic NO_x emission (Johansson, 1984; Slemr and Seiler, 1984; Williams et al. 1987) there is significant photochemical O_3 production that balances out most of the photochemical destruction and surface deposition of O_3 (Jacob and Wofsy, 1987; Trainer et al. 1987). Thus the region is probably only a weak net sink for O_3. We assume the sink strength is 12×10^{10} cm^{-2} s^{-1} for the NH in order to balance the net O_3 production in the free troposphere (5×10^{10} cm^{-2} s^{-1}) and the stratospheric flux (7×10^{10} cm^{-2} s^{-1}). Assuming 45% of the NH is clean continental region, a NH average sink of about 5.5×10^{10} cm^{-2} s^{-1} could be obtained.

Because biomass burnings have been estimated to be a significant source of NO_x, CO, and hydrocarbons, there may be substantial O_3 production associated with the burnings (Crutzen et al. 1979; Delany et al. 1985). Most of the burnings occur in the tropics and subtropics during dry season. An estimate of the amount of O_3 exported from the South America Cerrado boundary layer in the 3-month dry season amounts to about 2×10^9 cm^{-2} s^{-1} averaged over the SH (Delany et al. 1985). There has been no estimate of the O_3 production from biomass burnings on a global scale. If one takes the NO_x emission of 12×10^{12} gm(N) yr^{-1} from biomass burnings as an indication (Seiler and Crutzen, 1980; Logan, 1983), the O_3 production could be comparable to that from fossil fuel combustion. The production of O_3 occurs mostly in winter which usually coincides with the dry season. We assume 10% of the production occurs in summer. In addition, we assume the production is equally divided between the two hemispheres. Thus the O_3 production in the summer from biomass burnings is only about 0.2×10^{10} cm^{-2} s^{-1} for each hemisphere.

Table 1 summarizes the net O_3 production or loss in summer for each of the regions discussed above. These values are compared to the stratospheric flux. The two hemispheres are listed separately because of the gross asymmetry n the industrial areas. Values for the SH are derived by scaling to those of the NH. For the scaling we have made the following assumptions. (1) The NO_x concentrations in the free troposphere in the SH are half the values in the NH; (2) the O_3 concentrations in the SH are two third of the NH values; (3) the land area in the SH is 20%, the ocean area 80%; and (4), the industrialized areas in the SH are 5% of the NH. Obviously some of the assumptions are somewhat arbitrary. As a result, the values of net O_3 production and loss for the SH are more uncertain than the NH estimates.

Given the large uncertainties in estimating the net production and loss, the overall balance of the O_3 budget is surprisingly good, especially for the SH. The balance of budget for the NH is compromised to some extent because a major part of the net loss (i.e., the clean continental PBL) is forced to balance the flux from above. Nevertheless, several conclusions can be drawn by examining Table 1. First, photochemistry plays a major role in the O_3 budget even if one considers only the net photochemical production or loss. Second, the export of O_3 from industrialized areas has a significant but not dominant impact on the other regions. In this context, we note that the impact should be evaluated by comparing the amount of O_3 exported with the absolute production and loss terms rather than the net values of the other regions. In practice this is not a problem because the absolute values are on the same order of magnitude of the net values. Third, the free troposphere is the key region that determines the degree of anthropogenic impact on O_3 in the NH. If the NO_x in the free troposphere is mostly from natural sources, then the natural sources of O_3 would be about six times the anthropogenic source except in the industrialized areas. On the other hand, if the NO_x is mostly from anthropogenic sources, the natural O_3 sources would be about equal to the antrohopogenic source. Last but not the least is that the export and import fluxes among the five regions are small relative to the fluxes among the regions due to advection and turbulence. In other words, atmospheric transport processes should play a major role in determining the O_3 distributions among the regions.

O_3 Formation In Winter

The photochemical lifetime of O_3 in mid-latitude in winter is on the order of a few months (Liu et al. 1987). Surface deposition of O_3 in winter is also greatly diminished (Galbally and Roy, 1980; Wesely,

1983; Colbeck and Harrison, 1985). Furthermore, the lifetimes of O_3 precursors will also be considerably greater than the summer values. As a result, the boundaries that separate various regions discussed above become fuzzier and in certain cases hard to justify. Nevertheless, the separation of various regions will still be useful, especially in the tropics and subtropics.

In the following discussions it will become obvious that considerably greater uncertainty exists in the O_3 budget for winter season than that of summer. A quantitative evaluation as in Table 1 is not warranted. Furthermore, we will not explicitly discuss the budgets of O_3 for spring and fall seasons because the budgets probably lie between those of summer and winter and there is little information on seasonal variations of O_3 precursors.

There is little observational evidence suggesting any significant O_3 production in the mid-latitude during winter season. This is not surprising because the low solar insolation significantly slows down photochemical activities. However, because the photochemical lifetime of O_3 in the mid-latitude in winter is essentially the whole season, a significant accumulation of O_3 may occur even with a slow O_3 production. Furthermore, slow photochemical activities allow the procursors of O_3 to build up to higher concentrations and to be transported over greater areas than those in summer that tends to compensate for the decreased photochemical activities (Crutzen and Gidel, 1983). This is substantiated by model calculations showing that for a constant emission rate of O_3 precursors, the O_3 production integrated over the lifetime of a procursor, say NO_x , is almost independent of season (Fishman et al. 1986; Liu et al 1987).

Since the NO_x emissions from fossil fuel combustion is essentially constant with season while the NO_x emissions from lightning and biogenic sources are at their minima in winter, the anthropogenic NO_x will dominate the winter NO_x sources. This domination is enhanced by the addition of NO_x from biomass burnings. As a result the O_3 in the NH in winter may be dominated by O_3 produced from anthropogenic sources as hypothesized by Liu et al. (1987).

If the hypothesis is correct, there would be several important implications. First, with the long lifetime of O_3 in winter anthropogenically produced O_3 would be transported over most of the NH. Second, O_3 would accumulate over the winter months and contribute significantly to the spring O_3 maximum observed over many clean stations which has always been considered to be due to stratospheric O_3 injection (Junge, 1963; Fabian and Pruchniewicz, 1977). The hypothesis is consistent with the recent findings by Penkett and Brice (1986). They used PAN as an indicator for photochemical activity. Based on the observed correlation between PAN and O_3, and the springtime PAN maximum in background air, they concluded that photochemistry might contribute significantly to the spring O_3 maximum.

The seasonal invariability of O_3 production derived in the above model calculations depends on the seasonal variation of the lifetime of NO_x that compensates for the seasonal change in photochemical activity. In these model calculations a potential non-photochemical sink for NO_x, i.e., formation of HNO_3 or NO_3^- at night through NO_3 or N_2O_5, has been neglected. There is evidence that N_2O_5 or NO_3 is scavanged at night under humid conditions, presumed with the formation of NO_3^- or HNO_3 (Ehhalt and Drummond, 1982; Platt et al. 1984). If NO_3 at night is assumed to be totally removed, the O_3 production in winter would decrease by a factor of about 3. The factor would be doubled if N_2O_5 instead of NO_3 is removed (Liu et al. 1987). In the latter case the O_3 production from fossil fuel combustion in winter would be reduced to about one quarter of the stratospheric flux.

The free troposphere would remain a significant net O_3 production region in winter if the NO_x concentrations do not change with season. There is little data on the seasonal variability of NO_x in the free troposphere. A 1-dimensional calculation shows that formation of PAN in winter may diminish upper tropospheric NO_x concentration substantially (Kasting and Singh, 1986). If this is the case the O_3 production in the free troposphere maybe reduced by as much as a factor of 10.

A major O_3 source in winter may be from biomass burnings in low latitudes. If we assume 50% of the burnings occur in winter and evenly distributed over the two hemispheres, a hemispheric average net O_3 production of about 1×10^{10} cm^{-2} s^{-1} could be estimated.

The major O_3 sink region in winter is probably the PBL over clean tropical and subtropical altitudes. The sink strength would need to be about 10 to 20×10^{10} cm^{-2} s^{-1} to balance the O_3 sources in the NH. Assuming the area of the sink region covers 40% of the NH, the average sink strength over the area would be 25 to 50×10^{10} cm^{-2} s^{-1}. If we further assume that the height of the sink region is 2 km and the average O_3 mixing ratio is 20 ppb, a lifetime of 2.5 to 5 days for O_3 in this region could be derived. These values of O_3 lifetime are about a factor of 2 to 3 smaller than the photochemical lifetime calculated by models for the region (Liu et al. 1983; Liu and Trainer, 1987). Apparently, surface deposition is a major destruction mechanism for O_3 in the region.

Summary

Because CH_4 and CO are readily available throughout the troposphere, NO_x is usually the rate-limiting precursor for O_3. Observations of NO_x show that the distribution of NO_x is highly inhomogeneous and has large temporal variation, especially in the lower troposphere. Consequently, the troposphere should consist of various net O_3 production and net loss regions.

Based on observed NO_x distribution, we have divided the troposphere into five regions as shown in Table 1. The region with the largest volume and probably the most importance is the free troposphere. Model calculations based on observed NO_x and other key species suggest that the free troposphere is an important net production region for O_3. However, the sources of free tropospheric NO_x have not been definitely determined. Possible sources include lightning, stratospheric intrusion, aircraft emission, and near surface combustion emission that is transported upward either via PAN and/or by convective clouds. It is essential to determine the NO_x sources to evaluate the anthropogenic impact on the background O_3. Industrialized areas are another important net O_3 production region. Current estimates of the net production come from relatively simple models such as 1-dimensional model. However, because that the O_3 production is highly nonlinear relative to its precursors, models with realistic transport processes are needed. The clean continental boundary layer and the oceanic boundary layer are the two major net sink regions for O_3. Both photochemical destruction and surface deposition contribute significantly to the sink. Model calculations show that the relatively small biogenic NO_x emission in the clean continental region may contribute to substantial O_3 production that balances most of the photochemical and surface destructions of O_3. Areas under biomass burnings in the dry season of tropics and subtropics are also a large net O_3 production region. There are considerable uncertainties in the extent of burning, the temporal variation, and the NO_x and NMHC emitted.

In the summer, photochemistry plays a major role in the O_3 budget as evident by the fact that the sum of the net production is on the same order of the stratospheric O_3 flux. On the other hand, the export and import fluxs among the five regions are small compared to the exchange fluxes due to advection and turbulence at the interfaces of the regions, suggesting that transport processes play a major role in determining the O_3 distribution. Although it is firmly established that the O_3 production from anthropogenic sources dominates the O_3 budget in the industrialized region, it is not clear how important the export of O_3 from this region to the budgets in other regions.

There is very little information on the seasonal variability of NO_x distribution and thus the O_3 formation. Model calculations that neglect nighttime sink of NO_x show that O_3 production from a constant NO_x source is as large in winter as in summer because of the compensating seasonal variations in NO_x lifetime and photochemical activity. If this is true then in winter the anthropogenically produced O_3 would be a dominating source and would be transported over most of the NH because of the long lifetime of O_3. Furthermore, this O_3 source might contribute significantly to the spring O_3 maximum observed over clean stations.

A major source of uncertainty in the estimated O_3 production is the nighttime NO_x sink implied by the observed nighttime concentration of NO_3. If most of NO_3 or N_2O_5 is removed at night, the estimated O_3

production would be reduced by as much as factors of 2 and 6 for summer and winter seasons, respectively. Therefore, it is imperative to understand the mechanism and product of the nighttime sinks of NO_3 or N_2O_5.

Heteorogeneous removal of HO_2 and other radicals is another potential source of uncertainty for model calculated O_3 production due to its impact on the abundances of OH and HO_2. Measurements of the sticking coefficient of HO_2 with aerosols with various chemical and physical properties are needed. Furthermore, model calculated OH concentrations under various ambient conditions need to be tested against measurements by reliable instruments with adequate sensitivity.

From the above discussions, it is clear that modeling the tropospheric O_3 formation and distribution is an extremely complex problem that involves photochemical and transport processes of various temporal and spatial scales. Although relatively simple models such as 1 or 2-dimensional model will continue to contribute significantly to our understanding of the problem, 3-dimensional models from meso to global scales are needed to resolve some of the important issues raised in this study.

Acknowledgement

This research has been funded as part of the National Acid Precipitation Assessment Program by the National Oceanic and Atmospheric Administration.

Table 1. Net Ozone Production or Loss in Summer from Various Regions Averaged over each Hemisphere and Comparison with Stratosphere O_3 flux.

	Net Production or Loss ($cm^{-2}s^{-1}$)	
	NH	SH
Free Troposphere $\geq$ 2 km	5×10^{10}	2.5×10^{10}
Industrial Areas	$(1 \text{ to } 3) \times 10^{10}$	$(0.5 \text{ to } 1.5) \times 10^{9}$
Oceanic Boundary Layer	-4.5×10^{10}	-4.5×10^{10}
Clean Continental Boundary Layer	-5.5×10^{10}	-1×10^{10}
Biomass Burning Area	0.2×10^{10}	0.2×10^{10}
Stratospheric Flux	7×10^{10}	4×10^{10}

REFERENCES

Aldaz, L., 'Flux measurements of atmospheric ozone over land and water,' Geophys. Res., **74**, 6943-6946, 1969.

Barrie, L. A., R.M. Hoff, and S.M. Daggupaty, 'The influence of mid-lati pollution sources on haze in the Canadian Arctic,' Atmospheric Envir **15**, 1407-1420, 1981.

Beck, S. M., et al., 'Operational overview of NASA GTE/CITE 1 airborne instrument intercomparisons: Carbon monoxide, nitric oxide, and hydr instrumentation,' J. Geophy. Res., **92**, 1977-1985, 1987.

Bollinger, M. J., Chemiluminescent measurements of the oxides of nitroge in the clean troposphere and atmospheric chemistry implications, Ph.I Thesis, Department of Chemistry, University of Colorado, Boulder, Colorado, 1982.

Chameides, W. L., D.D. Davis, M.O. Rodgers, J. Bradshaw, S. Sandholm, G. Sachse, G. Hill, G. Gregory, and R. Rasmussen, 'Net ozone photochemi production over the eastern and central north Pacific as inferred fr GTE/CITE1 observations during Fall 1983,' J. Geophys. Res., **92**, 2131-2152, 1987.

Chatfield, R. B., and P.J. Crutzen, 'Sulfur dioxide in remote oceanic ai Cloud transport of reactive precursors,' J. Geophys. Res., **89**, 7111-7132, 1984.

Chatfield, R., and H. Harrison, 'Tropospheric ozone II. Variations alon meridional band,' J. Geophys. Res., **82**, 5969-5976, 1977.

Colbeck, I., and R.M. Harrison, 'Dry deposition of ozone: some measureme of deposition velocity and of vertical profiles to 100 meters,' Atmo Environ., **19**, 1807-1818, 1985.

Cox, R. A., A.E.J. Eggleton, R.G. Derwent, J.E. Lovelock, and D.E. Pack, 'Long-range transport of photochemical ozone in North-western Europe Nature, **255**, 118-121, 1975.

Crutzen, P. J., 'The role of NO and NO_2 in the chemistry of the tropospho and stratosphere,' Am. Rev. Earth Planet. Sci., **7**, 443-472, 1979.

Crutzen, P. J. and L. T. Gidel, 'A two-dimensional photochemical model o the atmosphere, 2, The tropospheric budgets of the anthropogenic chloracarbons, CO, CH_4, CH_3Cl, and the effect of various NO_x sources tropospheric ozone,' J. Geophys. Res., **88**, 6641-6661, 1983.

Crutzen, P. J., L.E. Heidt, J.P. Krasnec, W.H. Pollock, and W. Seiler, 'Biomass burning as a source of atmospheric gases CO, H_2, N_2O, NO, CH_3Cl, and COS,' Nature, **282**, 253-256, 1979.

Danielsen, E. F., and V.A. Mohnen, 'Project dust storm report: ozone transport, in situ measurements, and meteorological analyses of tropopause folding,' J. Geophys. Res., **82**, 5867-5877, 1977.

Davis, L. I., Jr., J.V. James, C.C. Wang, C. Guo, P.T. MOrns, and J. Fishman, 'OH measurement near the Intertropical Convergence Zone in t Pacific,' J. Geophys. Res., **92**, 2020-2024, 1987.

Delany, A. C., P. Haagensen, S. Walters, A.F. Wartburg, and P.J. Crutzen, 'Photochemically produced ozone in the emission from large-scale tropical vegetation fires,' J. Geophys. Res., **90, 2425-2429, 1985.**

Dickerson, R. R., 'Measurements of reactive nitrogen compounds in the fre troposphere,' Atmos. Environ., **18**, 2585-2593, 1984.

Drummond, J. N. D.H. Ehhalt, and A. Volz, 'Measurements of nitric oxide between 0-12 km altitude and 60°N - 60°S latitude obtained during STRATOZ IV,' J. Geophys. Res., submitted, 1987.
Ehhalt, D. H., and J.W. Drummond, 'The tropospheric cycle of NO_x, in Chemistry of the Unpolluted and Polluted Troposphere,' Eds. H. W. Georgii and W. Jaeschke, published by Reidel, Dordrecht, Holland, 1982.
Fabian, P., W.F. Libby, and C.E. Palmer, 'Stratospheric residence time and interhemispheric mixing of strontium 90 from fallout in rain,' J. Geophys. Res., **73**, 3611-3616, 1968.
Fabian, P., and P.G. Pruchniewiez, 'Meridonal distribution of ozone in the troposphere and its seasonal variation,' J. Geophys. Res., **82**, 2063-2073, 1977.
Fishman, J., S. Solomon, and P.J. Crutzen, 'Observational and theoretical evidence in support of a significant in-situ photochemical source of tropospheric ozone,' Tellus, **31**, 432-446, 1979.
Fishman, J., F.M. Vukovich, E.V. Browell, 'The photochemistry of synoptic-scale ozone synthesis: implications for the global tropospheric ozone budget,' J. of Atm. Chemistry, **3**, 299-320, 1986.
Fehsenfeld, F. C., M.J. Bollinger, S.C. Liu, D.D. Parrish, M. McFarland, M. Trainer, D. Kley, P.C. Murphy, D.L. Albritton, and D.H. Lenschow, 'A study of ozone in the Colorado mountains,' J. Atmos. Chem., **1**, 87-105, 1983.
Galbally, I. E., and C.R. Roy, 'Destruction of ozone at the earth's surface,' Quart. J. Royal Met. Sco., **106**, 599-620, 1980.
Gidel, L. T., 'Cumulus cloud transport of transient tracers,' J. Geophys. Res., **88**, 6587-6599, 1983.
Greenfelt, P., and J. Schjoldager, 'Photochemical oxidants in the troposphere: A mounting menace,' Ambio, **13**, 61-67, 1984.
Guicherit, R., and H. VanDop, 'Photochemical production of ozone in Western Europe (1971-1975) and its relation to meterology,' Atmos. Environ., **11**, 145-156, 1977.
Hov, O., 'Ozone in the troposphere: high level pollution,' Ambio, **13**, 73-79, 1984.
Hov, O., and R. G. Derwent, 'Sensitivity studies of the effects of model formulation on the evaluation of control strategies for photochemical air pollution formation in the United Kingdom,' J. of Air Pollution Control Assoc., **31**, 1260-1267, 1981.
Hov, O., J. Schjoldager, and B.M. Wathne, 'Measurement and modeling of concentrations of terpenes in coniferous forest air,' J. Geophys. Res., **88**, 10679-10688, 1983.
Isaksen, I. S. A., O. Hov, and E. Hesstvedt, 'Ozone generation over rural areas,' Environ. Sci. & Tech., **12**, 1279-1284, 1978.
Isaksen, I. S. A., O. Hov, S.A. Penkett, and A. Semb, 'Model analysis of the measured concentration of organic gases in the Norwegian Arctic,' J. of Atmos. Chem., **3**, 13-27, 1985.
Jacob, D. J., and S.C. Wofsy, 'Photochemistry of Biogenic Emissions over the Amazon forest,' J. Geophys. Res., in press, 1987.

Johansson, C., 'Field measurements of emission of nitrogen oxide from fertilized and unfertilized forest soils in Sweden,' J. of Atmos. Chem 1, 429-435, 1984.
Junge, C. E., Air Chemistry and Radioactivity, Academic Press Inc., New Y and London, 1963.
Kasting, J. F., and H. B. Singh, 'Nonmethane hydrocarbons in the troposphere: Impact on the odd hydrogen and odd nitrogen chemistry,' J. Geophys. Res., **91**, 13239-13256, 1986.
Kirchhoff, V. W. J. H., 'Ground based ozone measurements in an equatorial rain forest,' J. Geophys. Res., submitted, 1987.
Kley, D., J.W. Drummond, M. McFarland, and S.C. Liu, 'Tropospheric profile of NO_x,' J. Geophys. Res., **86**, 3153-3161, 1981.
Ko, M. K. W., M.B. McElroy, D.K. Weisenstein, and N.D. Sze, 'Lightning: A possible source of stratospheric odd nitrogen,' J. Geophys. Res., **91**, 5395-5405, 1986.
Lamb, B., A. Guenther, D. Gay, and H. Westber, 'A national inventory of biogenic hydrocarbon emissions,' Atmospheric Environ., in press, 1987.
Lenschow, D. H., R. Pearson, Jr., and B.B. Stankor, 'Measurements of ozone vertical flux to ocean and forest,' J. Geophy. Res., **87**, 8833-8837, 1982.
Levy II, H., J.D. Mahlman, W.J. Moxim, and S.C. Liu, 'Tropospheric ozone: The role of transport,' J. Geophys. Res., **90**, 3753-3772, 1985.
Liu, S.C., 'Possible effects on tropospheric $O_3$and OH cue to NO emissions, Geophys. Res. Lett., **4**, 325-328, 1977.
Liu, S. C., D. Kley, M. McFarland, J.D. Mahlman, and H. Levy II, 'On the origin of tropospheric ozone,' J. Geophys. Res., **85**, 7546-7552, 1980.
Liu, S.C., M. McFarland, D. Kley, O. Zafirious, and B. Huebert, 'Tropospheric NO_x and O_3 budgets in the equatorial Pacific,' J. Geophy Res., **88**, 1360-1368, 1983.
Liu, S. C., and M. Trainer, 'Tropospheric ozone response to column ozone change,' J. of Atmospheric Chemistry, in press, 1987.
Liu, S. C., M. Trainer, F.C. Fehsenfeld, D.D. Parrish, E.J. Williams, D.W. Fahey, G. Hubler, and P.C. Murphy, 'Ozone dproduction in the rural troposphere and the implications for regional and global ozone distributions,' J. Geophys. Res., **92**, 4191-4207, 1987.
Logan, J. A., 'Nitrogen oxides in the troposphere: Global and regional budgets,' J. Geophys. Res., **88**, 10785-10807, 1983.
Logan, J. A., 'Tropospheric ozone: Seasonal behavior, trends and anthropogenic influence,' J. of Geophys. Res., **90**, 10463-10482, 1985.
Logan, J. A., M.J. Prather, S.C. Wofsy, and M.B. McElroy, 'Tropospheric chemistry: A global perspective,' J. Geophys. Res., **86**, 7210-7254, 19
Mahlman, J. D., H. Levy II, and W.J. Moxim, 'Three-dimensional tracer structure and behavior as simulated in two ozone precursor experiments J. Atmos. Sci., **37**, 655-685, 1980.
McFarland, M., D. Kley, J.W. Drummond, A.L. Schmeltekopf, and R.H. Winkler 'Nitric oxide measurements in the equatorial Pacific region,' Geophys. Res. Lett., **6**, 605-608, 1979.
Mozurkewich, M., P.M. McMurry, A. Gupta, and J.G. Calvert, 'Mass accommodation coefficient for HO_2 radicals on aqueous particles,' J. Geophys. Res., **92**, 4163-4170, 1987.

Noxon, J. F., 'Atmospheric nitrogen fixation by lightning,' Geophys. Res. Lett., **3**, 463-465, 1976.

Noxon, J. F., 'NO_3 in the Mid-Pacific Troposphere,' J. of Geophy. Res., **88**, 11017-11021, 1983.

Parrish, D. D., B. Huebert, R.B. Norton, M.J.Bollinger, S.C. Liu, P.C. Murphy, D.L. Albritton, and F.C. Fehsenfeld, 'Measurements of HNO_3 and NO_3^- particulates at a rural site in the Colorado mountains,' J Geophys. Res. 91, 5379-5393, 1986.

Penkett, S. A., and K.A. Brice, 'The spring maximum in photo-oxidants in the Northern Hemisphere troposphere,' Nature, **319**, 655-6568, 1986.

Perner, D., U. Platt, M. Trainer, G. Hubler, J. Drummond, W. Junkermann, J. Rudolph, B. Schubert, A. Volz, and D.H. Ehhalt, 'Measurements of tropospheric OH concentrations: A comparison of field data with model predictions,' J. Atmospheric Chemistry, in press, 1987.

Platt, U., A.M. Winer, H.W. Biermann, R. Atkinson, J.N. Pitts, Jr., 'Measurement of nitrate radical concentrations in continental air,' Environ. Sci. Technol., **18**, 365-369, 1984.

Rahn, K. A., and R.J. McCaffrey, 'Long-range transport of pollution aerosol to the Arctic: a problem without borders,' in Proceedings WMO Symposium on the Long-Range Transport of Pollutants, Sofia (WMO-538), 1979.

Research Triangle Institute, Investigation of rural oxidant levels as related to urban hydrocarbon control strategies, EPA-450/3-75-036, Environmental Protection Agency, Research Triangle Park, N.C., 359 pp, 1975.

Ridley, B. A., M.A. Carroll, and G.L. Gregory, 'Measurements of nitric oxide in the boundary layer and free troposphere over the Pacific ocean,' J. Geophys. Res., **92**, 2025-2048, 1987.

Roberts, J. M., F.C. Fehsenfeld, S.C. Liu, M.J. Bollinger, C. Hahn, D.L. Albritton, and R.E. Sievers, 'Measurements of aromatic hydrocarbon ratios and NO_x concentrations in the rural troposphere: observation of air mass photochemical aging and NO_x removal,' Atmos. Environ., **18**, 2421-2432, 1984.

Sakamaki, F., M. Okuda, H. Akimoto, and H. Yamazaki, 'Computer modeling study of photochemical ozone formation in the propane-nitrogen oxides-dry air system, generalized maximum ozone isopleth,' Environ. Science and Tech., **16**, 45-52, 1982.

Schiff, H. I. D. Pepper and B. Ridley, 'Tropospheric NO measurements up to 7 km,' J. of Geophys. Res., **84**, 7895-7897, 1979.

Seiler, W. and P. Crutzen, 'Estimates of the gross and net flux of carbon between the biosphere and the atmosphere from biomass burning,' Climate Change, **2**, 207-247, 1980.

Slemr, F., and W. Seiler, 'Field measurements of NO and NO_2 emissions from fertilized and unfertilized soils,' J. Atmospheric Chemistry, **2**, 1-24, 1984.

Torres, A. L., and H. Buchan, 'Tropospheric nitric oxide measurements over the Amazon basin,' J. Geophys. Res., in press, 1987.

Trainer, M., E.Y. Hsie, S.A. McKeen, R. Tallamraju, D.D. Parrish, F.C. Fehsenfeld, and S.C. Liu, 'Impact of natural hydrocarbons on hydroxyl and peroxy radicals at a remote site,' J. Geophys. Res., submitted, 1987.

Trainer, M., E.J. Williams, D.D. Parrish, M.P. Buhr, F.C. Fehsenfeld, S.C Liu, E.J. Allwine, and H.H. Westberg, 'Impact of natural hydrocarbons rural ozone: Modeling and observations,' Nature, sumbitted, 1987.
U.S. EPA, Uses, limitations and technical basis of procedures for quantifying relationships between photochemical oxidants and precurso US Environmental Protection Agency, Research Triangel Park, NC, 1977; EPA 450/2-77-021a.
Vukovich, F. M., W.D. Bach, Jr., B.W. Crissman and W.J. King, 'On the relationship between high ozone in the rural surface layer and high pressure systems,' Atmos. Environ., **11**. 967-984, 1977.
Wesely, M. L., 'Turbulent transport of ozone to surfaces common in the eastern half of the United States,' Trace Atmospheric Constituents, Edited by S.E. Schwartz, J. Wiley and Sons, 345-370, 1983.
Wesely, M. L., D.R. Cook, and R.M. Williams, 'Field measurement of small ozone fluxes to snow, wet bare soil, and lake water,' Boundary Layer Meteorology, **20**, 459-471, 1981.
White, W. H., D.E. Patterson, and W.E. Wilson, Jr., 'Urban exports to the nonurban troposphere: Results from Project MISTT,' J. Geophys. Res., 10745-10752, 1983.
Williams, E. J., D.D. Parrish, and F.C. Fehsenfeld, 'Determination of nitrogen oxide emissions from soils: Results from a grassland site in Colorado, United States,' J. Geophys. Res., **92**, 2173-2181, 1987.
Worsnop, D.R., M.S. Zahniser, C.E. Kolb, L.R. Sharfman, J.A. Gardner, and Davidovits, 'Determination of HO_2 sticking coefficient on water droplets,' EOS, **68**, 270-270, 1987.

GLOBAL TRANSPORT OF OZONE

H. Levy II
Geophysical Fluid Dynamics Laboratory/NOAA
P.O. Box 308
Princeton University
Princeton, NJ 08542

ABSTRACT: The three principal mechanisms for large scale atmospheric transport of tropospheric ozone [injection from the stratosphere, transport from regions of net production in the boundary layer and distribution of O_3 precursors resulting from either stratospheric injection or surface emissions] are examined in the light of current observations. While the actual O_3 climatology may be much more complex than it currently appears, the limited data suggests that ozone in the Southern Hemisphere and the northern tropics and subtropics is strongly influenced by transport from the stratosphere. At this time, the major questions are in the southern tropics and the northern mid-latitudes. The high levels of ozone observed over South America appear to be either the result of local chemical production or transport from higher latitudes. Both the latitude gradient and the seasonal cycle in the northern mid-latitudes suggest a significant, if not dominant, role for the transport of both O_3 and its precursors from source regions in the boundary layer, though transport from the upper troposphere also pays a role.

1. INTRODUCTION

Recent observations suggesting long term increases in CO and CH_4 (e.g. Rasmussen and Khalil, 1984; Rinsland and Levine, 1985) have raised questions about the long term levels of tropospheric ozone. Before speculating on trends, the chemical and meteorological processes that control the current level of O_3 must be understood. We have known for many years that tropospheric ozone is both transported from the stratosphere and produced photochemically in the polluted boundary layer. In 1973 a photochemical theory was proposed for O_3 in the general troposphere (Crutzen, 1974; Chameides and Walker, 1973). The relative importance on a global scale of net photochemical production and direct stratospheric injection are still in question, though recent estimates of global photochemical production and loss exceed the estimates of stratospheric injection (e.g. Liu, this volume). Recently, net photochemical production of ozone in the upper

I. S. A. Isaksen (ed.), Tropospheric Ozone, 319–325.

troposphere driven by NO_x injected from the stratosphere has been proposed (Liu et al., 1980), and large net production in the more remote boundary layer has also been suggested (Liu, this volume).

After a brief discussion of the basic mechanisms of large scale transport and chemical production, we will consider the existing observations. Two regions in particular, the tropics and the northern mid-latitudes raise questions that are not easily explained by either local chemical production or large scale transport processes. Some future studies will then be proposed.

2. BASIC PROCESSES

2.1 Continuity Equation

The time dependent behavior of the ozone mixing ratio, R, is described by the following continuity equation,

$$\partial R/\partial t = - \backslash V_3 \backslash \nabla_3 R + P - LR - DR \qquad (1)$$

where P is the rate of chemical production, L is the rate coefficient for chemical destruction and D is the rate coefficient for surface destruction. The first term represents the 3-dimensional advection of R by the winds, the next two comprise the chemical tendency and the last one is surface destruction. Note that this equation is written for the mixing ratio of ozone, not the number density. The motions of the fluid parcels do not depend on the ozone concentration and, assuming no production or destruction, conserve the global ozone mixing ratio. Transport, in an Eulerian sense, depends on the existence of a gradient either horizontal or vertical, in the ozone mixing ratio. The rates of chemical production and destruction, while a direct function of number density, are easily converted to mixing ratio.

2.2 Transport

The largest gradients in the mixing ratio of O_3 are found between the stratosphere and the troposphere where the values drop from 7-10 ppm at 10mb to .06-.01 ppm at the ground. A simple description of the downward transport of O_3 from the stratosphere is provided by a diabatic circulation with rising motion in the tropics and sinking motion at higher latitudes, the same mechanism that carries heat poleward and maintains the latitudinal structure of the tropopause. The downward branch reaches a maximum at mid-latitudes and is strongest in the winter and spring. A detailed description is given by Mahlman et al. (1984). Growing extratropical cyclones tap the O_3 accumulating in the lower stratosphere and transport it into the troposphere. This process has been described as a "tropopause fold event" (Danielsen, 1968). These events carry the O_3 downward and equatorward along an anti-cyclonic path and are most prevalent at mid-latitudes in the spring. As a result, the stratospheric injection of O_3, with a global average flux in the range 3-12 x 10^{10} molecules $cm^{-2}sec^{-1}$, shows a maximum at mid-latitudes in the spring.

The transport of O_3 from the boundary layer is much more difficult to estimate. While the mean flow is toward the east(westerly) in the mid-latitudes and toward the west(easterly) in the tropics, the actual day to day winds are highly variable. Given both significant destruction and weak winds at the surface, effective long-range transport requires that the boundary layer air be lifted into the "free troposphere." This can be accomplished by processes ranging from a single convective cloud to a large synoptic-scale storm. However, the development of high O_3 levels in the boundary layer is normally associated with stable conditions implying limited vertical transport. An alternative is the transport of precursors for O_3 formation, rather than O_3 itself. The resulting O_3 formation becomes a product of both transport and chemistry.

2.3 Chemistry

While the magnitude of the O_3 chemical tendency in the clean "free troposphere", which ranges from 0.1-0.2 ppb/day in the winter to ~5.0 ppb/day in the summer, depends on many chemical species and physical parameters, the sign of the tendency depends most strongly on NO_x (Levy et al., 1985). When many non-methane hydrocarbons are available, the chemistry becomes much more complex, but the level of O_3 still remains most sensitive to the level of NO_x (e.g. Liu et al., 1987).

Since the vertical lifting is generally accompanied by condensation, the large scale transport of the key precursor, NO_x, becomes quite complicated. One of the reaction products, HNO_3, is very soluble, while another, PAN, is not. Thus, the indirect transport of O_3 via its precursors from source regions in the boundary layer requires an understanding of the transport and chemistry of the reactive nitrogen species.

3. OBSERVATIONS/GLOBAL TRANSPORT

Before attempting to assess the contributions of global transport to the climatology of tropospheric O_3, we will briefly examine the available observations. While there have been a number of measurement programs, some with time series of 10 years or more, the distribution of vertical profiles is highly skewed toward the populated northern mid-latitudes with only 3-4 stations located in the Southern Hemisphere (S.H.).

3.1 Global Tropospheric Climatology

To briefly summarize the detailed analysis of this limited data set by Logan (1985): O_3 increases poleward from the tropics and levels off at mid-latitudes where it shows a summer maximum in the Northern Hemisphere (N.H.); averaged over the year, O_3 appears to be more abundant in the N.H. than in the S.H. though the data is very limited; O_3 generally increases with height, except in the severely polluted boundary layer; throughout the "free troposphere," excluding continental northern mid-latitudes, the seasonal maximum occurs in that hemisphere's spring and the seasonal minimum in the fall or winter; the seasonal cycles in the boundary layer show no simple pattern and depend on the local sources and sinks. This relatively simple picture of the O_3 climatology may be due in large part to the sparse and highly skewed data sets.

Currently, there are two major exceptions: the large values found in the lower and mid-troposphere near South America in the southern tropics during October (Logan and Kirchhoff, 1986); the extension of the spring maximum into the summer up through 500 mb for continental mid-latitudes in the N.H. (London, private communication; Logan, 1985).

3.2 Stratospheric Injection

Excluding the two anomalies, the global climatology of tropospheric O_3 is quite consistent with stratospheric injection followed by equatorward/downward transport in the troposphere. The observed latitude gradients, interhemispheric differences, vertical gradients and seasonal cycles are all qualitatively consistent with a recent simulation of this process by a medium resolution general circulation/ transport model, though model defects are believed to exaggerate the interhemispheric gradient and to produce excessive levels of O_3 at northern high latitudes (Levy et al., 1985).

A more global data base may find a much weaker interhemispheric gradient. While stationary eddies and planetary wave activity are much weaker in the S.H., transient eddies are both more frequent and stronger. Therefore, both hemispheres may do an equivalent job of tapping their respective lower stratospheres for O_3. Any hemispheric difference would then be the result of differences in that hemisphere's diabatic circulation.

3.3 Tropical "Anomalies"

Until recently, the few available measurements of tropical O_3 resulted in a simple picture (small seasonal variability, low values, slight increase with height) that was consistent with stratospheric injection and the strong vertical mixing expected in the tropics from convection. However, Kirchhoff et al., (1984) and Gregory et al., (1984) found very high values over South America during October and further analysis by Logan and Kirchhoff (1986) found a strong September - October maximum over Natal, Brazil (6°S, 35°W) all the way up through 500 mb. These observations raise two questions: What is the source of the high O_3? What is the spatial and temporal extent of the maximum?

Dry season burning, which has been associated with an increase in the boundary layer concentration of both O_3 and O_3 precursors (Crutzen et al., 1985), has been proposed as a source of the observed maximum (Logan and Kirchhoff, 1986). How this produces a maximum off both coasts of South America is not clear, but the observed increase over Natal does grow during the burning season (Logan and Kirchhoff, 1986). Furthermore, Marenco (this volume) reports high O_3 throughout the troposphere over Africa in association with burning and Fishman (this volume) infers high values over a region running from western South America to eastern Africa during periods of extensive biomass burning. However, while some individual profiles over Natal may show a maximum at 700 mb, the monthly means increase with height up to 500 mb, and the largest excursions from "normal" tropical O_3 values appear to be highly correlated with sharp drops in relative humidity. These last two observations point to the excess O_3 being transported from above, and recent simulations with a high resolution GCM indicate equatorward transport from 60°S by the strong transient eddies (Mahlman, private communication).

Whether the seasonal maxima result from O_3 transported down from the upper troposphere at high latitudes or from O_3 and O_3 precursors transported from regions of biomass burning at the surface, large scale transport plays a major role. If the former, the maxima should be widely distributed and not necessarily so sharply seasonally dependent. If the latter, the seasonal maxima should be highly correlated with biomass burning and other products of combustion should be present.

3.4 Mid-latitude Summer Maxima

While the analysis of Logan (1985) shows a spring maximum at 300 mb for the multi-year average of continental ozonesondes in the N.H., at 500 mb and below this becomes a broad spring-summer maximum for continental stations at mid-latitudes. Recent studies of early O_3 measurements (Kley, this volume; Logan, 1985; Bojkov, this volume) find that the 1880 measurements in Paris were significantly lower than present values, that the seasonal maximum at Arosa previously occurred in the spring, and that a number of mid-latitude stations show increases over the last 10-20 years. It would seem that there has been an increase in tropospheric O_3 at mid-latitudes during the 1900's and that this increase has been most apparent in the summer. The coincidence of a rapid increase in the emission of O_3 precursors during this period and the largest observed increase in O_3 occurring during the summer, a time of maximum photochemistry, is hard to ignore. One should note, however, that the seasonal pattern at a station frequently varies with height in a given year and from year to year at a given height (Levy, 1985). This suggests that transport of the O_3 precursors is important.

4. PROPOSED STUDIES

4.1 Global

There is a great need for sufficient O_3 data to create a realistic global climatology, particularly in the S.H. where the only O_3 profile measurements are at Samoa, Aspendale, Natal and Antarctica. While the satellite studies reported by Fishman in this volume may point out regions where there is O_3 variability, the technique is limited to the low latitudes and is not sufficiently quantitative. A program to measure O_3 profiles in the tropics and over both oceans of the Northern Hemisphere is needed.

4.2 Tropics

It is clear that this region is far more complex than was thought previously. It is particularly important to determine the spatial and temporal extent of the O_3 maxima and their relation to the locations and times of biomass burning. We not only need more measurements of O_3, but we need to include measurements of O_3 precursors such as CO and NO_y and chemical tracers of the air parcel's past history.

4.3 Northern Mid-latitudes

Extensive O_3 measurements are needed to determine whether the yearly increase and the summer maximum are local, regional or hemispheric phenomena. Both surface and vertical profile data over the Atlantic and the Pacific would help. Measurements of reactive precursors as well as chemical tracers of atmospheric transport should be included with the O_3 observations.

5. REFERENCES

Chameides, W. and J.C.G. Walker, A photochemical theory of tropospheric ozone, J. Geophys. Res., 78, 8751-8760, 1973.

Crutzen, P.J., Photochemical reaction initiated by and influencing ozone in unpolluted tropospheric air, Tellus, 26, 58-70, 1974.

_______, et al., Tropospheric chemical composition measurements in Brazil during the dry season, J. Atmos. Chem., 2, 233-256, 1985.

Danielsen, E.F., Stratospheric-tropospheric exchange based on radioactivity, ozone and potential vorticity, J. Atmos. Sci., 25, 502-518, 1968.

Gregory, G.L., S.M. Beck, and J.A. Williams, Measurements of free tropospheric ozone: An aircraft survey from 44N to 46S latitude, J. Geophys. Res., 89, 9642-9648, 1984.

Kirchhoff, V.W.J.H., E. Hilsenrath, A.G. Motta, Y. Sahai, and R.A. Medrano, Equatorial ozone characteristics as measured at Natal, J. Geophys. Res., 88, 6812-6818, 1983.

Levy, II, H., J.D. Mahlman, W.J. Moxim, and S.C. Liu, Tropospheric ozone: The role of transport, J. Geophys. Res., 90, 3753-3772, 1985.

_____, Tropospheric ozone: Transport or chemistry?, in Proceedings of the Quadrennial International Ozone Symposium, edited by C.S. Zerefos and A. Ghazi, pp. 730-734, D. Reidel Pub. Co., Dordrecht, Holland, 1985.

Liu, S.C., et al., Ozone production in the rural troposphere and the implications for regional and global ozone distributions, J. Geophys. Res., 92, 4191-4207., 1987.

_____, D. Kley, M. McFarland, J.D. Mahlman, and H. Levy II, On the origin of tropospheric ozone, J. Geophys. Res., 85, 7546-7552, 1980.

Logan, J.A., and V.W.J.H. Kirchhoff, Seasonal variations of tropospheric ozone at Natal, Brazil, J. Geophys. Res., 91, 7875-7881, 1986.

_____, Tropospheric ozone: Seasonal behavior, trends and anthropogenic influences, J. Geophys. Res., 90, 10463-10482, 1985.

Mahlman, J.D., D.G. Andrews, D.L. Hartmann, T. Matsuno, and R.G. Murgatroyd, Transport of trace constituents in the stratosphere, in Dynamics of the Middle Atmosphere, edited by J.R. Holton and T. Matsuno, pp. 387-416, Terra Scientific Publishing Co., 1984.

Rasmussen, R.A., and M.A.K. Khalil, Atmospheric methane in recent and ancient atmospheres: Concentrations, trends and interhemispheric gradients, J. Geophys. Res., 89, 11599-11605, 1984.

Rinsland, C.P., and J.S. Levine, Free tropospheric carbon monoxide concentrations in 1950 and 1951 deduced from infrared total column amount measurements, Nature, 318, 250-254, 1985.

THE OZONE PROBLEM IN RURAL AREAS OF THE UNITED STATES

Jennifer A. Logan
Department of Earth and Planetary Sciences
Harvard University
29 Oxford Street
Cambridge, MA 02138, U.S.A.

ABSTRACT. A comparison of ozone measurements from rural sites in the eastern and western United States indicates that daily maximum values are 15-25 ppb higher in the east than the west in late spring and summer, and that high values (>80 ppb) are common in the east but rare in the west. We present the results of an analysis of ozone measurements in the east, which shows that high values of ozone occur frequently during episodes of large spatial scale (1000- >1600 km). Ozone episodes occurred 9 and 7 times, in 1978 and 1979, between May and September. They persisted for 3-4 days on average, with a range of 2-8 days, and were most common in June. Analysis of the tracks of anti-cyclones for each year indicates that ozone episodes occurred preferentially in the presence of weak, slow moving and persistent high pressure systems.

1. INTRODUCTION

It has been recognized for some time that air pollution episodes occur each summer, both in the eastern United States and in Europe, during which areas of large spatial scale (> 1000 km) may experience ozone concentrations in excess of 80 ppb for several days (e.g., Cox et al., 1975; Vukovich et al., 1977; Wolff et al., 1977; Guicherit and van Dop, 1977; Wolff and Lioy, 1980). These episodes are generally associated with slow moving high pressure systems, when meteorological conditions are particularly conducive to photochemical formation of ozone, with warm temperatures, clear skies and low wind speeds which allow accumulation of secondary products. Elevated concentrations of ozone are thought to be responsible for most of the crop damage caused by air pollution in the United States (e.g., Heck et al., 1982), and may be contributing to the observed decline of forests in Europe and the eastern United States (Skarby and Sellden, 1984).

Most investigations of ozone episodes in the U.S. have focussed on individual case studies (e.g., Decker at al, 1976; Vukovich et al., 1977; Wolff et al., 1977; Wolff and Lioy, 1980; Vukovich et al., 1985). More recently, Vukovich and Fishman (1986) analysed the influence of the paths of anticyclones across the U.S. on mean daily maximum concentrations of ozone in July and August of 1977-1981. They

I. S. A. Isaksen (ed.), Tropospheric Ozone, 327–344.

found that the spatial pattern of ozone in the eastern U.S. in a given month was influenced strongly by the passage of high pressure systems. Previous studies have shown that monthly average concentrations of ozone are rather similar in rural areas throughout the U.S., but that maximum concentrations are higher in the east than in the west (Logan, 1985; Evans, 1985).

This paper gives a brief review of the seasonal and diurnal behavior of ozone in the U.S. and then presents preliminary results of an analysis of the climatology of ozone episodes in the north eastern quadrant of the U.S. in 1978 and 1979. The focus is on the frequency of ozone episodes, their duration, and their spatial scale, rather than on characteristics of individual episodes.

2. DATA AND ANALYSIS

2.1 Sources of data.

The data discussed here were obtained in two measurement programs, the Sulfate Regional Experiment (SURE) and its continuation as ERAQS (Eastern Regional Air Quality Study), which consisted of nine stations operated from August, 1977, to December, 1979 (Mueller and Watson, 1982; Mueller and Hidy, 1983), and the National Air Pollution Background Network (NAPBN), run by EPA in cooperation with the U.S. Forest Service, which consisted of another nine stations operated from the late 1970s to 1983 (Evans et al., 1983; Evans, 1985). These programs both employed chemiluminescence ozone analysers; they overlapped only in 1979. The SURE data are given with resolution of 1 ppb, and the EPA data with resolution of 5 ppb. We also used data obtained concurrently at Whiteface Mountain, N.Y. (Mohnen et al., 1977; Lefohn and Mohnen, 1986). All of the data employed in this analysis were obtained at well characterized rural sites, although some of the sites in the east were unavoidably subject to local pollution; site locations are shown in Figure 1.

2.2 Seasonal and diurnal behavior.

It is well known that surface concentrations of ozone in the U.S. are highest in spring and summer (see review in Logan, 1985). Figure 2 shows monthly mean and maximum concentrations of ozone at four SURE sites in the east and at three NAPBN sites in the west. The figure illustrates that average maximum concentrations in the east in summer, 60-85 ppb, are higher than those in the west by 15-25 ppb, while mean concentrations are quite similar in the east and the west. The SURE sites are located in the more populated and industrialized part of the east; maximum concentrations at four of the five eastern NAPBN sites, which are in more remote and coastal locations, are somewhat lower (Logan, 1985).

The similarity between average concentrations of ozone in the east and the west suggests that minimum concentrations of ozone tend to be

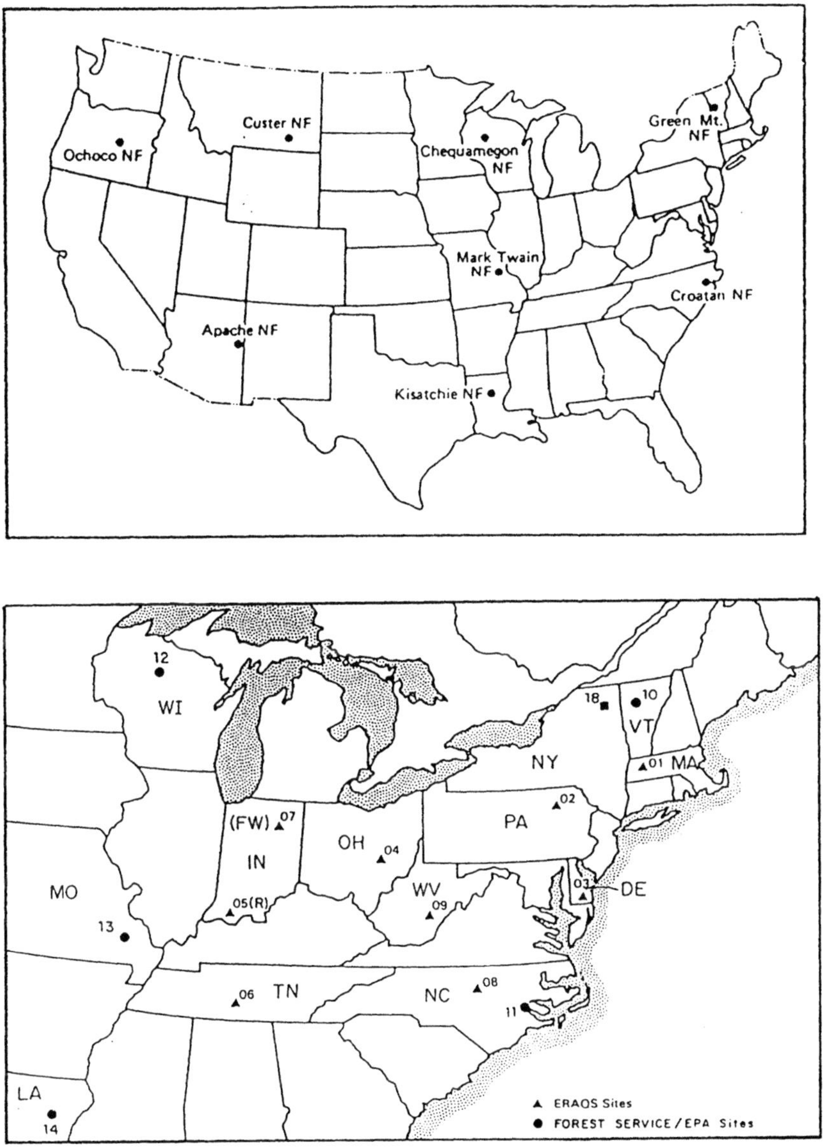

Figure 1. Locations of ozone measurement stations. The upper panel shows the sites of the National Air Pollution Background Network (Evans et al., 1983); the lower panel shows the SURE sites (▲; Mueller et al., 1982), the eastern NAPBN sites (●), and Whiteface Mountain, N.Y. (■).

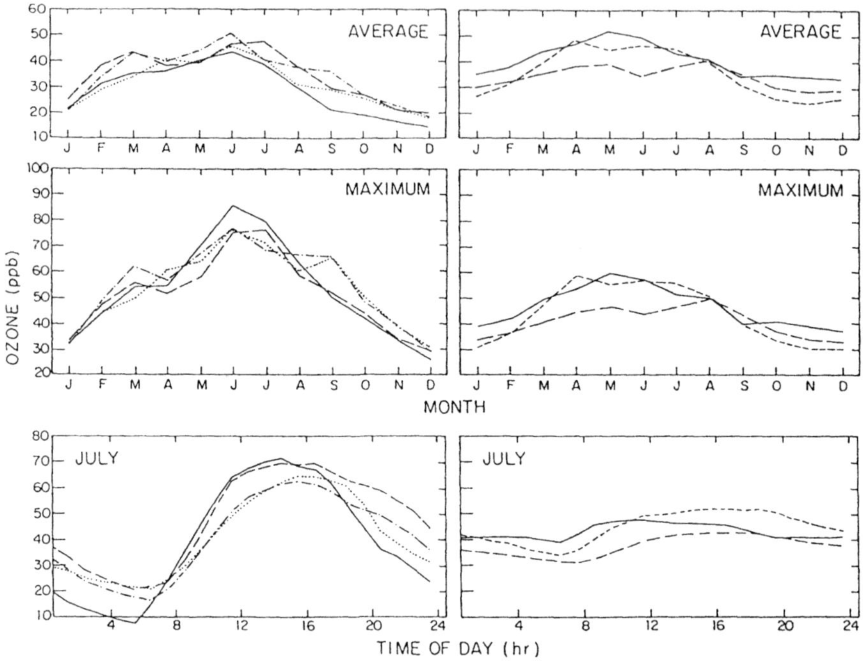

Figure 2. Seasonal and diurnal distributions of ozone at rural sites in the United States. The upper panels show the seasonal distribution of daily average values; the middle panels show monthly averages of the daily maximum values. The lower panels show the diurnal behavior of ozone in July. The left panels show results for sites 1, 2, 4 and 5 in the eastern U.S. (see Figure 1) from August 1, 1977 to December 31, 1979: Montague, MA (solid); Scranton, PA (dashed); Duncan Falls, OH (dot-dash); and Rockport, IN (dotted). The right panels show results for three sites in the western U.S. from four years of measurements: Custer, MT (1979-82, site at 1250 m, short dashes); Ochoco, OR (1980-83, 1350 m, long dashes); and Apache, AZ (1980-83, 2500 m, solid).

higher in the west, which is indeed the case. The lower panel of Figure 2 shows that the diurnal variation of ozone is much more pronounced at the eastern sites, with higher maxima, and lower minima. The lower minima may reflect nighttime titration of ozone by NO_x in the east, and more effective surface deposition to vegetation; the higher maxima are caused by photochemical production of ozone from anthropogenic sources of precursor emissions in the east. It is likely that the diurnal behavior at the western sites is dominated by the diurnal growth and decay of the mixed layer; we note that the three western sites are at altitudes of 1250-2500 m, but none are on isolated summits. The most probable time of occurrence of the daily maxima is between 1200-1400 at the western sites, and about three hours later, 1500-1700 at all of the eastern sites (see Figure 3), in accord with these concepts.

Another characteristic difference between ozone at the eastern and the western sites is illustrated in Figure 4 by cumulative frequency distributions for April 1 to September 30, for the hours of 1200-1800. Concentrations of ozone at the western sites almost never exceed 80 ppb, while values greater than 80 ppb occur more than 15% of the time at the eastern sites. Values above 80 ppb are most common in June and July in the east, as shown in Figure 5.

2.3 Analysis of ozone episodes.

Measurements made at the nine SURE sites and at Whiteface Mountain in 1978 and 1979 were used in an analysis of ozone episodes. Ozone on each day was characterized by the six hour mean value from 1000 to 1600. Time series of 6 hr ozone values from April 1 to September 30 are given in Figure 6 for three stations separated by up to 1300 km. The figure suggests that many of the high values of ozone occur concurrently at the three stations in events of a few days duration. For the purposes of this study, an ozone episode is defined as the simultaneous occurrence of the average value of ozone from 1000-1600 exceeding 70 ppb at four or more of the ten sites for two or more consecutive days; an episode-day is defined as the 6 hr average value of ozone exceeding 70 ppb at a single site. Inspection of Figure 6 indicates that the choice of 70 ppb should select periods when ozone was high. The upper quartile of the 6 hr averages exceeded 70 ppb at these ten sites during April 1 to September 30, while the mean was ~55 ppb. The fraction of episode-days which occurred during ozone episodes was 0.62 in 1978 and 0.49 in 1979; the fraction which occurred when the 6 hr average value of ozone was >70 ppb at three or more of the ten sites was 0.84 in 1978, 0.69 in 1979 (see Table 1). This suggests that the majority of episode-days occur over rather large areas.

The number of stations exceeding 70 ppb (6 hr average) on each day from May 1 to September 30 is given in Figure 7 for 1978 and 1979; April is not included because there was only one episode, on March 31 - April 1, 1978, during these two years. Inspection of Figure 7 indicates that there were 10 episodes in 1978 (including the one in early April) and 7 in 1979, lasting from 2 to 8 days, with a median duration of 3-4 days. Episodes were most common in May, June and July, and occurred with a

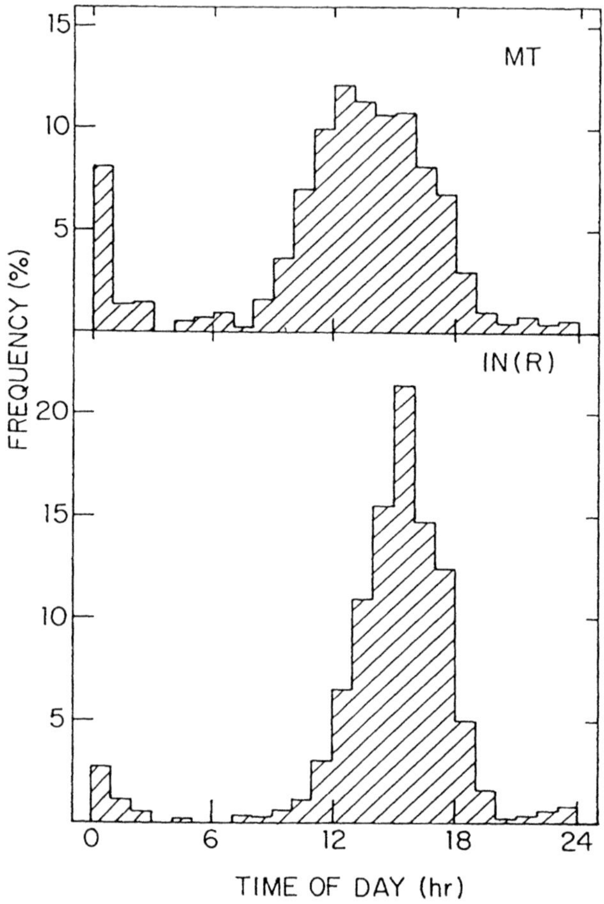

Figure 3. Frequency distribution of the time of occurrence of the daily maximum value of ozone. Results are shown for Rockport, IN, (lower) and Custer, MT, (upper) for April 1 to September 30. The distributions at these sites are typical for the eastern and western sites respectively.

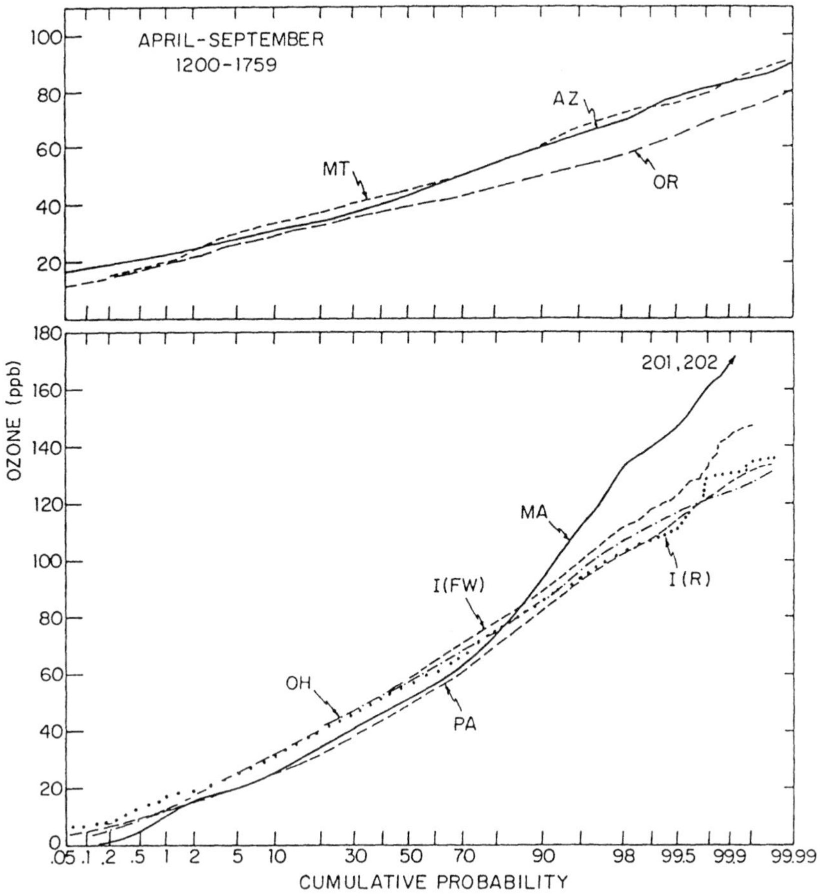

Figure 4. Cumulative probability distributions for ozone at sites in the western (upper) and eastern (lower) U.S. for April 1 to September 30, 1200-1800 h. Results are shown for the same sites as in Figure 2, with the addition of Fort Worth, IN, in the lower panel. Results are plotted on probability paper, such that normally distributed data define a straight line.

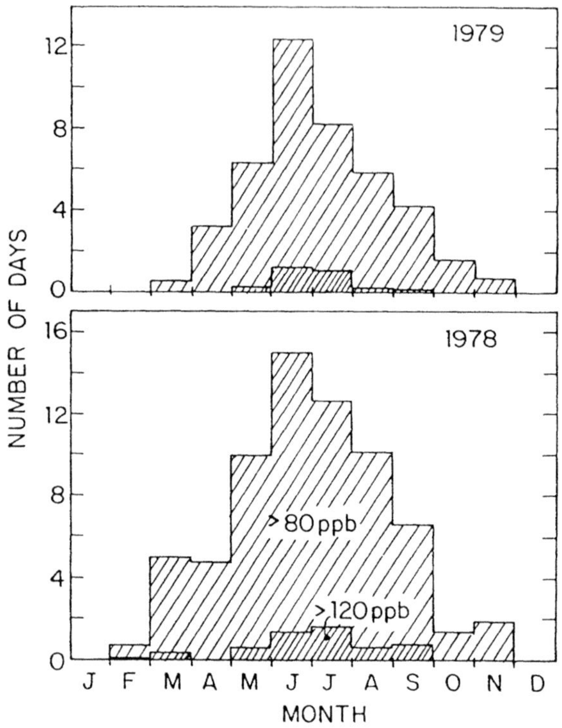

Figure 5. Number of days in each month when the daily maximum value of ozone exceeds 80 ppb and 120 ppb. Results are shown for an average over the nine SURE sites and Whiteface Mountain, N.Y., for 1978 and 1979.

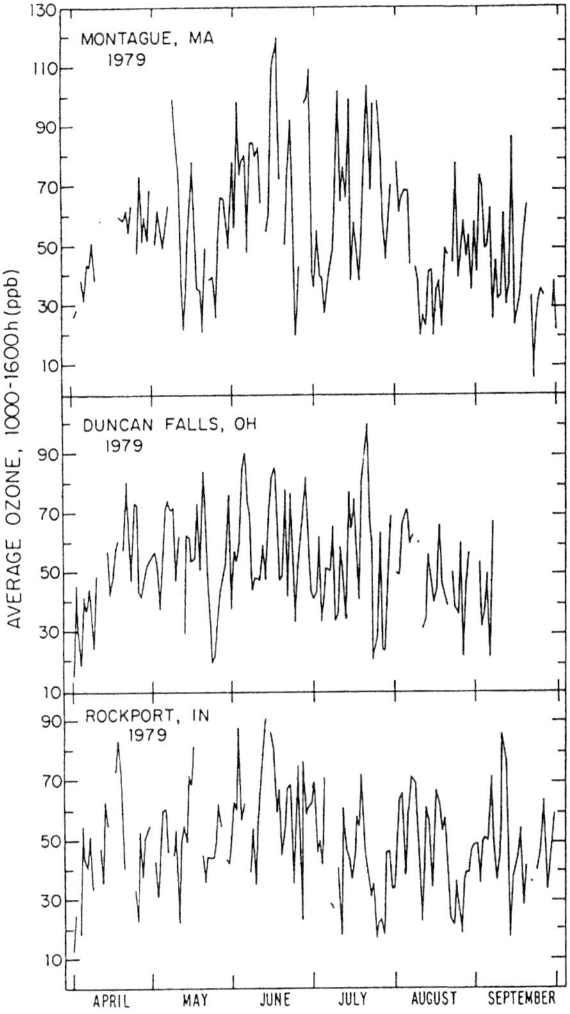

Figure 6. Time series of 6 hour mean values of ozone (1000-1600 h) for three sites in the eastern U.S. in 1979. Montague, MA, and Duncan Falls, OH, are separated by 810 km, Duncan Falls and Rockport, IN, by 470 km.

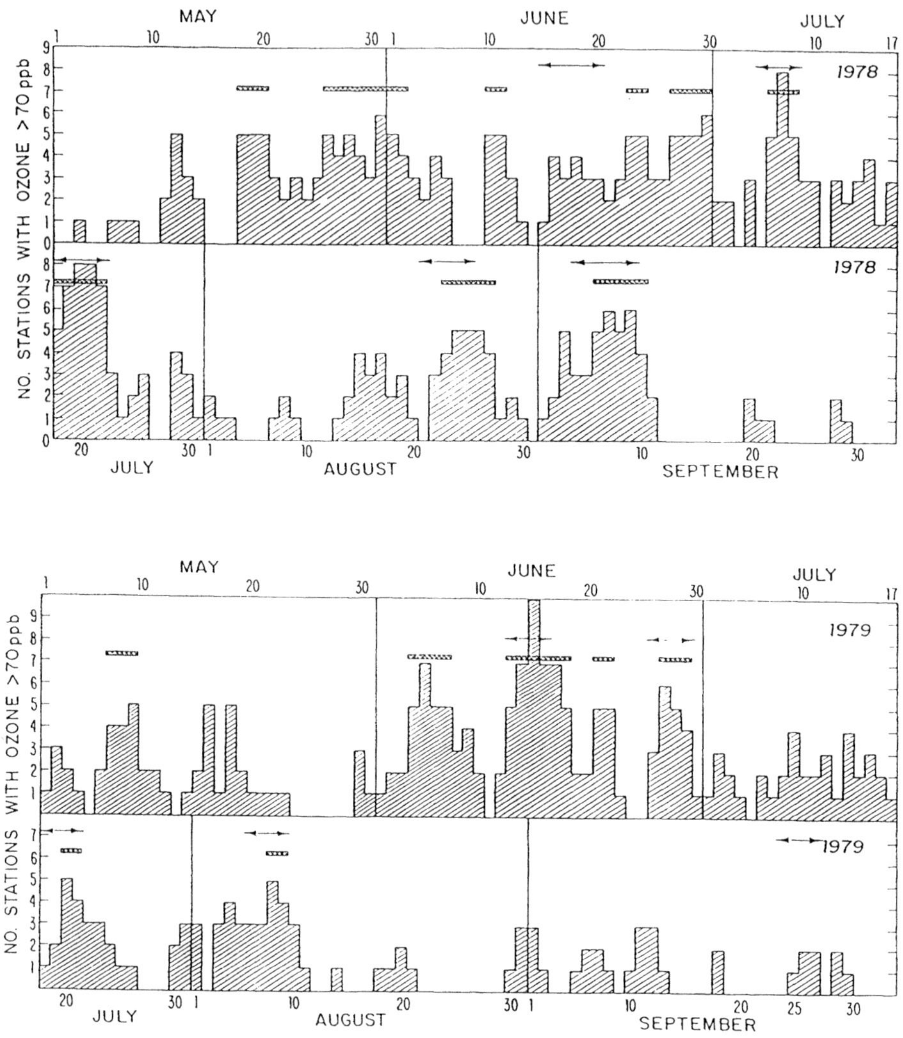

Figure 7. Time series of the number of stations with a 6 hour mean value of ozone (1000-1600) exceeding 70 ppb for 1978 and 1979. Results are show for the nine SURE sites and Whiteface Mountain, NY, (see Figure 1) for May 1 to September 30. The cross hatched bars show the times of ozone episodes, defined by the criterion described in the text. The double headed arrows show the times of air stagnation events in the region, defined according to Korshover's criterion (see text).

median separation of 7 days in these months. The criterion used here to define an episode may underestimate somewhat the actual number of episodes, since the fraction of stations reporting data was only 0.75 in 1978 and 0.86 in 1979; two stations were not operating for most of June, 1978.

TABLE 1

Ozone episodes in the eastern U.S. from May 1 to September 30

	1978	1979
No. of episode days	370	297
Fraction of episode-days which occur at ≥ 3 stations	.84	.69
Fraction of episode-days which occur at ≥ 4 stations	.62	.49
No. of ozone episodes	9	7
No. of days with an ozone episode	37	25

The selection of criteria for definition of an ozone episode is to some extent arbitrary, particularly in terms of the ozone concentration. We examined therefore the effect of an alternative definition, the simultaneous occurrence of the daily maximum value of ozone exceeding 80 ppb at four or more stations for two or more consecutive days. The 17 periods classified as episodes with the first definition (a 6 hr average value of ozone >70 ppb) were also classified as episodes with the second definition, although their starting and ending dates sometimes varied by a day; in addition, there were 5 more "episodes" in 1978, and 4 more in 1979. The extra "episodes" were shorter (2-4 days), and of much smaller spatial scale, 300-600 km in an east-west direction; the daily maximum values were also smaller. We are satisfied, therefore, that our original definition selects the appropriate periods as ozone episodes of large spatial scale.

Detailed analysis of the individual episodes is beyond the scope of the present paper and will be presented elsewhere. We give here a brief summary of results. Examination of daily maps of hourly maximum ozone concentrations indicates that the spatial scale of all the episodes exceeds 1000 km in an east-west direction; some of the episodes exceed 1600 km. Eight of the episodes occurred during air-stagnation events, according to the definition of Korshover (1976): areas where the surface geostrophic wind is less than 8 m sec^{-1} for at least four days, corresponding to a surface wind less than ~4 m sec^{-1} (see also Korshover and Angell, 1982). This is a very restrictive criterion for the type of meteorological conditions favorable for ozone production. We found that most of the ozone episodes shown in Figure 7 (12 of 17) occurred during the passage of anti-cyclones across the north-eastern U.S.. Maps showing the tracks of anti-cyclones are published for each month by the National Oceanic and Atmospheric Administration (NOAA, 1978;1979). Three episodes occured the day after an anticyclone dissipated over the region; for two of these episodes high pressure persisted in the region, and for one a warm front crossed the region. Inspection of daily weather maps showed that the remaining two episodes occurred when the eastern U.S. was under the influence of high pressure, even though anti-cyclones were not identified in the monthly summaries.

While the majority of ozone episodes were associated with tracks of anti-cyclones, the converse was not true. We identified a total of 34 anti-cyclones moving through the area of interest in 1978 and 1979, during May to August. All four of those that persisted for four or more days in the region were associated with ozone episodes; five of the ten that lasted for three days were associated with episodes, but only six of the twenty that lasted for two days. In some cases more than one anti-cyclone occured concurrently with an ozone episode. Considering anti-cyclones that lasted three or more days, the median pressure for those associated with ozone episodes was 1022 mbar, while it was 1028 mbar for those that were not. Evidently ozone episodes occurred preferentially in the presence of weak, slow moving and persistent high pressure systems.

The evolution of two major episodes, July 18-22, 1978, and June 13-18, 1979, is shown in Figure 8, which gives daily maximum values of ozone at each of the rural sites, the track of the anti-cyclone associated with the 1979 event, and the location of highs, lows, and fronts. Both of these episodes coincided with air-stagnation events, according to Korshover's criterion. In each case values exceeding 80 ppb occurred first in the southern and western halves of the region, and gradually spread to include the northeast states, New York, Massachusetts, and Vermont; values exceeding 120 ppb were found during each episode, with maximum values of 144 ppb and 161 ppb. Aircraft measurements made during one of the SURE intensive periods indicated that ozone values exceeded 100 ppb throughout much of the boundary layer on July 19-20, 1978 (Mueller and Hidy, 1983). For both the episodes shown, ozone maxima decreased again as a front moved from north west to south east across the region, bringing in cleaner air. (Note that the positions of the fronts are shown for for early morning, several hours before the daily maximum value of ozone occurs.) Ozone values were extremely low in the southeast in the middle of the 1979 event, as a warm front brought rain and clean air from the south Atlantic, with characteristically low values of ozone for this season (Logan, 1985).

3. DISCUSSION

Ozone concentrations exceeding 80 ppb are quite common in rural areas of the eastern U.S., but rare at remote western sites. The analysis presented here suggests that high values of ozone occur frequently during episodes of large spatial scale. There were 9 and 7 episodes in 1978

DAILY MAXIMUM CONCENTRATIONS OF OZONE (ppb), JULY 18-23, 1978

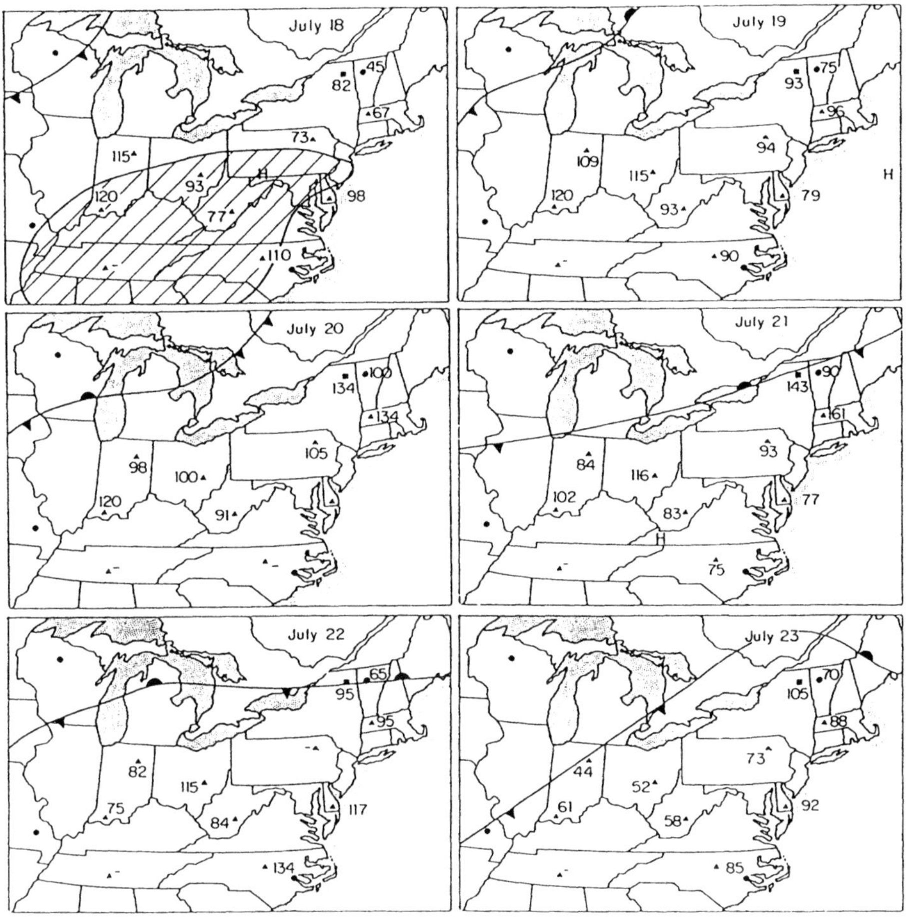

Figure 8a. Daily maximum values of ozone for July 18-23, 1978. The area of the corresponding 4 day air stagnation event is shown in the upper left hand panel, for July 18-21 (Angell, private communication). The locations of highs (H), lows (L), and fronts are taken from daily weather maps (National Weather Service, 1978), which give their positions at 7 am, some 8-10 hours earlier that the time of the daily maximum ozone values. Results are shown for the NAPBN sites when available, in addition to the SURE sites and Whiteface Mountain.

DAILY MAXIMUM CONCENTRATIONS OF OZONE (ppb), JUNE 13-18, 1979

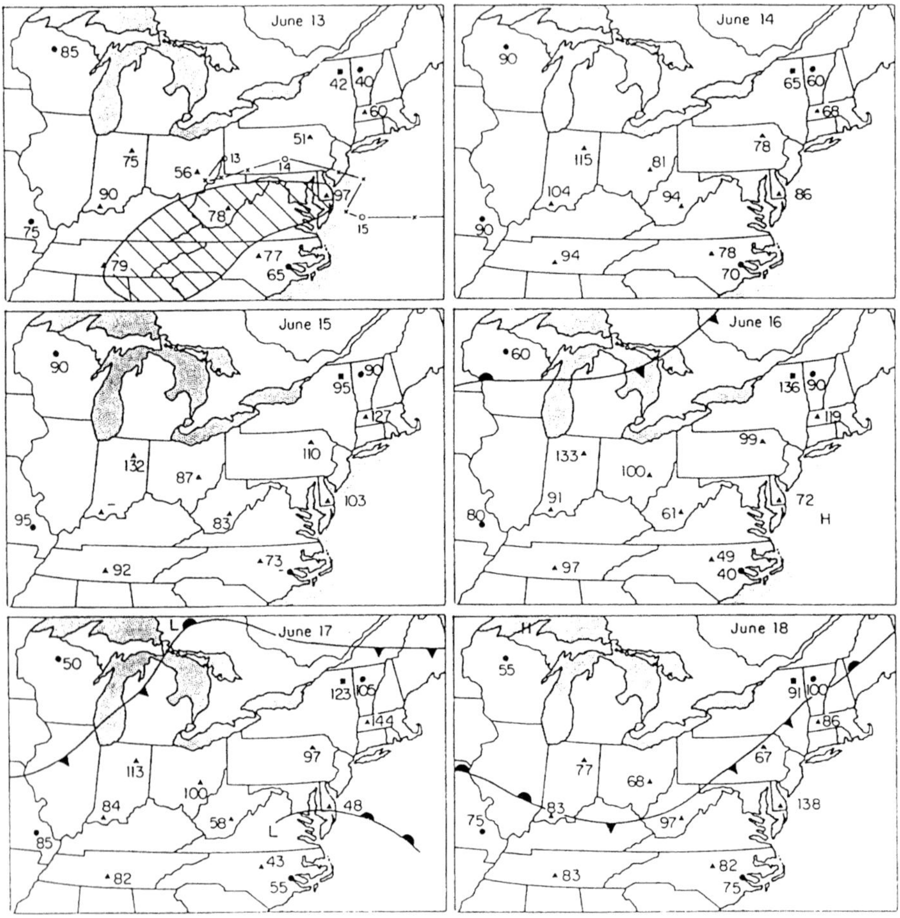

Figure 8b. Daily maximum values of ozone for June 13-18, 1979. The area of the corresponding 4 day air stagnation event is shown in the upper left hand panel for June 13-16 (Angell, private communication). The track of the anticyclone associated with the event is shown by the open circles (position at 7 am on the indicated day), with the crosses showing the intervening positions every 6 hours (NOAA, 1979). Additional highs (H), lows (L), and fronts are taken from daily weather maps, which give their positions at 7 am (National Weather Service, 1979). See Figure 8a for further details.

and 1979 respectively, between the months of May to September; they persisted for 3-4 days, on average, with a range of 2-8 days. Ozone episodes were present for 37 and 25 days in the two years, and were most common in June. They covered a spatial scale of 1000 km to more than 1600 km in an east-west direction, and more than 600 km from north to south.

These conclusions are based on only two years of data, and ozone is known to display considerable interannual variability. The five years of data available for the NAPBN sites on the periphery of the study region indicate that 1979 was a fairly typical year for ozone (Evans, 1985), while continued measurements at the SURE site in Tennessee suggest a similar conclusion for both 1978 and 1979 (Meagher et al., 1987). Measurements from Whiteface Mountain for 1974-85 indicate, however, that ozone values were higher than usual at this one site in 1978 and 1979 (Lefohn and Mohnen, 1986). Since photochemical formation of ozone is favored by particular weather patterns, we examined the meteorology for these two years using monthly summaries published at the end of each issue of Monthly Weather Review, and climatological data for stagnation events.

Temperature and precipitation were near normal for May-July, 1978, while August was generally 2°F warmer than usual, and somewhat wetter than usual. September was 2-4°F above normal, except in New York and New England which were -2-4°F below normal; most of the region was much dryer than usual. The weather was somewhat more anomalous in 1979. May was typical, but in June and July, temperatures were -2-4°F below normal; there were record lows in the region in the last week of June, the first week of July, and in the middle of July. Temperatures returned to normal in August, but it rained more than usual. September was an unusually wet month, with a large number of hurricanes.

A climatology of the number of stagnation events occurring each month east of the Rocky Mountains is available from 1936 to the present (Korshover, 1976; Korshover and Angell, 1982; Angell, private communication). We compared the frequency of occurrence of these events in June, July and August of 1978 and 1979 with average values for 1936-1975 (Korshover, 1976) and for 1976-1985, as shown in Table 2. (Comparisons were not made for May and September, because stagnation events occur primarily in the south east, outside the study region.) The two years 1978 and 1979 appear typical when compared to results for the same decade, but we note that the frequency of stagnation events from 1976-1985 was twice that of the previous four decades. The number of stagnation days shown in Table 2 are in some cases larger than those given in Figure 6, because the table relates to the entire east, while the figure shows only those events in the study region.

TABLE 2

Number of stagnation days east of the Rocky Mountains

	1978	1979	1976-85	1936-75
June	10	8	8.4	4.2
July	10	6	6.1	2.4
August	5	10	11.4	5.4
Total	25	24	25.9	12.0

The two years for which high quality rural ozone data are available appear to have had fairly typical summers by these measures. The somewhat cooler temperatures and wetter weather in 1979 versus 1978 is consistent with the smaller number of days with high values of ozone. There were more stagnation days between May 1 and September 30 in the study region in 1978, 26 compared with 21 in 1979. Four anticyclones spent 4-6 days crossing the region in 1978, while none lingered for more than three days in 1979. Given the strong association we found between ozone episodes and slow moving anti-cyclones, it is not surprising that high ozone values were more common in 1978 than in 1979. Our results are consistent with those of Vukovich and Fishman (1986) who showed that the path of anti-cyclones in July and August had a profound effect on the spatial distribution of ozone in these months.

Ozone episodes were present a significant fraction of the time during the warm months, on 23% of the days in May to August. Daily ozone maxima exceeded 80 ppb on most of these days. The value of 80 ppb was exceeded on almost half the days in June and one third of the days in July, as shown in Figure 5. Values greater than 120 ppb were much less common, but values up to 202 ppb were found at the rural sites. None of the sites in this study are immediately downwind of major cities, where even higher values of ozone may be found, although the sites in Massachusetts, Delaware and northern Indiana are clearly influenced by pollution from the New York conurbation, Washington-Philadelphia, and Chicago, respectively.

The results presented here indicate that both daytime average concentrations of ozone and daily maxima in the eastern U.S. frequently exceed values known to cause damage to vegetation and to impair human lung function (EPA, 1986). It is clear that ozone is severely impacted by anthropogenic emissions of NO_x and hydrocarbons. A number of modelling studies are in progress with the goals of enhancing understanding of this phenomana and devising appropriate control strategies, as described elsewhere in this volume. Ozone is no longer measured routinely at rural sites in the United States. A measurement program for ozone and its precursors is urgently needed in the United States to test these models and improve our understanding of this serious environmental problem.

ACKNOWLEDGEMENTS

This work was supported by CRC contract CAPA-22-83 and NSF grant ATM84-13153 to Harvard University.

REFERENCES

Cox, R.A., A.E.J. Eggleton, R.G. Derwent, J.E. Lovelock, and D.H. Pack, Long range transport of photochemical ozone in northwestern Europe, *Nature*, *225*, 118-121, 1975.

Decker, C.E., L.A. Ripperton, J.J.B. Worth, F.M. Vukovich, W.D. Bach, J.B. Tommerdahl, F. Smith, D.E. Wagoner, Formation and transport of oxidants along Gulf coast and in northern U.S., *EPA-450/3-76-003*, U.S. Environmental Protection Agency, Research Triangle Park, N.C. 27711, 1976.

E.P.A., Air Quality Criteria for Ozone and Other Photochemical Oxidants, *EPA 600/8-84-020-CF and EPA 600/8-84-020-EF*, U.S. Environmental Protection Agency, Research Triangle Park, N.C. 27711, August, 1986.

Evans, G.F., The National Air Pollution Background Network: Final Project Report. U.S. Environmental Protection Agency. *EPA 600/4-85-038*, 1985.

Evans, G., P. Finkelstein, B. Martin, N. Possiel, and M. Graves, Ozone measurements from a network of remote sites, *J. Air Pollut. Contr. Assoc.*, *33*, 291-296, 1983.

Guicherit, R., and H. Van Dop, Photochemical production of ozone in Western Europe (1971-1978) and its relation to meteorology, *Atmos. Environ.*, *11*, 145-155, 1977.

Heck, W.W., O.C. Taylor, R. Adams, G. Bingham, J. Miller, E. Preston, L. Weinstein, Assessment of crop loss from ozone, *J. Air Pollut. Contr. Assoc.*, *32*, 353-361, 1982.

Korshover, J., Climatology of stagnating anti-cyclones east of the Rocky Mountains, 1936-75. NOAA Technical Memorandum, ERL ARL-55, U.S. Dept. of Commerce, 1976.

Korshover, J., and J.K. Angell, A review of air-stagnation cases in the eastern United States during 1981-annual summary, *Mon. Wea. Rev.*, *110*, 1515-1518, 1982.

Lefohn, A.S., V.A. Mohnen, The characterization of ozone, sulfur dioxide, and nitrogen dioxide for selected monitoring sites in the Federal Republic of Germany, *J. Air Pollut. Contr. Assoc.*, *36*, 1329-1337, 1986.

Logan, J.A., Tropospheric ozone: seasonal behavior, trends, and anthropogenic influence, *J. Geophys. Res.*, *90*, 10,463-10,482, 1985.

Meagher, J.F., N.T. Lee, R.J. Valente, and W.J. Parkhurst, Rural ozone in the southeastern United States, *Atmos. Environ.*, *21*, 605-615, 1987.

Mohnen, V.A., A. Hogan, P. Coffey, Ozone measurements in rural areas, *J. Geophys. Res.*, *82*, 5889-5895, 1977.

Mueller, P.K., and G.M. Hidy, The sulfate regional experiment: Report of findings, *Rep. 1901* (3 vols.), Electric Power Research Institute, Palo Alto, CA, 1983.

Mueller, P.K., and J.G. Watson, Eastern Regional Air Quality Measurements. EPRI EA-1914, Volume 1, Project 1630-1. Electric Power Research Institute, Palo Alto, CA, 1982.

National Weather Service, National Meteorological Center, Climate Analysis Center. *Daily Weather Maps*, Weekly Series, Published by NOAA, 1978,1979.

NOAA. Climatological Data - National Summary - Annual Summary. Vol. 29,30. National Climate Center, Ashville, N.C., 1978,1979.

Skarby, L., and G. Sellden, The effects of ozone on crops and forests, *Ambio 13*, 68-72, 1984.

Vukovich, F.M., W.D. Bach, B.W. Crissman, and W.J. King, On the relationship between high ozone in the rural surface layer and high pressure systems, *Atmos. Environ.*, *11*, 967-983, 1977.

Vukovich, F.M., and J. Fishman, The climatology of summertime O_3 and SO_2 (1977-1981), *Atmos. Environ.*, *20*, 2423-2433, 1986.

Vukovich, F.M., J. Fishman, and E. V. Browell, The reservoir of ozone in the boundary layer of the eastern United States and its potential impact on the global tropospheric ozone budget, *J. Geophys. Res.*, *90*, 5687-5698, 1985.

Wolff, G.T., and P.J. Lioy, Development of an ozone river associated with synoptic scale episodes in the eastern United States, *Environ. Sci. Technol.*, *14*, 1257-1261, 1980.

Wolff, G.T., P.J. Lioy, G.D. Wight, R.E. Meyers, and R.T. Cederwall, An investigation of long range transport of ozone across the mid western and eastern United States, *Atmos. Environ.*, *11*, 797-802, 1977.

A PROGRAM OF TROPOSPHERIC OZONE RESEARCH (TOR)

S.A. Penkett, I.S.A. Isaksen* and D. Kley+,
School of Environmental Sciences,
University of East Anglia,
NORWICH NR4 7TJ,
United Kingdom.

ABSTRACT. There are several indications that the concentration of tropospheric ozone at mid-latitudes in the northern hemisphere is on the increase. The reason is probably increased formation in the background troposphere but an alternative explanation is enhanced transfer of ozone-rich air from the stratosphere. TOR (Tropospheric Ozone Research) is an experiment which is designed to investigate the potential for year-round tropospheric ozone formation on a sub-hemispheric scale. TOR has three main components. These include a ground-based network of stations measuring ozone, its precursor molecules and other peroxy species characteristic of tropospheric chemical activity, a vertical ozone sonde program capable of distinguishing tropospheric from stratospheric contributions, and an integrated theoretical modelling program producing specific models to accept and interpret TOR data. This paper presents details of the TOR experiment including the experimental philosophy and the specific objectives to be attained within the overall 8-year framework of the Eureka Environment Program EUROTRAC.

1. INTRODUCTION

Ozone is one of the most important atmospheric trace gases. This has been emphasized recently in the substantial report WMO 16,[1] which focusses its attention mainly on the behaviour of ozone and related trace gases in the stratosphere. Tropospheric ozone has received much less scientific attention, which is unfortunate since it plays a crucial role in tropospheric chemistry both as an oxidant and more importantly by acting as the primary source for the hydroxyl radical. This radical is responsible for most of the chemical transformations leading to removal atmospheric pollutants and thus acts to conserve the

* Institut For Geofysik, Universitetet I Oslo, Postboks 1022, Blindern, OSLO 3, Norway.
+ Institut fur Chemie, Atmospharische Chemie, KFA, Postfach 1913, D-5170, JULICH, West Germany.

I. S. A. Isaksen (ed.), Tropospheric Ozone, 345–363.

composition of the atmosphere.

Troposphere ozone also plays an important role in climate since it can absorb infrared radiation reflected up from the earth's surface. Temperature changes in the troposphere are thus sensitive to ozonechanges particularly those occurring in the upper troposphere.

Finally ozone is also a toxic gas. Regional pollution has reached levels which cause damage to plants in rural environments and concentrations are often large enough in highly populated areas for human health to be adversely affected.

Ozone is produced photochemically in the stratosphere, and a small fraction leaks into the troposphere to form a natural background. In addition man-made and natural emissions of nitrogen oxides and hydrocarbons cause photochemical production of ozone in the sunlit troposphere. There is a considerable body of evidence to testify that tropospheric ozone levels at the mid-latitudes of the northern hemisphere have increased substantially over the last thirty years (see Figure 1). The reason is not known at present.

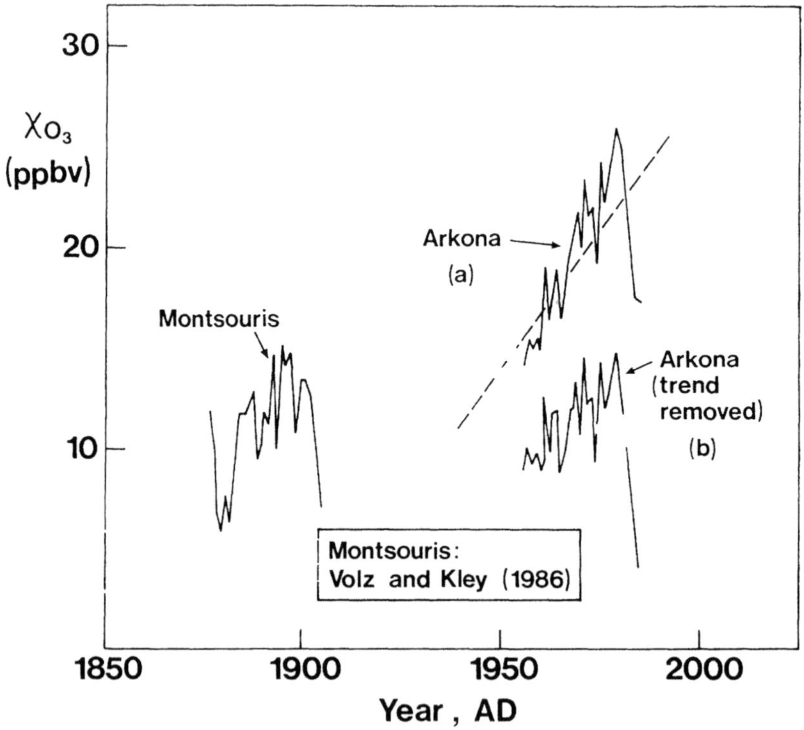

Figure 1 Ground level ozone concentrations compared from Montsouris, Paris 1876-1905 and Cape Arkona, Baltic Sea. Volz and Kley, Nature (in the press (1987)).

Because of the potential effects on chemistry, vegetation and climate, this increase is a matter of serious concern. Unfortunately, for various reasons, the tropospheric budget of ozone is not well established. Firstly, the dynamics of air exchange between the stratosphere and the troposphere can only be quantified in an approximate manner, such that the exact flux of ozone from the stratosphere into the troposphere is not known. Secondly, ozone interacts chemically in many ways and thus its tropospheric chemistry is complex. Ozone is both produced and destroyed in the troposphere and the preponderance of one or the other process depends on the presence or absence of other trace gases, as outlined below.

1.1 Photochemical Ozone Destruction

The major destruction process is photolysis by ultraviolet light.

$$O_3 \xrightarrow[<310\ nm]{h\nu} O_2 + O(^1D)$$

$$O(^1D) + H_2O \longrightarrow OH + OH$$

The amount of destruction depends on the photon flux at wavelengths less than 310 nm and on the water vapour content. These both maximise over the tropical ocean.

Ozone is also destroyed by reaction with the free radicals OH and HO_2, with the latter being more efficient.

$$O_3 + OH \longrightarrow HO_2 + O_2$$

$$O_3 + HO_2 \longrightarrow OH + 2O_2$$

1.2 Photochemical Ozone Production

Ozone can only be formed in the troposphere from the photolysis of nitrogen dioxide

$$NO_2 \xrightarrow[<\ 400\ nm]{h\nu} NO + O\ (^3P) \qquad (j_1)$$

$$M + O(^3P) + O_2 \longrightarrow O_3 + M \qquad (k_2)$$

This reaction produces ozone but

$$NO + O_3 \longrightarrow NO_2 + O_2 \qquad (k_3)$$

destroys ozone. Peroxy radicals produced in the course of hydrocarbon oxidation by hydroxyl radicals provide an additional route for NO oxidation to NO_2.

$$HO_2 + NO \longrightarrow OH + NO_2$$

$$RO_2 + NO \longrightarrow RO + NO_2$$

This allows further production of ozone by NO_2 photolysis and shifts the ratio of NO_2 to NO in the expression below in favour of ozone.

$$O_3 = \frac{j_1 \, [NO_2]}{k_3 \, [NO]}$$

In the course of hydrocarbon oxidation in the presence of NO_2, peroxy-acetyl nitrate (PAN) can be produced.

$$HC \xrightarrow{OH} CH_3CHO + HO_2 \text{ etc.}$$

$$CH_3CHO + OH \longrightarrow CH_3CO + H_2O$$

$$CH_3CO + O_2 \longrightarrow CH_3COO_2$$

$$CH_3COO_2 + NO_2 \longrightarrow CH_3COO_2NO_2 \text{ (PAN)}$$

Similarly peroxides are produced from radical termination reactions.

$$HO_2 + HO_2 \longrightarrow H_2O_2 + O_2$$

$$RO_2 + HO_2 \longrightarrow RO_2H + O_2$$

The presence of these various peroxide species is a strong indicator of free radical chemistry in the lower atmosphere.

1.3 Background to the Establishment of TOR.

Knowledge of the basic chemical processes which lead to ozone production in the troposphere has increased substantially over the last two decades but a number of outstanding questions remain in terms of the overall tropospheric ozone budget. These questions can be summarised as follows:

(i) How much ozone today is exchanged between the stratosphere and the troposphere?

(ii) How much ozone is produced in the troposphere?

(iii) How much ozone is destroyed by photochemistry in the troposphere and by deposition on the ground?

Only when these questions have been answered quantitatively rather than qualitatively can a true assessment be made of the ozone increase which may have occurred in the lower atmosphere. In order to provide more knowledge of the basic photochemical behaviour and transport properties

TABLE I Proposers of TOR

Name	Organization	Country
John Burrows	MPI-C	Fed. Rep. of Germany
Paul J. Crutzen	MPI-C	Fed. Rep. of Germany
Adolf Ebel	AMUK	Fed. Rep. of Germany
P.I. Grennfelt	ERI	Sweden
Robert Guicherit	TNO	Netherlands
Geoff Harris	MPI-C	Fed. Rep. of Germany
Oystein Hov	NILU	Norway
Ivar Isaksen	UO	Norway
Sylvain Joffre	FMI	Finland
Dieter Kley	KFA	Fed. Rep. of Germany
Alain Marenco	UPS	France
Gerard Megie	CNRS	France
Tom C. O'Connor	UG	Ireland
P. Oyola	NEPB	Sweden
Stuart A. Penkett	UEA	United Kingdom
Dieter Perner	MPI-C	Fed. Rep. of Germany
Pascal Perros	LPCE	France
Hans Puxbaum	TUW	Austria
Henning Rodhe	MISU	Sweden
Wolfgang Seiler	FIU	Fed. Rep. of Germany
Jean Servant	UPS	France
Peter Simmonds		United Kingdom
Gerard Toupance	LPCE	France
Nelly Tsalkani	LPCE	France
Andreas Volz	KFA	Fed. Rep. of Germany

of ozone in the troposphere a group of interested scientists decided to set up a coordinated program of Tropospheric Ozone Research (see table I) to be carried out on a large scale over a period of at least 5 - 8 years. It will utilise the fact that Europe is one of the major sources of precursor gases to ozone in the troposphere and thus has a strong regional flavour. In addition however, it is also hoped that the program will reach to greater understanding of perturbations produced in the ozone budget of the whole of the northern hemisphere.

1.4 Objectives of the TOR Experiment

The TOR experiment will set up a measurement and a modelling program with the object of finding answers to the following questions.

1. What is the seasonal, latitudinal and vertical variation of ozone and precursor molecules within the latitude band investigated by TOR?

2. How much higher is the mean ozone concentration in the boundary layer over Europe than that averaged over northern mid-latitudes?

3. How much of the excess ozone in the boundary layer over Europe spills over into the background atmosphere?

4. How much ozone is produced in the background atmosphere adjacent to source regions from the transport of precursor molecules?

5. Is there a secular trend in the concentrations of ozone and precursor molecules in the boundary layer or in the background atmosphere?

6. Is it possible to quantify, by making co-measurements of ozone and tracers of both tropospheric and stratospheric air, the proportion of ozone produced in the troposphere at our location?

7. How much ozone and its precursors are transported across regional boundaries?

8. Can we model the observations, and how well do the model calculations agree with the observations?

The problem of ozone uptake on surfaces will not be a primary objective of TOR but it is known that this problem will be addressed in other programs to be carried out over the same time scale as TOR within the EUROTRAC program.

2. DESCRIPTION OF THE TOR EXPERIMENT

2.1. General

The TOR experiment has three main components which are intended to be continued over a period of at least five years. The first component is the operation of a comprehensive suite of air sampling apparatus at various ground based sites in Europe over a wide latitudinal band, 80°N to 40°N (figure 2). The second is the collection of vertical profile data for a limited number of components, including ozone, at selected sites. The third component is a model development program specifically designed to formulate models to interpret the data being produced in components one and two. In addition it is intended that the program will be supplemented on occasion by aircraft experiments designed to increase the number of chemical components being sampled above ground level. The aircraft experiments will also have specific objectives concerned with the study of chemical change within well-defined synoptic situations.

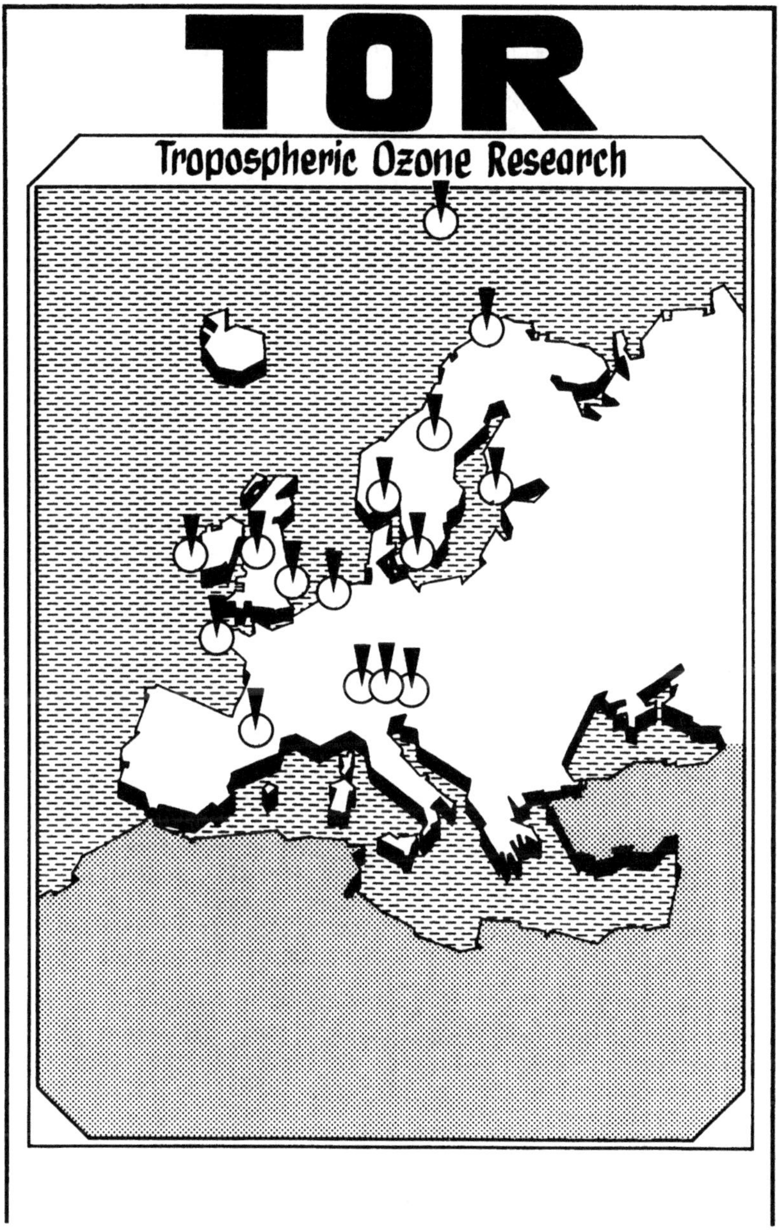

Figure 2 Distribution of TOR measurement sites

The experimental philosophy is such that research grade instruments will be operated by highly qualified scientists who will pay close attention to detail. There will be continuous dialogue between experimentalists and modellers both to sharpen the focus of the experiment as it proceeds and to ensure that relevant models are developed. These should be able to identify ozone sources and to predict the transport of ozone on regional and hemispheric scales at different times of the year. They also should be capable of testing whether a large scale ozone increase has occurred in the troposphere. Ultimately it is hoped that TOR can be incorporated into other ozone research programs operated outside Europe and active participation of non-European scientists will be strongly encouraged.

2.2. Site Location of Ground-based Measurement Program

Sites will be located over a 40° latitudinal range between 40°N and 80°N in order to study seasonal and latitudinal changes in the concentration of ozone and precursor gases. Some sites will be located close to the Atlantic shore and will be exposed to "background" air for much of the time. No doubt instances of pollution will also be observed and this will provide information on the chemical lifetime and of the transport of pollutants out of the source areas. Other sites will be located in regions close to source areas and will provide data on the extent of boundary layer air pollution. The elevation of the sites is such that some will provide boundary layer data and some will provide data from the free troposphere.

Names, locations and operators of sites for ground-based measurements are given in table II.

Stations exposed to the most polluted air will be Nos. 7 and 9. It is also possible that station No. 11 will receive highly polluted air on occasion, although it is at a considerably higher altitude than Nos. 7 and 9.

The cleanest boundary layer stations are expected to be Nos. 1 and 8 although it is very likely that clean air will dominate the sampling patterns at all stations except Nos. 7 and 9. (It is very simple to determine the percentage of time a station is exposed to clean air from a consideration of the frequency distribution of the CFC tracer record).

Stations Nos. 12, 13, and 14 will provide information on the composition of air in the free troposphere. This is essential to determine the extent to which pollution is affecting the lower atmosphere and to determine whether ozone perturbation is purely a boundary layer phenomenon.

The distribution of the stations in table II is not ideal. The most southerly boundary layer station is located at approximately 50°N. It is highly desirable to have such a station at 40°N and ideally one should be operated at 30°N to examine as large a latitudinal range as possible. Hopefully additional stations will become operational as the TOR program proceeds and in principle it is possible to run a measurement station for ozone on Ascension Island, which is in the southern hemisphere. A comparison of the potential of

TABLE II Site Location for Ground Based Measurements

No.	Station Name	Location	Coordinates	Altitude
1	Ny Alesund[1,2]	Spitzbergen	78°N : 4°E	sea level
	or Jergul	Finmark	69°24'N:24°E	225 masl
2	Birkenes	near Kristiansand	58°23'N:8°15'E	119 masl
3	Utö	Finnish Archipelago	59°47'N:21°23'E	7 masl
4	Areskutan	Central Sweden	63°26'N:13° 6'E	1250 masl
5	Rorvik	Swedish West Coast	57°23'N:11°55'E	sea level
6	Great Dun Fell	North West England	54°40'N: 2°30'W	800 masl
7	Norwich	East Anglia	52°40'N: 1°20'E	sea level
8	Mace Head	Ireland	53°15'N: 9°30'W	sea level
9.	Delft[3]	Holland	52° : 4°30'E	<sea lev.
10.	Brittany	[4]	[4]	
11.	Schauinsland	Black Forest	47°54'N: 7°48'E	1100 masl
12.	Garmisch-P	German/Austrian border	47°29'N:11°04'E	740 masl
	Wank	"	47°40'E:11°09'E	1776 masl
	Zugspitze	"	47°25'N:10°59'E	2962 masl
13.	Sonnblick	Central Austr. Alps	47°03'N:12°57'E	3000 masl
14.	Pic du Midi	French Pyrenees	42°56'N: 0°09'E	3000 masl

[1] Location not shown on Figure 1
[2] Station may be moved to higher altitude
[3] A station at a Dutch coastal site is also proposed by TNO
[4] Exact location to be determined

the atmosphere to create ozone in the two hemispheres would be a very useful additional goal for TOR.

2.3. Station Instrumentation

All stations will be equipped with the same high quality research type instrumentation. This guarantees inter-changability of instruments and easier comparability of the data. Also it may be necessary because commercial instrumentation is not available in many cases.

The instrument coverage shown in table III will ensure that many of the precursor molecules responsible for ozone formation in the troposphere are measured. Methane and carbon monoxide are present in much larger concentrations than the non-methane hydrocarbons (nmhc's), but the carbon flux through these latter can exceed that through the

TABLE III Details of Instrumentation

Species	Technique
Ozone	UV-absorption
NO, NO_2, NO_y	Chemiluminescence
CH_4, CO, CO_2	GC
C_2-C_5 Hydrocarbons	Automated - GC or Flask-sampling - GC
> C_5 Hydrocarbons	as above
J_{NO_2}; $J_{O(^1D)}$	photometry
Meteorology	Standard instrumentation
CFC-tracers	GC(ECD)
PAN	GC(ECD)
H_2O_2	UV-fluorescence in solution
Aldehydes and ketones	HPLC

former two molecules at certain times of the year. It is also more likely that higher NO_x concentrations necessary for ozone production will correlate with the nmhc's rather than with methane and CO.

Hydrogen peroxide can be formed as a result of oxidation of both the single and multiple carbon compounds listed above and theory suggests that ideal conditions for its formation involves low NO concentration, mainly because of the competition from the following reactions

$$HO_2 + NO \longrightarrow OH + NO_2$$

$$HO_2 + HO_2 \longrightarrow H_2O_2 + O_2$$

PAN is only formed from compounds with more than one carbon atom and its presence in clean air is strongly indicative of tropospheric free radical chemistry including nmhc's. Co-measurements of these various peroxy species are therefore considered to be an essential requirement of a comprehensive tropospheric ozone chemistry measurement program.

Aldehydes and ketones are important intermediates in the oxidation of hydrocarbons to carbon monoxide and thus provide useful measurements for tests of models with complex chemical schemes. This remark also applies to PAN and H_2O_2 of course.

The tracer molecules, such as CFC 11 and 12 and many halocarbons including CH_3CCl_3, C_2HCl_3 and C_2Cl_4 are very useful in establishing the nature of the air mass being sampled. Clean and polluted air masses can be distinguished without difficulty.

2.4 Comments on Instruments shown in table III

A limited amount of detail is presented here, much more information relating to details of calibration, etc. are provided in the full TOR proposal.

Ozone Measurements

These will be made using a simple UV absorption instrument which requires minimum attention. Standard commercial instruments are available at reasonable cost with adequate sensitivity levels (1 ppbv).

NO_x and NO_y Measurements

(1) NO Detector

The instrument for the measurement of the oxides of nitrogen must be very sensitive and therefore requires a state-of-the-art chemiluminescence detector. It is anticipated that ultimately a commercial NO detector will be available with a detection limit of 50 pptv. This is less sensitive than some instruments but should prove easier to operate over long time periods.

(2) NO_y Detector

$$NO_y = NO + NO_2 + NO_3 + N_2O_5 + HNO_3 + PAN + \text{nitrates}$$

NO_y is an important conservative quantity and there is experimental evidence that NO_y correlates positively with O_3. It can be measured by using the NO detector, described under (i) above with the addition of a CO catalysed gold tube at 300°C to convert NO_y to NO. This is a technique that was recently perfected at the Aeronomy Laboratory in Boulder. The sensitivity for NO_y will be equal to that for NO.

(3) NO_2 Detector

NO_2 can be converted to NO by UV photolysis and it may therefore be possible to again use a chemiluminescent detector of the type described above. An alternative may be a surface-chemiluminescence detector (luminol) with a theoretical detection limit of 20 pptv. This should be evaluated against the photolytic converter technique at a few key stations.

Hydrocarbon Measurements

The technology is available for measuring concentrations of many individual hydrocarbons down to approximately 20 pptv in a 2 litre sample of air. Subsequent analysis is performed with an FID gas chromatograph using an inert capillary column. Tenax tubes for sampling of C_6 C_{15} hydrocarbons have been successfully used by NILU and can possibly be employed at stations if a central laboratory for analysis is established. This should increase the sampling frequency.

CFC Tracers/PAN Measurements.

Apparatus has been developed to measure PAN and the common CFC's at all concentrations below 20 pptv in 5 ml air samples. The main problem with the PAN measurement is calibration but several techniques have now been reported in the literature for on-site PAN calibration. Calibration of the more stable halocarbons can be performed without much difficulty.

Aldehydes and Ketone Measurements.

One technique for the measurement of aldehydes and ketones is based on the derivatisation of the 2,4-dinitrophenylhydrazine followed by liquid chromatography analysis of the hydrazone formed. However, this method cannot be used in remote stations. Another possibility, involving direct GC analysis for formaldehyde and acetaldehyde is promising.

Carbon Monoxide.

Carbon monoxide can be measured by gas chromatography using either a hot HgO detector or an FID after conversion of the carbon monoxide to methane with a hot nickel catalyser. The analysis system is readily automated since it requires only a few CO's of air.

Hydrogen Peroxide.

Measurements of gaseous H_2O_2 will be made using an enzyme-specific fluorescence method developed at NCAR by Lazrus and his co-workers. Its sensitivity is such that 20 pptv of H_2O_2 can be detected and it is expected that a positive signal will be measured throughout much of the year. Calibration can be performed easily on site using liquid standards.

3. STANDARDISATION AND CALIBRATION EXERCISES

Standardisation of all instruments used in the experiment is essential and must be carried out at least at yearly intervals. It may be necessary to convey instruments to a central laboratory for this purpose although if standard procedures can be agreed upon it may be possible to standardise in-situ or at a laboratory close to the measurement site.

It is very likely that long-term standards can be made for HC's and CFC's and possibly for NO. It is most unlikely that any long term standards can be made for the other species.

In order to assure the good quality of the data, each and every proposer must be involved in a rigorous calibration and cross calibration exercises. This will be an essential prerequisite for TOR to succeed.

4. VERTICAL SOUNDINGS OF OZONE, WATER VAPOUR, TEMPERATURE

Long-term vertical soundings of ozone at Hohenpeissenberg in Southern Germany have indicated that free tropospheric ozone has seen a dramatic rise over the last twenty years. This could be due to either tropospheric production or to a change in the amount of ozone that is exchanged annually from the stratosphere. A well-designed vertical ozone sounding program should be well-suited for investigations of the origin of free tropospheric ozone, its budget, and trends.

4.1. Measurements

Standard RS 80 radiosondes will be modified to include an electrochemical ozone instrument of either the ECC or the Brewer-Mast type. This combination will provide pressure, temperature, humidity and ozone data. In addition to this experimental data, calculations can be made of the potential vorticity. A combination of ozone with humidity and temperature data will give information on ozone coming from the lower troposphere, including the boundary layer, whereas the ozone/potential vorticity combination will give information on ozone of recent stratospheric origin. This latter will be associated with extremely dry air and identifiable, therefore, by more than one criterion.

4.2. Latitudinal and Longitudinal Coverage of Sonde Sites

Concentrations of molecules such as nmhc's, carbon monoxide and oxides of nitrogen are probably largest between 45°N and 55°N. Consequently, these latitudes have the largest potential for photochemical ozone formation in the free troposphere. An ozone sounding network should therefore be evenly positioned, centered at 40° to 50° but extending somewhat to the south of 40° and to the north of 50°. Site selection will be a compromise between optimizing each site in the center of a triangle of radiosonde sites and the logistically and administratively feasible sites. Five ozone sonde stations seem to be a reasonable initial number in addition to the Hohenpeissenberg station.

In the west-east direction across Europe there is a strong potential for ozone generation on a regional scale. Sonde stations should therefore be positioned at the Atlantic coastline and a substantial distance away from the Atlantic coastline to pick up this regionally produced anthropogenic signal. One or two additional continental sonde sites should be positioned in Germany and another sonde station should be selected in the eastern part of western Europe (i.e. Greece). This would be capable of detecting long range transport of ozone in the free troposphere.

Names and locations of the initial network of stations for ozone sonde operations are given in table IV.

TABLE IV Site Location for Ozone Sondes

No.	Site	Location	Altitude
1	Tromso	North Norway	sea level
2	Jülich/Köln or Schauinsland	West Germany Black Forest	60 m asl 1100 m asl
3	Garmisch-Partenkirchen	South Germany	700 m asl
4	Observatoire de Haute Prov.	France	~ sea level
5	Pic du Midi	French Pyrenees	~3000 m asl
6	Hohenpeissenberg	South Germany	~1000 m asl

4.3. LIDAR Profiles of Ozone

The LIDAR technique for generating tropospheric profiles of ozone has now reached operational status. It is proposed that at a few key stations, LIDAR-generated ozone profiles will be compared to those obtained by the classical balloon soundings.

5. AIRCRAFT FLIGHTS

Aircraft flights will be performed to determine the spatial distribution of O_3 in the troposphere between the balloon launching stations, both in vertical and horizontal direction. The aircraft will be equipped with a set of instruments which allow measurements of O_3 and also those substances which are related to tropospheric O_3 formation and/or which may be used as tracers for stratospheric O_3 (see table III). The flight frequency would probably not exceed 4 times per year and needs to be carefully controlled by the prevailing meteorology.

In addition various distinct phenomena will be investigated such as the influence of injections of stratospheric ozone on atmospheric reactivity, and the extent of increased ozone concentrations in anticyclonic situations. Ideally the ozone build-up in these situations should be tracked until it disappears. In this way it may be possible to establish the extent to which ozone found in boundary layer photochemistry during the summer months mixes into the surrounding clean troposphere. The efficiency of destruction processes such as ground removal and photochemical removal may be quantifiable in these aircraft experiments.

6. MODELS

The models will be constructed to study the spatial and temporal behaviour of atmospheric trace gases which lead to ozone production in the troposphere. Specific examples include

a) Episodes of length comparable to the duration of synoptic weather situation giving rise to a layered structure in the tropospheric composition and regional pollution in the PBL.

b) Seasonal variation.

c) Trends on a global, long term scale.

The episodic behaviour of the trace substances is an important boundary condition for the description of the behaviour of atmospheric trace gases in the free troposphere. Models under part (a) will involve the description of the chemistry and transport in the PBL and the exchange with the free troposphere when the PBL breaks up (convection, large scale synoptic disturbances, nocturnal "pumping", etc.).

The measuring sites are located and equpped to monitor the composition of air coming into and out of Europe. Models to investigate the episodic behaviour (in the PBL and as revealed from sonde measurements) would include a Lagrangian trajectory model with a chemical description of clean and polluted air for Europe extending from the Mediterranean into the Arctic and a two-dimensional "channel model" extending in a zonal direction from the U.S.A. to Europe.

A three-dimensional Eulerian model for the PBL over Europe will provide useful input (boundary conditions) for a 2- or 3-d model of the northern hemisphere or global atmosphere. It is planned to adapt NCAR's Regional Eulerian Acid Deposition Model to the European conditions. The construction of numerical models should be done in close relation to the development of the measurements programs. This is expected to strongly enhance the scientific output of the joint project.

6.1. Model Input

The day-to-day variability of atmospheric trace gases will be calculated on the basis of the synoptic data from the European Centre for Medium Range Weather Forecast (ECMWF) and use input data based upon emission inventories for many molecules. The horizontal and vertical resolution of the models will be compatible with that of the meteorological data.

The chemical turnover of the emissions of CH_4, the most common nmhc's, NO_x and CO, and the formation of O_3, PAN, aldehydes and other secondary products will be described in the models. Clouds can influence the chemistry, the transfer of UV-radiation and the vertical exchange of material. These are important processes to include in the models.

The location, time variation and strength of the sources of NO_x, hydrocarbons and CO are important input data. Model calculations of seasonal and long-term composition changes may also be based on climatological data (temperature, wind and pressure fields).

An important contribution to model input is the determination of rate constants for many chemical reactions in the laboratory experiments. Coordinated programs of these experiments are therefore highly desirable, if possible with a close link to the TOR experiment. In a similar manner a coordinated program of deposition studies should be closely associated with TOR.

6.2. Present Activity

Models for the calculation of the episodic behaviour of trace gases in the PBL of Europe have been developed through EMEP (Lagrangian model). Several 3-d photochemical pollution models are under development (PHOXA, TADAP). Of more relevance to the present proposal are the models developed at the University of Oslo, and MPI-C, for the description of the chemistry and transport of trace gases out of the boundary layer and in the free troposphere. Work on the parameterisation of cloud processes in connection with tropospheric chemistry models are under development at the University of Stockholm.

7. METEOROLOGICAL MEASUREMENTS

The following meteorological parameters should be measured at all ground level stations:

(i) Wind velocity and direction at one level (10 m),
(ii) Temperature at two levels (2 and 10 m) if possible,
(iii) Humidity at one level (10 m),
(iv) Pressure at ground level,
(v) Extent and elevation of clouds,
(vi) Radiation (spectral measurements).

Nowadays, reliable automatic weather stations providing this data are available on the market. The sampling frequency should be comparable to the chemical observation frequency (60 minutes) and data should be centralized in the chemical data archive.

Surface meteorological fields can be constructed with data available from synoptic and/or climatological stations and upper air data provided by aerological stations. Upper air measurements (height, wind, temperature, humidity) will provide information on the structure of the atmosphere and large-scale transport conditions. Moreover, tropopause folding processes can be traced back giving information on the exchange of mass between the stratosphere and the troposphere. Specific analyses of quantities such as potential vorticity, can give additional information on the vertical and horizontal transport in the free atmosphere. Satellite data can provide additional information on cloudiness and solar fluxes particularly. Total ozone data provided by the existing ozone-network and/or measured at the ground-based TOR stations can give valuable additional input to the interpretation of observations and model experiments.

Episodic studies may require specific meteorological setups expecially for the exploration of PBL processes (many vertical levels of observations, dense spatial network of sensors). They also need accompanying synoptical analyses in order to determine ongoing dynamical processes and to provide short-term forecasts for better planning of field experiments.

8. JOINT DATA BASE

It is intended to combine the measurements into a joint data base at a central location. This data base will be capable of being integrated by TOR scientists within the first two years from the commencement of the experiment. Ultimately the data base will be made available to other interested scientists.

9. WORK PLAN AND TIMETABLE

On the assumption that this study is considered a suitable one for joint European cooperation, it is hoped that proposals for funds to augment this program can be made in early 1987. These proposals will be submitted separately to individual research councils in the respective countries but with a common end in view and a common timetable.

1987 February: workshop on calibration, standardisation and joint data base.
Purchase and calibrate instruments.
Inter-calibration exercise.
Start measurement program.
Begin model development program.
Start measurement program for vertical profiles.

1988 Assessment of one year's data in conjunction with model work (PI's workshop).
Modification incorporated into measurements and modelling work.
Aircraft programs designed and started.
Ideas put forward for incorporation of TOR into US program.

1989 Completion of 3 years of ground based measurements.
Availability of good working model.
Further aircraft flights at different seasons.
Workshop.

1990 Assessment of data.
Workshop meeting of all PI's to present data and prepare basis of reports concerned with measurement and modelling phases. Write

up of papers for publication in European Atmospheric Chemistry Journal.
Review papers prepared for JGR, etc.
European Symposium.

1991 Major effort to collect episodic data.

1992 Large latitudinal and longitudinal range experiment with U.S. workers.
Possibility of extension to cover 90° latitude.

1993 Joint workshop with U.S. and European workers.
Preparation of major report on tropospheric chemistry.

REFERENCE

1. Atmospheric Ozone 1985: Assessment of our Understanding of the Processes Controlling its Present Distribution and Change. World Meteorological Organisation Global Ozone Research and Monitoring Project Report No. 16.

LIST OF ABBREVIATIONS

NILU Norwegian Institute for Air Research
KFA Kernforschungsanlage Jülich
CNRS Centre National de la Reserche Scientifice
UEA University of East Anglia
FMI Finnish Meteorological Institute
MISU Meteorological Institute Stockholm University
TNO Netherlands Organisation for Applied Scientific Research
FIU Fraunhofer Institut für Umweltforschung
MPI-C Max-Planck-Institut für Chemie, Mainz
LPCE Laboratoire de Physico-Chimie de l'Environment
TUW Technische Universität Wien
UPS Universite Paul Sabatier
ECMWF European Center for Medium Range Weather Forecasting
PBL Planetary Boundary Layer
NMHC Non-Methane Hydrocarbons
EMEP
PHOXA
TADAP
NCAR National Center for Atmospheric Research
AMUK Albertus-Magnus-Universität zu Köln
PI Principal Investigator
NEPB National Environment Protection Board (Sweden)
ERI Environmental Research Institute (Sweden)

ACKNOWLEDEMENT

The authors wish to thank other proposers of the TOR experiment who contributed to various sections of this manuscript at various times. It is also necessary to acknowledge the support of many funding agencies in Europe who are providing financial support for the TOR experiment to proceed under the scientific coordination of EUROTRAC.

THE AEROCE PROJECT

H. Levy II
Geophysical Fluid Dynamics Laboratory/NOAA
P.O. Box 308
Princeton University
Princeton, NJ 08542

ABSTRACT. The Atmospheric/Ocean Chemistry Experiment (AEROCE), a coordinated multi-institutional atmospheric and marine chemistry research program, will establish a network of four stations in the North Atlantic (Barbados; Bermuda; Mace Head, Ireland; and Tenerife). Using proven measurement technologies, this program will address three major scientific issues: the climatology of background tropospheric ozone in the northern mid-latitudes; the export of combustion sulfur and nitrogen from North America; the flux of mineral aerosol and trace metals to the North Atlantic from the continents. As instruments are developed and proven for continuous field use, we hope to examine a range of atmospheric chemistry issues including the importance of chemistry in the budget of tropospheric ozone, the role of biogenic sources of sulfur in sulfate deposition to the ocean, and the flux of continentally derived organics to the surface of the North Atlantic. Furthermore, this network of chemical observations will provide a framework for developing intensive studies of such specific processes as atmospheric oxidation, the sulfur cycle, and heterogeneous removal.

1. INTRODUCTION

The Atmospheric/Ocean Chemistry Experiment (AEROCE) will measure ozone, radon, and a range of chemical species in particulate aerosol and precipitation at four locations (Barbados; Bermuda; Tenerife; and Mace Head, Ireland) in the North Atlantic. If sufficient funds and personnel become available, it is hoped that the network will expand to include Sable Island, Sal, Azores and Iceland. Current plans are to operate the network long enough to produce realistic chemical climatologies at the sites. The sites and their data sets will provide a foundation for the design and implementation of field experiments which employ state-of-the-art measurement technology and focus on a specific process or chemical cycle.

I. S. A. Isaksen (ed.), Tropospheric Ozone, 365–370.

2. SCIENTIFIC OBJECTIVES

Initially the climatological data sets will be used to study three major scientific issues: the climatology of tropospheric O_3 in the remote Northern Hemisphere and the relative importance of boundary layer and stratospheric sources; the fate of combustion S and N emissions not accounted for in acid deposition over North America; the impact of continental sources of mineral aerosol and trace metal on surface seawater chemistry and biological activity. In the future, as the necessary measurement techniques are developed, we hope to study the role of chemistry in the behavior of tropospheric ozone at remote locations and to determine the impact of biogenic sources on the sulfur cycle over the North Atlantic.

2.1. Tropospheric O_3

Recent observations in the northern mid-latitudes (see Levy, this volume) strongly suggest a major and possibly dominant role for chemical production and destruction in the budget of tropospheric O_3. However, the current data are from surface and ozonesonde measurements over the polluted continents. The continuous measurements of O_3, in combination with tracers of atmospheric transport from the upper troposphere (^{7}Be) and the continental boundary layer (mineral aerosol, ^{222}Rn) and with the isentropic trajectory analyses, will be used to establish the actual climatology of tropospheric O_3 away from continental sources and to estimate the relative importance of boundary layer and stratospheric sources. When they become available, the continuous measurements of CO and reactive nitrogen compounds will permit a detailed examination of the role of tropospheric chemistry in the budget of O_3.

2.2. Nitrogen and Sulfur Deposition

Current measurements of acid precipitation over North America account for no more than 30% of present emissions (e.g. Levy and Moxim, 1987). While considerable dry deposition is expected, the extent of export to the North Atlantic and beyond remains a significant issue. The four stations will generate local climatologies of deposition which, combined with realistic transport/chemistry models, should both provide an accurate picture of deposition over the North Atlantic and help determine the extent of export from North America and its impact on the global budgets for S and N. In addition, ^{222}Rn and ^{210}Pb provide simplified and important surrogates for S and N which can be used to test a model's simulation of precipitation deposition.

2.3. Continental Fluxes of Mineral Aerosol and Trace Metals to the North Atlantic

Recent studies indicate that atmospheric input can be the dominant source for a number of trace species in surface seawater (Settle and Patterson, 1982; Prospero et al., 1986; Gagosian, 1986): mineral material; trace metals such as aluminum, lead and tin; vital nutrients such as iron and nitrogen; organics such as chlorinated hydrocarbons, polycyclic aromatic hydrocarbons and heavy alkanes. Of particular importance is the biological impact of large sporadic pulses of mineral material. It now appears that most of the micronutrient iron in surface seawater is atmospherically derived (Duce, 1986), but little is known about the climatology of such atmospheric inputs. Thus both the accumulated deposition and the climatology of deposition must be measured for these continental materials.

2.4. Future Studies

Recent studies suggest that a major fraction of non-sea-salt (nss) SO_4 in the remote oceanic regions may come from dimethyl sulfide (DMS), a product of marine organisms (Saltzman et al., 1986). Very little is known about the spatial and temporal characteristics of this source and its reaction pathways. We hope to establish a network measurement program for DMS, methanesulfonic acid (MSA), SO_2 and OCS in the near future.

While net chemical production of ozone is thought to be very important in the continental regions (see Logan, this volume), its role is not clear in the more remote locations of the northern mid-latitudes. The key ingredient in this chemistry is NO_2 whose measurement currently requires state-of-the-art instrumentation. As soon as devices suitable for remote locations and unsupervised operation become available, we plan to include them in the network. The measurements of NO_y and SO_2 will also provide more information about the export of combustion N and S.

3. FACILITIES

The four surface measurement sites are located at Daniels Head on western Bermuda; Mace Head, Ireland; Ragged Point, Bermuda; and Izania, Tenerife. Preliminary deployment as secondary sites is planned for late 1987. They will have 15 m folding towers, limited daily supervision and minimal on-site laboratory facilities. Beginning in 1988, the sites will be upgraded to primary status with permanent walk-up towers, daily supervision by a trained technician, and a small air-conditioned laboratory. Each site will have automated sampling systems for particulate aerosol, precipitation, selected trace gases and meteorological variables. The data will be automatically logged by a computer.

Barbados lies in the trade winds and will sample air with a long over-ocean trajectory. North Africa will be the dominant continental source and anthropogenic sources, with the possible exception of biomass burning, should have little impact relative to the other stations.

Izania, Tenerife is located at 2300 m and the winds are generally from either the North Atlantic or Europe. The station lies above the boundary layer 90% of the time except in the winter when it is 70%. It is the only location which should routinely sample "free troposphere" air and will provide an interesting comparison with the observations from Mauna Loa in the Pacific.

Bermuda is affected by a wide range of meteorological conditions and will allow us to sample air masses originating in North America and, in some instances, in Europe and Africa. The degree of continental and anthropogenic impact as well as the length of the trajectories over the ocean should vary widely.

Mace Head lies in the westerlies though the air trajectories are highly variable (see Chatfield, this volume). While we will often sample air masses with long over-ocean trajectories, the influence of European and North American sources will frequently be present.

We would prefer more stations, particularly in the middle of the ocean, but funding constraints and geography make this impossible. The four stations described above provide the best network possible in the North Atlantic, given the facilities and logistical support required and the financial support available.

4. MEASUREMENTS

It is our plan to start the program with instrumentation that has already been successfully used for long-term continuous measurements at remote locations. Later as new techniques are developed and proven for network use, they will be included.

4.1. Particulate Aerosol

Particulate aerosol will be continuously filtered from the surface air at all four sites and collected daily. These filter samples will be analyzed for mineral aerosol, trace metals, NH_4^+, $nssSO_4^=$, NO_3^-, MSA, ^7Be, ^{210}Pb and the sea salt components.

4.2. Precipitation

The accumulated precipitation will be collected daily from samplers with lids that open only during precipitation events. Most of the same analyses carried out for the aerosol samples will be performed with certain exceptions: ^{210}Pb in precipitation will be determined monthly rather than daily; fewer trace metals will be measured in the precipitation; standard rain chemistry measurements will be made.

4.3. Gases

Both O_3 and ^{222}Rn will be measured continuously at the four sites and grab samples of long lived trace gases including CFM's, methane and other hydrocarbons will be collected weekly for analysis by NCAR and NASA/LRC. In the near future we hope to begin continuous measurements of CO and some of the long lived trace gases. When the techniques have been developed and proven for use in the field with minimal supervision, continuous measurements will be started for the reactive nitrogen compounds (NO, NO_2, NO_y) and the sulfur compounds (SO_2, DMS, CS_2, OCS).

4.4. Meteorological Observations

Continuous measurements of the standard meteorological variables (wind, temperature, pressure and humidity) will be automatically recorded. Furthermore, daily isentropic trajectories will be calculated for each of the four sites.

5. COLLABORATION

AEROCE will provide numerous opportunities for interaction and collaboration with other groups and individual investigators. The four sites are designed to provide state-of-the-art chemistry sampling facilities for the atmospheric chemistry community. New groups could test new equipment, initiate scientific programs that are complementary or independent of AEROCE objectives, and carry on intensive studies of specific processes such as atmospheric oxidation, the sulfur cycle and heterogeneous removal. The network data will be evaluated by a panel composed of AEROCE and non-AEROCE experts, held as proprietary information of the PIs for a predetermined time, and then released to the community as a whole. We estimate that the elapsed time from sample collection to data release will not exceed two years. Active collaboration with individuals and groups interested in data analysis and numerical simulation will be sought.

The AEROCE network will overlap with a number of existing or planned international programs. AEROCE and ALE/GAGE will both be measuring at Barbados and Mace Head, Ireland. The chemical and meteorological tracer data from AEROCE should be useful to ALE/GAGE investigators, and their continuous measurements of stable gases will supplement the initial AEROCE flask data. When the continuous measurement system for CO and the more stable trace gases comes on line in the AEROCE network, there will be an excellent opportunity for intercalibration and expansion of the network coverage. In the same way, the NOAA surface ozone network can expand to the Atlantic Ocean, and, through an overlap of ozone measurements at Mace Head, connect with the European project TOR (Tropospheric Ozone Research). The overlap between AEROCE and TOR at Mace Head will also provide a connection to the detailed ozone chemistry data that TOR plans to gather in Europe. Additional interactions are planned with BAPMON through measurements at their station in Tenerife and with the Global Ocean Flux Study (GOFS) through the study of continental fluxes of mineral aerosol and trace metals to the surface North Atlantic.

6. BIBLIOGRAPHY

Gagosian, R.B., The air-sea exchange of particulate organic matter: The sources and long range transport of lipids in aerosols, in The Role of Air-Sea Exchange in Geochemical Cycling, edited by P. Buat-Menard, in press, Reidel, Dordrecht, 1986.

Levy II, H., and W.J. Moxim, Fate of US and Canadian combustion nitrogen emissions, Nature, in press, 1987.

Prospero, J.T., R.T. Nees, and M. Uematsu, Deposition rate of particulate aluminum derived from Saharan dust in precipitation at Miami, Florida, J. Geophys. Res., in press, 1987.

Saltzman, E.S., D.L. Savoie, J.M. Prospero, and R.G. Zika, Methanesulfonic acid and non-sea-salt sulfate in Pacific air: Regional and seasonal variations, J. Atmos. Chem., 4, 227-240.

Settle, D., and C.C. Patterson, Magnitudes and sources of precipitation and dry deposition fluxes of industrial and natural leads to the North Pacific at Enewetak, J. Geophys. Res., 87, 8857-8869, 1982.

SAMPLING MID-TROPOSPHERIC AIR AT AEROCE SURFACE SITES

Robert B. Chatfield and Philip L. Haagenson
National Center for Atmospheric Research
P.O. Box 3000, Boulder, Colorado 80307
U.S.A.

ABSTRACT. Air chemistry sampling stations at the earth's surface, like one proposed for Mace Head, Eire, appear to be frequently and predictably exposed to air of recent origin 1500 to 4000 meters above the surface. A suite of backward isentropic trajectory analyses for this and other North Atlantic sites suggests that significant subsidence almost invariably occurs when an air parcel originates from the northwest. A simple practical model demonstrates how boundary-layer mixing usually smoothes out the impact of any small-scale variations in upper-air composition, acting like a time filter with a 6- to 18-hour time constant. Examination of the trajectories also reminds us of the slanting and stochastic nature of tropospheric air motions, and delimits the types of simulations appropriate in understanding the chemical measurements. An attempt to describe a geographical domain of influence which instruments at the site sample reveals some difficulties facing climatologically oriented air chemistry models, and suggests a concept of "complex influence."

1. FREQUENT AND SYSTEMATIC EXPOSURE TO MID-TROPOSPHERIC AIR

1.1. Introduction

It is important to develop a climatology describing the abundance and correlations of gas-, aerosol-, and liquid-phase compounds in the atmosphere. The AEROCE and TOR programs are two current attempts to describe the degree of cleanliness or pollution of a significant portion of the global troposphere which lies above the North Atlantic. One main concern is the description of tropospheric oxidants and the processes that determine them. The nitrogen oxides, nitrates, and formaldehyde, PAN, and carbon monoxide are closely tied to the levels of hydroxyl radicals and ozone, and observations of these concern this symposium. However, observations about many element cycles should come from these studies, and studies of one cycle will inform studies of the others. Sites near the on islands and borders of the North Atlantic which are nearly unaffected by any local pollution are ideal for measurements made in a continual, partially automated mode.

I. S. A. Isaksen (ed.), Tropospheric Ozone, 371–380.

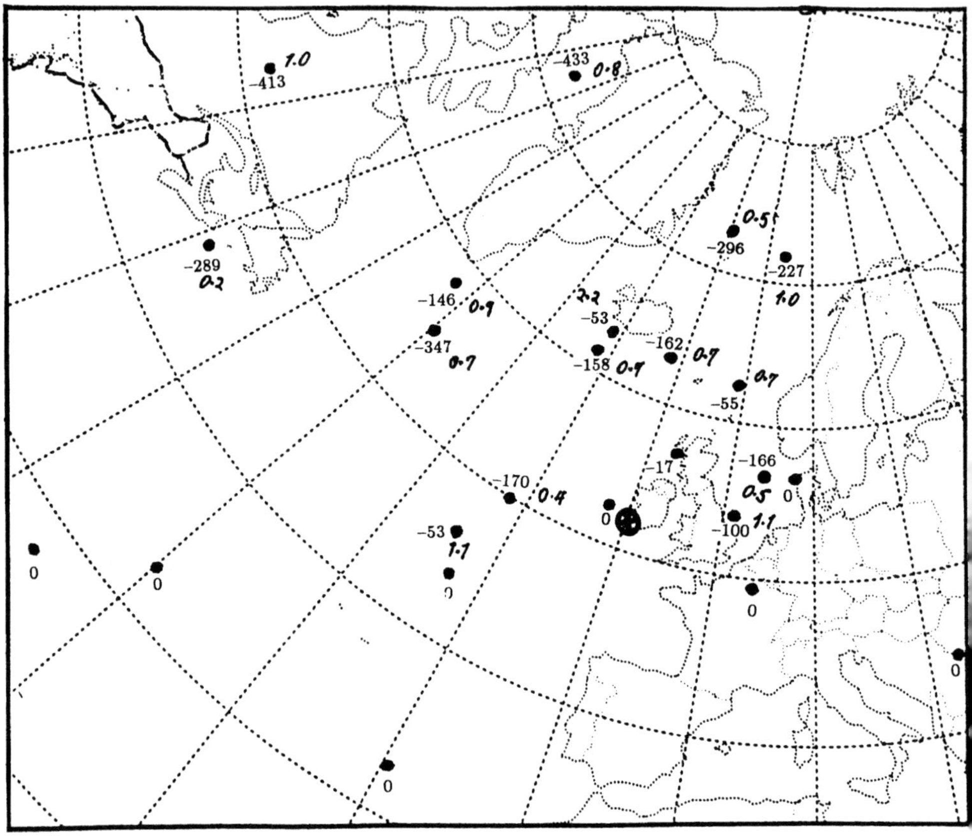

Figure 1. Starting points of 72-hour isentropic trajectories leading to the spot on the west coast of Eire marked with a cross. Each point is marked with a number indicating the change in pressure altitude of the parcel along the trajectory. Those trajectories marked with a large negative pressure change (subsidence) also have a small figure indicating the fraction of the subsidence that occurred during the final 36 hours. (See text)

One difficulty with such measurements is that they appear to miss most of the troposphere. Ozone or non-sea-salt sulfate may be measured at a certain local concentration, but what were the concentrations of precursors when these products were made at another altitude? Aircraft or balloon measurements appear to be necessary to answer such questions.

We initiated a study of transport in order to guide airborne measurements and help relate them to measurements at the surface sampling stations. The analyses suggested that the sampling stations appear to be frequently and predictably exposed to air of recent origin 1500 to 4000 meters above the surface. It was the consistency of this exposure that surprised us and suggested that estimates of upper-air concentrations from surface measurements was a worthwhile endeavor, but one that requires some care.

1.2. Beginnings of a Trajectory Climatology: Eire

We calculated twenty-four back trajectories ending at each of several sites that have been proposed for AEROCE, TOR, and other measurement networks. The isentropic trajectory cases were selected from four spring seasons at time intervals of two weeks, so that many fundamentally independent weather situations were analyzed. This paper focuses on one site on the western shore of Eire that is especially interesting for European and American air chemists. We chose as our endpoint the 925 mb (600m) level above the rawindsonde station at Valentia, so as to have the most accurate estimates of winds near the surface at the trajectory endpoint. This is a sensitive point in the trajectory calculations, and it is most important that these wind derive from observations rather than an interpolating analysis. We expect the results to be equally valid for the Mace Head site, if allowances are made for local wind systems like land-sea breezes.

The calculated origins of the air parcels three days earlier are shown in Figure 1. Beside each origin point is a number noting the change in pressure altitude the parcel experienced on the trajectory (in millibars), and another number describing what fraction of the subsidence occurred in the final 36 hours. Occasionally that figure is greater than one, indicating previous rising motion. The numbers do come close to one, suggesting that the trajectories have essentially captured the latest vertical motion event.

Analyses for this station showed almost without exception that there had been substantial (53 to 433 mb) subsidence over 72 hours *whenever* the trajectory originated from the sector northwest of the station. Approximately 70 percent of the subsidence occurred within the last 36 hours, meaning that the mean subsidence rate over that period was approximately 1.5 cm/s. Air with origins in sectors distinctly away from the northwest showed very little vertical motion according to the isentropic analysis, but since the air stays near the surface, diabatic effects are likely important for these air parcels. For the starting points marked zero, the back trajectory hit the surface, and followed the surface, crossing isentropes. It would seem that these boundary-layer trajectories are not to be trusted. However, Haagenson *et al.* [1987] have shown that isentropic trajectories perform as well as any trajectory considered when compared to an analysis of a perfluorocarbon release in the Cross-Appalachian Tracer Experiment. We do not know of systematic errors that would make the marked beginning points too close or too distant from Mace Head.

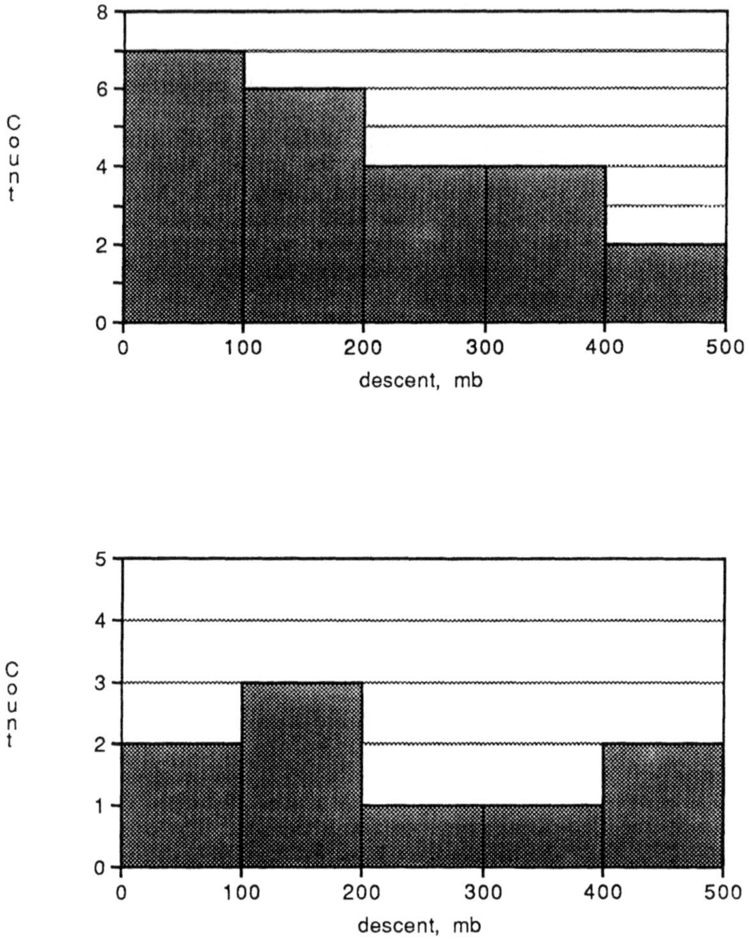

Figure 2. (a) Histogram showing the amount of subsidence in mb for trajectories leading to Irish and to St. John's, Newfoundland. (b) Histogram for the Irish trajectories only. The Irish trajectories are as described in the text, while for Nova Scotia, all trajectories originating between 250° and 360° were accepted. The three cases with origins between 250° and 270° showed strong subsidence, with pressure changes of 204, 330, and 305 mb. For all stations, there were many trajectories from all directions. Those with origins between 90° and 245° had isentropic trajectories with pressure change near zero, and did not make a very interesting histogram.

1.3. Corroboration at Other Sites and by Other Data

That equatorward motion is generally associated with subsiding air in mid-latitudes is well-known in synoptic meteorology [Palmen and Newton, 1969]. Merrill *et al.* [1985] have reported several instances of subsiding transport of Asian air to the Marshall Islands. However, real weather is complex, and so we think it worthwhile to report how common, coordinated, and intense this phenomena is in the North Atlantic. The phenomena is of course not strictly limited to the northwest quadrant, and some of the cases strongest sinking in our analyses occur when the wind is generally between 225° and 45°. However, there are also more cases of little motion for these trajectories.

Similar analyses for other stations gave very similar results. We composited trajectory data for a Nova Scotia site with the Eire site to make the histogram shown as Figure 2. All the data included show parcels with origins in the northwest sector. Data for Iceland and Azores sites are generally similar. Iceland has a more complex meteorology than other sites in several respects, a behavior which we ascribe to the frequency of low pressure systems and Iceland's proximity to mountainous Greenland. That proximity greatly affects trajectories from the northwest. There were cases of very strong indicated subsidence and cases of very little. Even the Azores showed a very similar pattern when transport was from the northwest or north.

The statistics of the trajectories do fit rather well with general notions of the climatology of high pressure systems and air motions within them [Palmen and Newton, 1969]. We suggest that subsiding trajectories will be more common from the southwest quadrant toward more northern latitude stations like Iceland, and more common from the northeast quadrant toward stations at the Azores latitudes and further south. Nearer the tropics and at the height of the summer season, we expect that true parcel motions will be less slanted, and essentially vertical convective downdrafts will be a major mechanism bringing free tropospheric air to the surface.

However, it is interesting to compare our meteorological results to a report of sudden, strong ventilation of the mixed layer at 23 N which was noticed in chemical measurements. Andreae *et al.* [1985] reported a very rapid change in air composition during shipboard dimethyl sulfide (DMS) and aerosol sampling off the coast of the Bahamas. This followed the passage of a November cold front, when both the local and synoptic surface winds were out of the northwest. A rapid drop in measured dimethyl sulfide concentrations was attributed to ventilation by low-DMS air and the introduction of unspecified oxidants originating from the United States. It took six to eight hours after the frontal passage for the concentrations to drop to near-constant values.

It is not likely that all the trajectories with substantial downward transport leading to Eire were within a few hours of a cold front, but we are investigating whether the strongest vertical motions were associated with these fronts, as might be expected. It does seem very likely that the subsidence was associated with undisturbed weather or shallow, suppressed convection at the endpoint. One conclusion is that scientists should beware of a fair-weather bias in air composition measurements in rainy coastal environments! This is especially true if one wishes an observation in which the mixed-layer air composition has reached some kind of steady state with the ocean surface.

One of us (P.L.H.) is examining correlates which would allow a description of trajectory directions and subsidence in terms of variables readily obtainable from a weather analysis or forecast. We hope that these will provide a more useful tool for categorizing data.

2. ACCOUNTING FOR PARCEL MIXING NEAR THE SAMPLING SITE

2.1. Isentropic Parcel Trajectories are Meaningful above the Mixed Layer

Isentropic trajectory analyses of air parcel motion have known difficulties, but many need not concern us in describing subsiding motion. One potential difficulty is that the analysis is not valid if heat is added to or removed from the parcel. Another is turbulent mixing, which disperses the parcel and makes the trajectory much less informative. The fact of subsidence suggests that air is stable and that cloud transport, undetectable in the synoptic rawindsonde collection used for our analyses, will not be very deep. Latent heat release effects on the large scale are similarly ruled out. Radiational cooling of the air is another matter. When a nominal one degree per day cooling is taken into account, the estimated subsidence may well be *increased*, but only by 50 mb over our three day period. Turbulent mixing may break out in clear air, even away from mountainous regions, but its effect is difficult to estimate, and, we believe, small. [Pao and Goldburg, 1969, Rosenberg and Dewan, 1974]. The advantages over isobaric trajectories should be clear. Clearly, the subsidence shown in Figure 1 could not be discovered from isobaric trajectories; Danielsen [1961] and many others have shown that directional and distance errors for the origin points would also be consididerable.

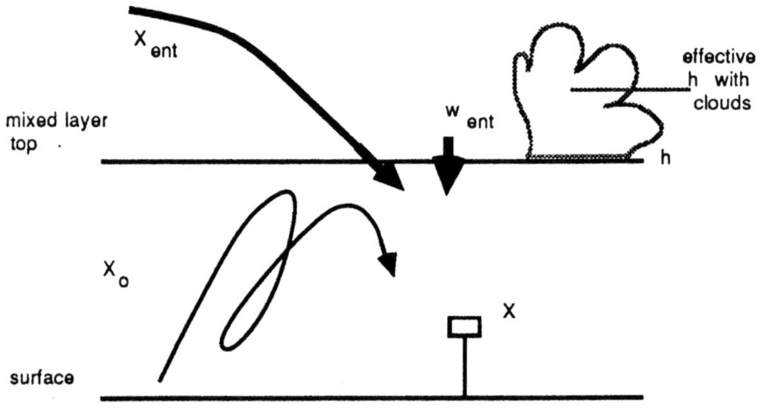

Figure 3. A simple conceptual model used to describe the effect of subsiding motion upon concentrations χ measured over time at a surface site. The entraining parcel concentration χ_{ent} is brought in with an entrainment velocity w_{ent}, but its effect on the sensor is diluted by turbulent mixing with the original mixed layer concentration χ_o. An effective mixing height h may be estimated as cloud base, or may be somewhat higher due to partial mixing of higher layers by clouds. This model suggests a 6 to 18 hour relaxation time of concentrations toward χ_{ent} if chemical deposition to the surface is not too strong.

2.2. Parcels Mixing into the Boundary Layer Have Diluted Impact

When the air parcel reaches the atmospheric boundary layer, mixing processes have a major effect. If air parcels aloft are contained within isolated blobs or streams of pollution [Ehhalt *et al.*, [1986], J.Greenberg, P. Zimmmerman personal communication, 1987], then these will usually influence the sampling site only gradually, as the air parcel mixes with and displaces the previous air. Figure 3 shows a simple conceptual picture we will use in describing the process.

Imagine a parcel just becoming entrained into the boundary layer, with a mixing ratio of some important chemical species χ_{ent}. This parcel is at the leading edge of air with distinctly different composition. It has "caught up" to the boundary layer air due to the substantial wind shears that are typical at the top of that layer. The entrainment velocity into the boundary layer is w_{ent}. (Assume that this is the same as the subsidence velocity, since strongly maritime sites show little diurnal change in boundary layer height). Idealize the situation upwind within the boundary layer as horizontally homogeneous for air upwind of the station, with mixing ratio of χ_o. The boundary layer has a depth h, which we take to be 500 to 800 m. A 770 m depth can be estimated from the average of reports from weather stations on the Irish west coast [Warren, *et al.*, 1986], while a 500 m depth is estimated from the mean cloud base when clouds are reported over the ocean upwind [S. Warren and C. Hahn, personal communication, 1986]. Alternatively, a one-kilometer height would make allowances for partial mixing of boundary layer air into a low cloud layer.

When material enters the boundary layer, it mixes throughout rapidly, typically with a time scale of a half hour or more. These two time scales, an entrainment time scale and a deposition time scale, are natural ones and simplify our quantitative discussion:
Let

$$\tau_{ent}^{-1} = \frac{w_{ent}(\rho_h/\rho_m)}{h} \quad \text{and} \quad \tau_{dep}^{-1} = v_{dep}/h \tag{1}$$

where v_{dep} is a bulk deposition velocity for the species applicable to the whole mixed layer and ρ_h and ρ_m describe the atmospheric density at the entrainment height and the mean density of the mixed layer. The situation of Figure 3 is described by this linear differential equation:

$$\frac{d\chi}{dt} = \chi_{ent}\tau_{ent}^{-1} - \chi\left(\tau_{ent}^{-1} + \tau_{dep}^{-1}\right) \tag{2}$$

Notice that the second occurrence of τ_{ent}^{-1} describes the gradual loss of material out the side of the box towards some distant storm system. The concentration χ of material within the boundary layer then evolves in this manner, the equation of adjustment:

$$\chi(t) = \chi_o e^{-(t/\tau_{ent}+t/\tau_{dep})} + \frac{\tau_{ent}^{-1}\chi_{ent}}{\tau_{ent}^{-1} + \tau_{dep}^{-1}}\left(1 - e^{-(t/\tau_{ent}+t/\tau_{dep})}\right) \tag{3}$$

This expression shows the effect of mixing as χ_o is gradually transformed in time towards a quasi-steady state in which introduction of new material is balanced by deposition to the surface. For materials like ozone, which are relatively unreactive with the surface, the entrainment process dominates $((\chi_{ent} - \chi_o)\tau_{ent}^{-1}$ is rather larger than $\chi_o \tau_{dep}^{-1}$). For very reactive materials like gas-phase nitric acid, the two time scales may be quite similar, since a 1 cm/sec entrainment velocity matches a 1 cm/sec bulk deposition velocity. (Nitric acid may also have other reactions besides those at the earth's surface, and their rate may be added to the deposition rate. Sources within the boundary layer need to be treated like the first entrainment term on the right side of Equation 2)

The characteristic adjustment time of weakly deposited materials may then be estimated as $\tau_{ent} = 500\text{m}/1.5\text{x}10^{-2}\text{m s}^{-1}$ or 9 hours. Here we assumed that the entrainment rate is equivalent to the average subsidence rate for the parcel over the last 36 hours, and a shallow mixed layer. This compares well with the adjustment time noted by Andreae *et al.* [1985] for the wintertime cold-front situation near the Bahamas. In fact, the mixed-layer height *and* the entrainment velocity are likely to be somewhat higher in both situations. Shallow convective clouds are very likely, both near the Irish coast and in the Bahamian seas. These mix on a much slower time scale than the subcloud layer (on the order of half a day to a day) and show substantial gradients of tracers [Betts and Albrecht, 1987, Ferek *et al.* 1986]. We surmise that an equivalent mixed layer height of 800 to 1200 m is a reasonable approximation: if all the material were redistributed to match the sub-cloud layer concentration, that would be the appropriate height. On the other hand, we expect that the vertical velocities of entraining parcels should be higher than average near the end of the trajectory, and should be especially high after a cold front. (Quantitative estimates are shown in Fleagle and Nuss, [1985].) It is the common situation that the subsiding air first begins to affect an area after the passage of a cold front, so that relatively rapid change of concentrations may occur. It is important to remember that the concentrations of reactive species in the upper air source region are frequently not homogeneous, so a continual adustment process should occur.

Practically speaking, the preceding discussion and equations can be used as a guide in an preliminary analysis of remote station data, so as to get a semi-quantitative description of influence by the upper air. It may be tempting to take concentration differences and solve the equations for the original upper-air concentrations. The main difficulty we expect would be in the assumption that the boundary layer air continuing to blow over the station is horizontally homogeneous in concentration. Inhomogeneities in boundary layer concentrations or chemical processes could be mistaken for features of the upper air. Very important species like the reactive nitrogen compounds can be studied only as elemental groups, like "total reactive nitrogen". Nevertheless, we believe that careful interpretations will be useful in illuminating the climatology of trace species in the global troposphere.

3. TRAJECTORIES AND MODELS: SOME CONCLUSIONS

3.1. Models Appropriate for Estimating a Region of Influence

The trajectories of Figure 1 also make a point which should be repeated. They show that the region of influence for the Mace Head remote site is quite large; including much of the North Atlantic and Europe. By domain of influence we mean the area or volume of the atmosphere that has a significant effect on measurements at the sampling site. A naive definition might be that volume which encloses half or two-thirds of the air parcels that are sampled at the site. Now, the points on the map of Figure 1 do not indicate that domain exactly, for they are the points that are associated with the origin and properties of χ_{ent}. But they do indicate the breadth of influence on the sampling station. We have seen that local air within the boundary layer also has a strong influence, and this air moves much more slowly.

Another way to estimate a domain of influence is to use a simple mean-motion and eddy-diffusion description of transport, such as is used in global two-dimensional tropospheric models [*e.g.*, Derwent and Curtis, 1977, Crutzen and Gidel, 1983] or "traveling" one-dimensional models [Gidel, 1984] which are successively exposed to oceanic and continental influences. These models have proven useful in describing general features of global oxidant chemistry. However, if one uses this approach, selecting local vector mean winds applicable for the area of the Irish coast (2 m/sec or so) and horizontal diffusivities used in global models ($3x10^6 m^2/sec$), the calculated region of influence has a diameter of only 1200 km, or approximately 15 degrees. This would suggest that samples should predominantly reflect the air composition determined for a short distance upwind into the Eastern Atlantic.

Clearly, Figure 1 shows that a significant part of the influence is from distinctly further away. That figure suggests a domain of influence for three days would be perhaps 3500 km wide and centered perhaps 1000 km upwind, although these figures represent only one part of a complex influence. It also suggests that the site is frequently influenced by air recently over the North American continent, but that air is also likely several kilometers above the surface, and may experience no or diluted influence from the continent. When low levels of continentally derived compounds are sampled at Mace Head, that may be due to the high altitude origin of air parcels influencing the site, rather than extensive decay during weeks of low-altitude over-water transport.

We speculate that simple advection/diffusion models would do better if they could incorporate the interaction of horizontal and vertical transport more intimately.

3.2. Conclusions

The air sampled at remote stations can be used to help describe the composition of mid-tropospheric air. These stations can contribute toward a relatively inexpensive continuing climatology of the middle troposphere. It appears likely that simple weather analysis and forecast data can indicate when upper-air influence will be strong. More definitive isentropic trajectory analyses will be an important part of the AEROCE data collection strategy. However, airplane and balloon measurements are necessary to sample the upper middle troposphere and also to discover if that air which subsides strongly may have biases in its composition compared to other air from the origin level. The interpretations we have presented require that local boundary-layer perturbations upon the station be small, *i.e.*, it is even more important to escape or screen out local sources than for estimation of mean concentrations. Quantitative description of these influences, improving on the formulas given here, can magnify the value of remote air sampling.

However, the idea of a domain of influence for a station is a complex one: frequently a substantial portion of the influence is local, and another substantial portion is quite distant in altitude and location. Trajectories extending much of the way across the Atlantic are common, but many of them trace back to high altitudes. The correlation of vertical and horizontal transport effects must be considered in modeling the air chemistry of reactive species affecting ozone levels in the mid-latitudes.

ACKNOWLEGEMENTS: The National Center for Atmospheric Research is funded by the United States National Science Foundation. (NCAR Air Chemistry Division publication number 1905-87-6). We thank Richard Brost and Rolando Garcia for commenting on a draft of this work, and several attendees of this meeting for discussions.

REFERENCES

Andreae, M.O., R.J. Ferek, F. Bermond, K.P. Byrd, R.T. Engstrom, S. Hardin, P.D. Houmere, F. LeMarrec, H. Raemdonck, and R.B. Chatfield., 1985. 'Dimethyl sulfide in the marine atmosphere,' *J. Geophys. Res*, **7**, 12891-12900.

Betts, A.K, and B.A. Albrecht, 1987. 'Conserved variable analysis of the convective boundary layer structure over the tropical oceans,' *J. Atmos. Sci.*, **44**, 83-99.

Crutzen, P.J., and L.T. Gidel, 1983. 'A two-dimensional photochemical model of the atmosphere: The tropospheric budgets of the anthropogenic cholrocarbons, CO, CH_4, CH_3Cl, and the effects of various NO_x sources on tropospheric ozone,' *J. Geophys. Res.*, **88**: 6641-6661.

Danielsen, E.F., 1961. 'Trajectories, isobaric, isentropic, and actual,' *J. of Meteorol.*, **18**, 479-486.

Derwent, R.G., and A.R. Curtis, 1977. 'Two-dimensional model studies of some trace gases and free radicals in the troposphere,' Report AERE-R8853 of AERE Harwell, U.K., (H.M. Royal Stationery Office).

Ehhalt, D.H., J. Rudolph, and U. Schmidt, 1986. 'On the importance of light hydrocarbons in multiphase atmospheric systems,' *Chemistry of Multiphase Atmospheric Systems*, ed. W. Jaeschke, Springer Verlang, New York, 321-350.

Ferek, R.J., R.B. Chatfield, and M.O. Andreae, 1986. 'Vertical distribution of dimethyl sulphide in the marine atmosphre,' *Nature*, **320**, 514-516.

Fleagle, R.G., and W.A. Nuss, 1985, 'The distribution of surface fluxes and boundary layer divergence in midlatitude ocean storms,' *J. Atmos. Sci*, **42**, 784-789.

Gidel, L.T., 1984. 'The role of clouds in micro and macro-scale transport of atmospheric constituents,' in *Gas-Liquid Chemistry of Natural Waters*, Report BML 51757, Vol. **I**, of Brookhaven National Laboratory, Upton, N.Y., U.S.A.

Haagenson, P.L., Y.-H. Kuo, M. Skumanich, and N.L. Seaman, 1987. 'Tracer verification of trajectory models,' *J. Clim. Appl. Meteor.*, **26**, *in press*.

Merrill, J.T.,, R. Bleck, and L. Avila, 1985. 'Modeling atmospheric transport to the Marshall Islands,' *J. Geophys. Res.*, **7**, 12927-12936.

Palmen, E. and C.W. Newton, 1969. *Atmospheric Circulation Systems: Their Structure and Physical Interpretation*, Academic Press, New.York.

Pao,. Y.H., and A. Goldburg, 1969, *Clear Air Turbulence and its Detection*, Plenum Press, New York.

Rosenberg, N.W., and E.W. Dewan, 1974. 'Stratospheric turbulence and vertical effective diffusion coefficients,' *Proceedings, Third Conference on the Climatic Impact Assessment Program, eds.*, A.J. Broderick and T.H. Hard, United States Department of Transport DOT-TSC-OST-74-15, 91-101.

Warren, S.G., C.J. Hahn, J. London, R.M. Chervin, and R.L. Jenne, 1986. 'Global Distribution of Total Cloud Cover and Cloud Type Amounts Over Land,' NCAR Technical Notes NCAR/TN-273+STR.

Part II

WORKING GROUP SUMMARIES

ATMOSPHERIC TRANSPORT

G.Brasseur (chairman), F.Stordal (rapporteur), H.Van Dop (rapporteur), R.Chatfield, S.Joffre, J.Jonson, H.Kelder, H.Levy, V.Vaughan.
(WORKING GROUP I)

1. INTRODUCTION

Trace species with chemical lifetimes significantly longer than the local characteristic time for horizontal and vertical exchanges may be transported far away from their emission sources before they react with other chemical species. It is therefore of primary importance to understand the basic mechanisms governing the transport of chemicals and to formulate these processes as accurately as possible in numerical models.

The study of trace species dispersion in the atmosphere requires a detailed understanding of the basic transport processes on a wide range of spatial and temporal scales especially between the different atmospheric layers. Shown in Figure 1 is a schematic representation of various transport mechanisms below 30 km. Of particular importance are the exchanges between the atmospheric boundary layer (ABL) and the free troposphere, the ABL and the surface, including wet and dry deposition processes, and between the stratosphere and the troposphere. Several papers in this volume deal with some of these questions. The purpose of this report is to identify the important problems to be resolved in the next decade or so and to suggest research strategies in this context.

1.1. Spatial and Temporal Scales in Atmospheric Transport

The type of transport affecting chemical species varies with the time scale involved. For some applications it may be desirable to know hourly values of concentrations; for other only monthly, seasonally or yearly averaged concentrations are required. Also the spatial resolution of observations vary widely. In the modelling of transport, therefore, it makes sense to distinguish these various scales.

The area of concern in this study ranges from the atmospheric boundary layer through the tropopause, i.e. the vertical scale ranges from 1 to 20 km. Numerical models which are able to resolve the structure of transport over the vertical range should have a grid which is at least an order of magnitude smaller. In a consistent three-dimensional model, horizontal scales should be compatible with vertical scales and can be inferred from the vertical scale by equating their corresponding time scales. Vertical transport velocities vary from $\sim$ 0.01 m/s

I. S. A. Isaksen (ed.), Tropospheric Ozone, 383–392.

Fig. 1 A schematic view of transport processes in troposphere and strat

1. Earth/ocean-atmosphere exchange processes
2. dry-convective ABL-mixing (Ekman-layer pumping)
3. Mid-latitude vertical exchange processes associated with front systems and moist convection
4. Deep convection in tropical storms
5. Mean tropospheric circulation
6. Horizontal exchange by breaking of Rossby waves
7. Troposphere-stratosphere exchange (tropopause folding)
8. Horizontal eddy diffusion in the stratosphere

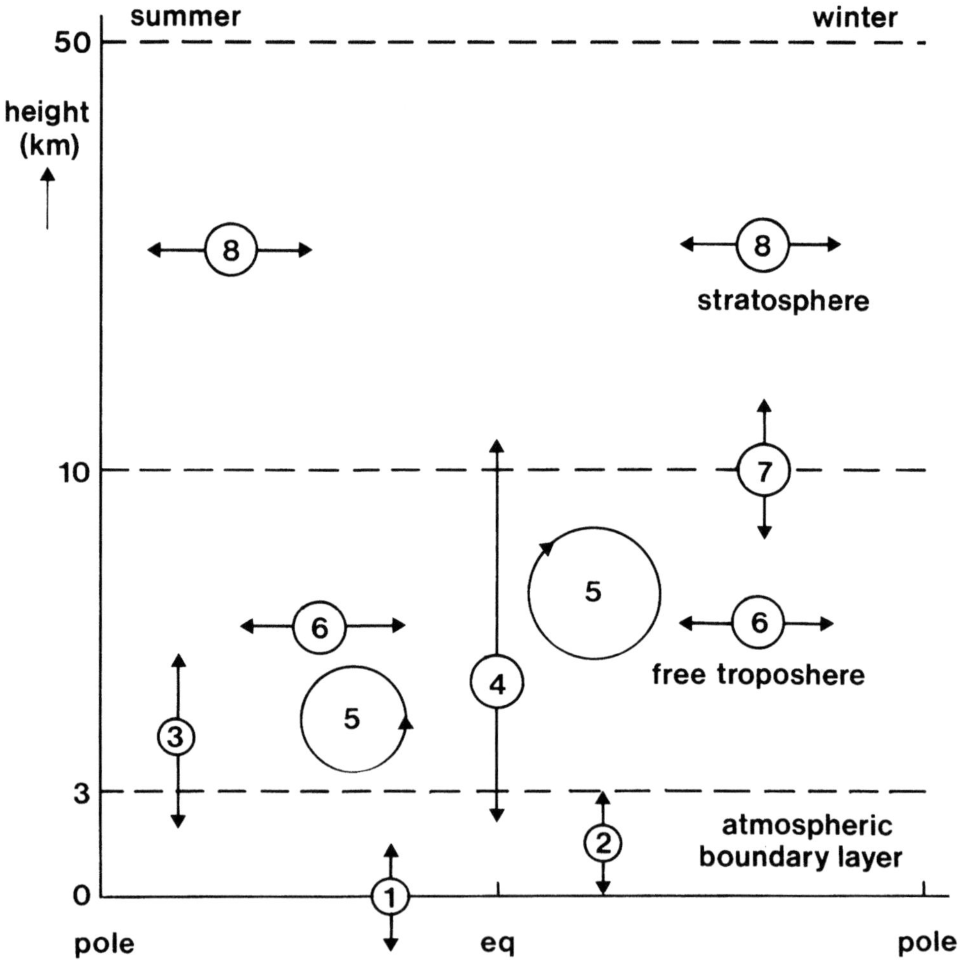

(subsidence) to $\sim$ 1 m/s (convection), which leads to time scales of 1000 - 100.000 s. Realizing that horizontal transport velocities are of the order of 1-100 m/s, the horizontal mesh size should be somewhere between 0.1 and 1000 km. (It should be noted here that chemical processes have their own timescales which may set different requirements. This problem, however, is not addressed here).

A typical model with a grid size of 100 km, a vertical resolution ranging between 100 and 1000 m and a timestep of three hours should resolve the (horizontal) Rossby planetary waves, most circulations associated with cut-off lows, orographically induced vertical motions by the large mountain ridges (Alps, Andes, Himalya, Rocky mountains), large scale subsidence, cyclones and anticyclones.

They however do not resolve small scale turbulent disturbances (boundary layer and clear air turbulence),cumulus cloud convection and associated precipitation, local circulations such as nocturnal jets, valley circulation, land-sea breeze and drainage flow, incidental cumulonimbus entrainment of stratospheric air in the tropics and fronts.

Features that are partly resolved include tropopause folds, mesoscale flow over mountains and mesoscale organization of convection in the tropics.

1.2. Potential Applications of Transport Modelling

The modelling of transport in the troposphere may contribute to our understanding of the distribution and trends of atmospheric constituents. In particular, it may contribute to the following fields:

- regional modelling of ozone

Causal relationships may be found between various sources of primary pollutants and the occurrence of high ozone concentrations.This can be done by episodic transport models, i.e. models which describe transport (and chemistry) over a period of a few days only.

Models which have proved their ability to predict concentration fields can be used either to supply or to interpolate data in regions where measurements are sparse.

Simpler and faster means to identify sources are trajectory analyses. Such analyses (and forecasts) may be helpful in guiding experiments and examining chemical transformations during field experiments.

- global modelling of ozone

For the study of the global distribution and trends of ozone and other relevant trace constituents, measurements will provide only a very limited and incomplete picture. They are, especially at elevated levels, difficult to carry out and relatively expensive. Models provide global fields and may realistically represent the spatial and temporal variability of trace gases.

2. FUNDAMENTAL PROCESSES IN THE TROPOSPHERE

2.1. Vertical Stratification

Away from the boundary layer, the potential temperature of the atmosphere increases with height. Its vertical gradient determines the stability against convective mixing and therefore the potential for laminar, stratified flow. In the stratosphere, where the potential temperature increases rapidly with height, laminar flow along isentropic surfaces dominates, and any trace gas released as a parcel into the lower stratosphere spreads rapidly while remaining on the surface. The stratification is not as pronounced in the troposphere, but is nevertheless present, particularly in high pressure systems where subsidence inhibits convection.

Laminar flow of this kind has several implications for atmospheric chemistry which are not well understood. Most obviously, if reactive tracers lie in adjacent thin laminae a few hundred meters thick, their chemistry will be most active at the interface - which can be very sharp. Thus, in stable flows the inhomogeneous mixing of tracers of different origin suppresses chemistry and permits long-range transport of reactive trace constituents. Modelling the transport of reactive species becomes very hazardous under these conditions - the familiar expression of the law of mass action,

$$d(A)/dt = -k\,(A)\,(B)$$

with (A) and (B) averaged over some model box, will clearly be a misrepresentation. Fundamental research into the extent and the properties of laminar flow, as well as its impact on chemistry, is required.

2.2. Wind Shear and Parcel Integrity

Related to this problem is the effect of wind shear on tracer distributions. A parcel of air, initially of a regular shape, perhaps 100 km square, will be extruded by wind shear over a period of a few days to a long thin, coiled parcel. In doing so, parcel integrity is eventually lost, since the surface area increases enormously and diffusion permits exchange of tracers with adjacent air parcels. Again, Eulerian type models are sensitive to this process of differential advection since it can act to increase gradients by bringing together air parcels of very different origins. Furthermore, when the extruded air parcel becomes thinner than the grid of the model, its identity within the model is lost and its contents immediately mixed into its surroundings. Thus Eulerian models may be expected to underestimate the long-range transport of trace constituents.

Isentropic trajectory calculations based on numerical forecast models are a powerful tool for calculating air transport in the free atmosphere. But these too are sensitive to wind shear - both from the standpoint of parcel integrity, and because of the sensitivity of the calculations to initial conditions. Wind shear is very large near jet streams, and trajectory calculations for longer than two or three days in the free troposphere should be undertaken only for specific meteorological situations.

2.3. Stratosphere-Troposphere Exchange

The problems of describing exchange of air between the troposphere and the stratosphere are magnified in situations with tropopause foldings, where stratospheric air flows in a lamina beneath a northwesterly jet stream as a response to frontogenesis. The fraction of the ozone flowing into the fold which remains in the troposphere (rather than returning isentropically to the stratosphere) is not known. It is important to know this fraction, since any air returning to the lower stratosphere is likely to contain a tropospheric component introduced by the vigorous mixing at the base of the fold. This would then re-enter the troposphere in the next fold downstream, so that all the air entering the troposphere in a folding event would not be of stratospheric origin.

Calculations of the global flux of ozone from stratosphere to troposphere, based on experimental data, have so far consisted of extrapolation from a number of case studies, mostly in the U.S. For the Southern Hemisphere, in particular, the flux is not known, even in the mean, let alone in its variability. Global estimates of stratosphere-troposphere exchange have been calculated with GCMs, and agree quite well with those based on case studies, but the results need to be confirmed by global studies using real data before confidence can be expressed in flux estimates. An uncertainty of a factor of two, particularly in the Southern Hemisphere, is likely.

2.4. Fronts

The mesoscale dynamics of frontal systems are not well understood by meteorologists. Some structure is well-known - e.g. the conveyor belt mechanism ahead of a cold front and the extensive slow ascent of air above a warm front. Embedded in these are mesoscale rainbands where enhanced precipitation and vertical transport is found, often in a highly organised structure. Fronts also behave erratically over land, especially in mountainous regions where they occasionally disappear, perhaps reforming on the other side of the mountains.

These fronts are not regions of uniform precipitation and ascent, and must be treated especially carefully in chemical models. By their nature, they are important for exchanges from the boundary layer to the free troposphere in the extratropics - so that boundary layer trajectory calculations, for instance, should terminate when a front is encountered. Trajectory calculations through frontal regions are also difficult to perform in the free troposphere. Experiments with releases of non-reactive chemical tracers in the vicinity of fronts would greatly aid chemists and meteorologists in studies of the transport in these regions.

2.5. Cloud Sheets

Extensive stratocumulus sheets are a feature of winter anticyclones in the high midlatitudes. These layer clouds often interact very little with the rest of the atmosphere during daytime, so that trace constituents trapped in them can be transported up to 1000 km. Wet chemistry occurs in these clouds, but there is seldom any precipitation- the clouds usually evaporate.

2.6. Cumulus Cloud Transport

One area of fundamental uncertainty concerns the transport properties of convective clouds. There are strong indications that the inflow, outflow and trajectories of molecules depend on the depth and breadth of the cloud, the environmental wind shear, the rainfall process, etc... Observations and modelling studies are now only beginning to define quantitatively the answers to these fundamental questions. Carbon monoxide and hydrocarbons provide tracers of air motions that are free of the measurement and interpretative difficulties associated with conventional meteorological analyses. Cloud water and cloud air may take different trajectories in cumulonimbi. Soluble species with low vapor pressures are expected to follow the water droplets, but many important species that are soluble have a high vapor pressure and may move from drop to drop.Soluble conserved tracers may illuminate these processes.

Nitric oxide may have important sources within thunderstorms. The impact on NO distribution in the global troposphere depends on the transport through and out of the clouds as well as complex heterogeneous chemistry. Currently, all of these processes are poorly understood.

Cloud droplets are very efficient for scavenging particles and soluble gases. Complicated cloud models including interactions between dynamical, chemical and microphysical processes in the gaseous, liquid, ice and particle phase are difficult to incorporate into even the highest resolution mesoscale models. Further complications arise from the complexity of radiative transfer and heterogeneous chemistry in this environment. Nucleation is the main in-cloud scavenging process whereas impaction dominates below cloud. The stochastic nature of precipitation fields makes it difficult to compare model results with rain chemistry observations because individual rain gauge data are not representative enough. In fact large discrepancies are often observed.

2.7. Transport in the Atmospheric Boundary Layer

Processes within the lowest part of the ABL, say 5-50 m (the surface layer), are well represented by th Monin-Obukhov similarity theory which provides a framework for parameterizing the aerodynamic resistance to deposition of particles and weakly reacting gases. Closer to the surface ($\sim$1 mm), molecular and laminar resistance is generally parameterized on the basis of models tested by laboratory measurements. How well these laboratory conditions represent the real world remains to be demonstrated.

The case of particle deposition over water is more difficult due to the growth of particle size as they approach the water surface. Effects of bubbles, spray and foam should also be clarified since these mechanisms may apply to a large fraction of the earth's surface.

More data (surface type, moisture content, chemical/biological state) are required in order to assess the surface resistance to transfer for trace gases over a wide variety of surface conditions. It should be mentioned that some species have their main resistance to transfer in the water, so that mixing and molecular diffusion in the upper layer of the ocean should be coupled to the meteorological model.

The horizontal transport of species is performed by the advecting wind (longitudinal diffusion being at most 10 % except in stagnant wind conditions). Under unstable conditions, the upward mixing of ground-emitted species together with the relatively large gridsize of models make the assumption of a vertically homogeneous wind field reasonable in the ABL. However, the 925-mb surface should be more representative of ABL-wind than the often used 850-mb surface generally located above the ABL top. The interfacial layer at the top of the ABL is often characterized by a strong jump in wind speed and direction.

In the case of the stable ABL, there is a large velocity and directional shear throughout the ABL so that the horizontal transport of species should be height dependent close to the source as they rise slowly (due to weak turbulence) from the surface.

The collapse of the continental daytime well-mixed ABL at sunset leaves boundary layer air in a residual layer aloft, subsequently transported elsewhere. This type of exchange has not been properly addressed in many existing transport models.

Surface emission can be treated in the same way as deposition. However, all models suffer from the absence of temporal distribution in the emission data (both anthropogenic and biogenic). Incorporation of time-dependent emission input would significantly improve transport models capability.

3. USING METEOROLOGICAL/DYNAMICAL MODELS TO SIMULATE TROPOSPHERIC TRANSPORT

High resolution global general circulation (climate) and weather prediction models (100 km grids and 20-40 levels in the vertical) can simulate realistic synoptic meteorology down to and including such important mechanisms for free tropospheric transport as extra-tropical cyclones and tropical waves. Their precipitation patterns are accurate on a regional scale, but not quantitatively correct on the mesoscale. Medium resolution meteorological models ($\sim$300 km grids and 10-30 vertical levels) are also able to simulate such important features but show only qualitative agreement with observation. Meteorological fields from global models can simulate the global climatology of distribution and deposition of chemical trace species and provide 3-5 day forecasts for particular experiments. Unfortunately, at their highest resolution, these models tax the limits of current computers. We can reasonably expect that the next generation of computers (approx.1990) will support high resolution global transport/chemistry models which transport 1-5 chemically reactive species or groups. We should be planning to develop the medium resolution transport/chemistry models (1-5 transported species) which are consistent with current computer resources and, at the same time, begin to plan for the future.

Current regional and meso-scale (50-100 km sq) meteorological models have only been used to study specific event on a 1-2 day time scale though their use has been proposed for climatological studies. Such models produce more realistic simulations of synoptic scale phenomena as they evolve over 1-2 days. Beyond that time, the results rapidly deteriorate. The precipitation fields are more realistic than the highest resolution global models, but they do not yet capture the observed local variability.

Models which transport 10-20 reactive species during a particular meteorological event are currently under development.While these models simulate the gross distribution and deposition features for reactive trace chemicals, there are serious deficiencies in the detailed structure. Beyond 2 days the current model meteorology rapidly degrades. The continued improvement of such meteorological models and their extension in space and time is clearly needed.

An alternative approach is the nesting of a mesoscale model in a coarse global model. This is just in the development stage and the key scientific question is whether the result will be a regional meteorology that has been seriously degraded by the deficiencies of the coarse global model or a coarse global model that has a meteorology in a region that is of meso-scale quality. If it works, this may provide a framework for generating a realistic meso-scale chemical climatology, not just a 2 day forecast.

All these models exhibit two major common difficulties: the inability to resolve the important boundary layer turbulent transport processes and convective vertical transport in the free troposphere. While realistic meteorological models of these processes do exist, it is not possible, as discussed in 2., to include them explicitly in the transport/chemistry models. Therefore it is necessary to parameterize their effects. Development of realistic parameterizations for both boundary layer turbulence and cloud transport and precipitation is a critical need for both meteorology and atmospheric chemistry.

4. MODELLING SUB-GRID TRANSPORT PROCESSES

4.1 Parameterization of Subgrid-Scale Convective Transports

The underlying transport functions describing the transport by clouds of material from each level of the troposphere to every other level are not clearly defined by theory or experiment. Consequently, we cannot be very confident of our parameterizations. Traditional parameterizations for convection have errors in their conceptual underpinnings. New parameterizations for weather-forecast and climate models will probably await more fundamental understandings of cloud-scale and mesoscale meteorology. Air chemical applications cannot wait.

There are good reasons why air chemistry makes different demands on transport models. The parameterization of convection in meteorological models addresses the problems of latent heat release and momentum generation, but does not pay particular attention to material transport. This problem will remain a very difficult one that requires the cooperation of atmospheric chemists and dynamical meteorologists.

4.2. Parameterization of Subgrid-Scale ABL Transport

Exchange between the ABL and the free atmosphere occurs by entrainment in the case of a well-mixed ABL (warm continental daytime conditions or cold air outbreaks over warmer seas). Since this is a small-scale process, it has to be parameterized by using resolved or calculated quantities. In the case of a stable ABL (nocturnal continental case or spring/summer case over mid- and high latitude seas), the weaker entrainment, resulting from internal wave breaking or downward bursts of turbulence driven by strong velocity shear, has not been properly parameterized. Synoptic scale downward motion (subsidence) modulates the growth of the ABL, both in the stable and unstable cases.

Convective and mechanical production of turbulence have been mainly observed over the sea together with the occurrence of an extended capping layer of stratocumulus clouds. Radiative destabilisation at the top of the cloud layer, enhancing mixing inside the cloud, should be included.

Most parameters required to describe surface and ABL- top processes in transport models (surface heat flux, friction velocity, convective velocity and ABL- height) are provided by modern meteorological models (e.g. European Centre for Medium Range Weather Forecasts).

5. MODEL STRATEGY

Numerical models are important tools for studying the behaviour of trace-species in the atmosphere and for assisting in ozone control strategy formulation. Despite much progress made in recent years, a number of difficulties in model formulation remain to be solved.

The type of model to be used for regional or global studies should be governed by the particular problem to be addressed. For example, the chemical behaviour of fast reacting species for a given solar illumination can be derived from a simple zero-dimensional model if the concentration of the long-lived species, especially the source gases is specified either from field measurements at the given location or from a comprehensive transport model. Trajectory models (Lagrangian models) with a coupled chemical code may consider the changes of chemical composition within an air parcel which encounters different solar and meteorological conditions. These models are useful for the interpretation of trace species observations at selected measurements sites or for the assessment of the distribution of pollutants emitted at a fixed point. They are only reliable if the flow is not disturbed by wind shear, small-scale turbulence, strong convection or fronts. Trajectory calculations have shown that the origin of air parcels above a give location can vary strongly with altitude. Such simulations can only be performed over a limited period of time (a few days).

Regional or global distributions of trace species calculated over long integration times are usually provided by Eulerian models. The simplest approach involves one-dimensional models in which the vertical exchanges are parameterized by eddy diffusion. This method oversimplifies the representation of the transport processes which are advective rather than diffusive in nature and therefore other empirical parameterizations have been used. One-dimensional models can still be useful to give order of magnitude estimates of global budgets and to study the local details of chemical processes. Because of large horizontal inhomogeneities in both dynamical and chemical processes in the troposphere, the interpretation of the results given by such models is severely limited and the calculated vertical distributions cannot be regarded as global average conditions.

Two-dimensional models are in principle capable of accounting for latitudinal and seasonal dependence of solar illumination, and boundary conditions and allow a rudimentary representation of meridional advection. The limitation of such models arises from the difficulty of representing zonal averages of mass, momentum and energy fluxes, which are the result of complex dynamical processes involving large zonal assymmetries. The net fluxes resulting from these processes are usually specified through empirical parameters. However by using transport coefficients based on observed meteorological variability and dispersion of tracers, one may simulate the meridional distribution of chemical tracers.

A detailed representation of the transport of trace species on any scale (e.g. ABL turbulence, convective cloud, front, cyclone or interhemispheric) should be based on a full three-dimensional representation. The success of such models depends on the spatial and temporal resolution adopted in the dynamical formulation as well as on the completeness of the chemical scheme included. Despite the fact that the simulation of the real atmosphere is significantly improved by such models, some parameterization is still required to account for sub-grid transport and eventually to ensure numerical stability. With the rapid development of more powerful computers, the resolution of three-dimensional models can be expected to improve in the next five to ten years. Interaction with observations is important in model development, both for validating model predictions and for designing complex field experiments. Chemical tracer measurements have an important part to play in this interaction.

The most comprehensive chemical transport models of the troposphere should be linked to general circulation models, have the highest possible resolution, extend to the stratosphere and be capable of eventually studying the climatic impact of natural and anthropogenic perturbations.

A practical approach to the three-dimensional transport/chemistry problem is to use the assimilation fields from numerical forecast models to represent atmospheric dynamics. Chemical calculations may then be conducted in an "off-line" mode, but care should be taken to extract the full range of information required from the forecast model, not just temperature and wind fields.

Although less complex models remain quite useful for addressing a wide range of specific questions and for trying out new ideas, a major effort should be undertaken to develop high resolution three-dimensional chemical transport models. Because of the magnitude of this task, such projects require international cooperations and a long-term committment.

NOx IN THE TROPOSPHERE

F.Fehsenfeld (chairman), Ø.Hov (rapporteur), G.A.Ancellet,
R.A.Cox, D.Ehhalt, H.Hakola, M.Legrand, S.Liu.
(WORKING GROUP II)

1. GOALS AND RATIONAL

Present estimates indicate that much of the ozone in the earth's troposphere is produced photochemically in situ within the troposphere.- It is generally believed that this production is NOx limited.Thus, in order to estimate the ozone production in the lower atmosphere and to determine whether it can be perturbed, it is necessary to know:

a) the origin and distribution of tropospheric NOx.

b) the details of the chemical mechanisms that transform NOx toother members of the reactive nitrogen oxide family ($NOy = NOx + PAN + HNO_3 + NO_3^- + NO_3 + N_2O_5 + ...$), and of those leading to ozone formation.

c) the processes responsible for the eventual removal of NOy from the atmosphere.

To highlight the important areas of investigation required to understand these processes, the discussion is divided into four sections: 1) sources and sinks, 2) distributions and trends, 3) gas phase and heterogeneous chemistry and 4) instrument development. This research requires laboratory studies, modelling and field measurements.

2. SOURCES AND SINKS

The sources of NOx include emissions from fossil fuel combustion, biomass burnings, lightning, biogenic emissions, and stratospheric intrusion. Depending on the reaction scheme assumed, photochemical reactions of NH_3 may lead to either production or destruction of atmospheric NOx. Except for fossil fuel combustion, large uncertainties exist in the total emission rates and temporal and spatial distributions of the NOx-sources.

In order to attain an adequate understanding of the sources and sinks of NOx, the following programs should be undertaken:

I. S. A. Isaksen (ed.), Tropospheric Ozone, 393–401.

2.1. Source inventory and emission trends

An accurate inventory of the temporal and spatial distribution of NOx emission for all major sources, including fossil fuel combustion, should be developed. The data base should have adequate spatial and temporal resolution to satisfy mesoscale, as well as, global scale modeling requirements at all seasons. In this connection, it is recommended that measurements of emissions in plumes from well defined anthropogenic sources such as power plants and large urban areas be undertaken to determine fluxes from these sources in order to verify emission inventories.

In addition to an adequate inventory of current NOx emissions, a knowledge of the trends of NOx emissions are essential for evaluating the anthropogenic impact of the regional and global ozone distribution. Currently the data on the historical trends of NOx emissions exist for only fossil fuel combustion and in only a few industrial countries. The NOx emission data base must be capable of defining the trends of NOx emission from all major sources. In this regard it is recommended that a NOx monitoring network be developed that is capable of demonstrating how trends in NOx emissions are reflected in atmospheric NOx concentrations.

2.2. Nitrate Deposition

Measurements of nitrate deposition are carried out regularly in many industrial areas. However, there are few systematic observations of nitrate deposition at relatively clean and remote sites. However, these observations are required to help characterize the spatial distributions and temporal variations of the major NOx emission sources and to verify the loss process of NOx from the atmosphere.

3. DISTRIBUTION AND TRENDS

Since the transport processes and the chemistry of NOx and NOy molecules are not fully understood, we presently cannot translate a given source distribution into an atmospheric distribution of NOy with any certainty. Research is required:

a. To study the transport of NOy from the planetary boundary layer, where most of the sources are located, into the free troposphere.

b. To study the atmospheric distribution of the various NOy molecules in order to initialize and validate model calculations.

c. To determine any trend in concentration pattern which only result from long term changes in NOx emissions.

3.1. NOy-Distribution

Initialization and validation of model calculations require a knowledge of the distribution of NOy molecules throughout the troposphere. To provide the most useful data set the following parameters should be measured:

a. The sum of NOy - this is the conserved quantity and best reflects the relations between sources, sinks and atmospheric distribution.

b. The sum of NOx - this determines the local O_3 production and thereby the potential for photochemical ozone production.

c. PAN - this indicates past photochemical activity in an airmass and has been proposed as an important vehicle of NOx transport away from high emission areas.

Economically, it is advisable to measure these parameters simultaneously The techniques to measure the parameters in-situ are currently available. Thus both ground based and aircraft measurements are needed to provide the lateral and vertical distribution. Measurements should be carried out on both local and regional scales over low and high, natural and anthropogenic emission areas to study the transport of NOy. They also need to be done on a global scale and in all seasons to define the hemispheric and global O_3 production in the troposphere. In considering the implementation of this study it must be noted that the technology required to successfully undertake the measurements of these parameters are limited to a very small number of research groups.

3.2. Transport From the Planetary Boundary Layer

The boundary layer is a buffer zone for heat, moisture and pollution between the sources at the surface and the free troposphere. Typically in temperate latitudes the boundary layer breaks down every 2-4 days,

a. due to large scale convective instability,

b. at fronts where upsliding motion occurs,

c. in mountainous regions with considerable vertical stirring, and

d. through the diurnal pumping effect of the varying height of the PBL.

In the tropics, the typical exchange time between the surface and the free troposphere over land may be shorter than 2-4 days, due to stronger convective activity, while at high latitudes atmospheric stability is higher such that the PBL-free tropospheric exchange may on the average take longer than 2-4 days. Due to the similarity in typical PBL residence times of an air mass, and the chemical and physical decay times of NOx and NMHC-species, the net flux of these species into the free troposphere can be considerably smaller and with different speciation, than the emissions at the surface.

Model calculations of the net flux from the PBL need to address the inhomogeneous nature of the vertical exchange processes. Account should be taken of transport both on a synoptic and mesoscale (convection) in the calculation of fluxes between the PBL and the free troposphere.

Thus both theoretical and experimental studies should be made to assess vertical fluxes of NOx, and NMHC-species and to characterize the mechanisms found in:

a. convective cells

b. vertical motion in frontal regions

c. vertical motion set up by the land topography

d. diurnal pumping of the PBL due to day/night-changes in radiation and due to synoptic weather

e. PBL over maritime areas

When the transport out of the PBL from these different mechanisms is characterized, a parameterization for longer times and larger regions may be arrived at.

3.3. The Historic Nitrate Trend in Polar Ice

Nitrate measurements in polar snow and ice allow the determination of the atmospheric response at high latitudes to natural (solar activity variations, climatic change) and anthropogenic perturbation in the past. Such records have been obtained in several locations, both in Antarctica and Greenland, and they do not confirm the earlier proposition of a modulation of nitrate deposition flux in polar ice caps by the 11-year solar cycle.

In order to use these data to test calculations by a global model, more knowledge on the physical (gas or aerosol) and chemical form of nitrate in the polar atmosphere is required. At the present time this question is partly solved by a comprehensive study of soluble species in antarctic ice to establish the ionic balance. The deposition processes of nitrate in these areas remain poorly understood and therefore the relation between atmospheric concentration of NO_3^- and deposition is not well defined, and should be investigated. In addition, the role played by supercooled water droplets and the conversion of NOx to NO_3^- by ice particle aerosols needs to be investigated.

4. GAS PHASE AND HETEROGENEOUS CHEMISTRY

A detailed description of the mechanisms and time scales for the interconversion of NOy molecules is necessary for an understanding of the distribution and budget of NOx in the PBL and in the free troposphere. Advances in knowledge of the relevant chemical transformations require a combined approach using laboratory studies of elementary chemical and photochemical reactions, laboratory simulation studies of atmospheric processes and field measurement campaigns designed to test and measure the transformations in situ. We recommend that the following program areas of NOx chemistry should be emphasized.

4.1. NOx Reservoir Study.

Most of the NOx is emitted to the atmosphere as nitric oxide, which is rapidly interconverted with NO_2 on a time scale of minutes. Conversion of NO_2 to HNO_3 occurs on a longer timescale and is usually a sink for NOx. Peroxyacetyl nitrate (PAN) is an important temporary reservoir for NOx since it is relatively stable with respect to thermal decomposition with a lifetime of about one day at 275 K. However, there are important gaps in our understanding of these fundamental processes. For example, there is an apparent difference in the magnitude of the conversion of NOx to HNO_3 between North America and Europe as deduced from the reported data.To identify the interconversion mechanisms, one needs the seasonal and diurnal patterns of the relative ratios of the nitrogen oxide molecules. This can be derived from simultaneous measurements of NO, NO_2, PAN, HNO_3 along with the total NOy concentrations. These measurements will have to be made at carefully chosen locations. Measurements at several altitudes are needed since the evolutions of the different reservoirs are not alike in the planetary boundary layer and the free troposphere (e.g. HNO_3). In addition to this, the determination of the difference (NOy) -[(NOx) + (PAN) + (HNO_3) + (NO_3^-)+ N_2O_5 should reveal the existence of any missing NOy molecules. It should be noted that this study could be conveniently coupled with the NOy distribution study.

4.2. Fast Photochemistry of NO and NO_2

Peroxy Radicals (RO_2 and HO_2) are formed in the troposphere during the oxidation of CO and hydrocarbons. The oxidation of NO by the peroxy radicals are presently thought to be the primary chemical production mechanism for ozone in the troposphere. However, chemical schemes presently used in model studies of O_3 formation suffer from imperfect knowledge of RO_2 production during decomposition of the more complex hydrocarbons. To test the understanding of this ozone formation mechanism it is recommended that the concentrations of peroxy radicals be measured simultaneously along with a determination of the balance of the photochemical stationary state between NO and NO_2.

The photochemical stationary state between NO and NO_2 is determined by photolysis of NO and the oxidation of NO by O_3. Additional oxidation of NO to NO_2 by peroxy radicals would cause an imbalance in the stationary state and the degree of imbalance would provide an indirect determination of the peroxy radical concentration. The measurements should be coupled with laboratory studies of the identity and yields of peroxy radicals formed by hydrocarbon oxidation and of the rates of reaction of these radicals with NO.

The field experiments should cover a range of NOy concentrations and be undertaken in areas where anthropogenic, as well as biogenic, sources of hydrocarbons are predominant. The measurements should be compared with model predictions of the local ozone mixing ratio attendant to this production mechanism.

4.3. Plume Study

A complementary approach to study ozone production is provided by the measurement of the evolution in a plume of NOy compounds (NOx, PAN, HNO_3) and of the more abundant and reactive NMHC, in parallel with O_3 measurements. An urban plume drifting over the ocean would be the ideal case since it will minimize the further addition of NOy or NMHC (natural or anthropogenic) from surface sources. It will also be important to conduct such a lagrangian plume study at nighttime so that the evolution of NOy can be studied above the nocturnal inversion layer and in the dark.

4.4. Laboratory Studies to Expand and Improve Knowledge of the Chemical Transformations in the NOy Family

The underlying mechanisms and kinetics of the reactions leading to ozone formation from coupled NOx - hydrocarbon chemistry are now established. However there remain a number of gaps in our knowledge. For example the rate constants for the reactions of the organic peroxy radicals with NO, which play a fundamental role in O_3 formation in the troposphere, have only been measured for a few simple peroxy radicals. Data for other peroxy radicals of atmospheric significance need to be obtained and a reactivity pattern established. The importance of the alternative channel in this reaction forming organic nitrates needs to be established for molecules other than the alkanes.

Kinetic data for the formation and stability of peroxynitrates, formed through the reactions of organic peroxy radicals with NO_2, is also needed. For example there are no comprehensive measurements of the pressure and temperature dependence of the rate constants for the formation reaction of the most abundant peroxynitrate i.e. PAN.

The photochemical dissociation and reactions with OH of the peroxynitrates and also the rates of these reactions of the organic nitrates are not well established. These reactions may be important in determining the interconversion of NOx and its organic reservoir molecules in the background troposphere.

In addition to the study of elementary reactions we recommend the development of integrated experiments where selected aspects of NOx-interconversion mechanisms are investigated under simulated atmospheric conditions, e.g. the uptake of NO_3 and N_2O_5 in aerosols; formation of organic nitrates from peroxy radicals and formation of HONO.

4.5. NO_3 Chemistry and the Conversion of NOx, HNO_3 and NO_3

NOx is lost from the atmosphere by conversion to HNO_3 or nitrate in aerosol, which are both removed by dry or wet deposition. The most important transformation in this respect is the oxidation of NO_2 to HNO_3, which occurs by two routes, i.e. directly via reaction of NO_2 with OH or indirectly via reaction of NO_2 with O_3 to form NO_3 which may be converted to HNO_3 by one of several mechanisms. The understanding of the NO_3 chemistry is a major challenge at the present time.

In daytime NO_3 undergoes rapid photolyses or reaction with NO, reforming either NO or NO_2 so balancing the rate of NO_2 oxidation; reaction of NO_3 with NO has the same effect. At night, NO_3 loss by reaction with organics to give HNO_3 or nitrates, and with NO_2 to give N_2O_5, become the major NO_3 loss routes. The N_2O_5 route is potentially the most important but is complicated by rapid equilibration between NO_3 and N_2O_5 and the heterogeneous nature of the hydration of N_2O_5.

$$NO_3 + NO_2 \longrightarrow N_2O_5 \xrightarrow{\text{aerosol or droplet}} 2HNO_3$$

The possibility also exists for direct loss of NO_3 to droplets with formation of NO_3 in the liquid-phase.

Experimental studies to simulate these processes in the laboratory are required, as well as studies to establish the kinetics and products of the elementary reactions of NO_3 with H-containing molecules and radicals.

4.6.Ammonium Nitrate Formation

The travel distance of NOy in the PBL depends on the partitioning between HNO_3 and nitrate aerosol, because of the different lifetimes of these NOx carriers with respect to deposition. Ammonia can have an important effect on the partitioning between the gas and aerosol phases through the equilibrium

$$(NH_3)_g \ + \ (HNO_3)_g \rightleftharpoons (NH_4NO_3)_s$$

The equilibrium constant for this reaction is known and models describing the influence of relative humidity, liquid water and sulphate content of the aerosol on the partitioning of NOy in this process, have been developed. However, there are discrepancies between model results and atmospheric observations, which need to be clarified.

5. INSTRUMENT DEVELOPMENT

In order to facilitate the field studies described above it will be necessary to develop and test several key measurement techniques. In terms of new technique development, it is critical to our understanding of photochemical ozone production, and the chemical transformation of the NOy molecules to measure the atmospheric concentrations of OH, HO_2 and RO_2. In addition, real-time techniques with improved sensitivity and response time are needed for in-situ measurements of ambient concentrations of HNO_3 and particulate nitrate. Finally, the measurement of the emission of NOx molecules into the troposphere requires the development of fast response detectors (response time < 0.1s).

5.1. Instrument Validation

The capability to measure trace concentrations of a photochemically active compound in the ambient atmosphere is an extremely difficult goal to achieve. Severe operating conditions and a multitude of potential interfering compounds that are present in ambient air may serve to limit the accuracy, sensitivity or dependability of many promising techniques. To ensure the integrity and field reliability of the instruments to be used for tropospheric measurements, a carefully controlled field intercomparsion is essential. The ability of two or more detectors with different operating principles to simultaneously measure concentrations of compound in ambient air over a wide range of operating conditions and obtain satisfactory agreement, constitutes the basis for instrumentation certification. To ensure credibility the intercomparsion should be done with a predetermined protocol. The results should be evaluated by an independent third party and published in the open scientific literature. It should also be noted that the quantification of systematic differences in measured concentrations by the intercompared techniques allow data sets previously obtained by the techniques to be reevaluated and merged.

5.2. Techniques to Measure OH, HO_2, AND RO_2

There is a need for high-sensitivity instruments that are capable of measuring OH, HO_2 and organic peroxy radicals. To allow meaningful measurements to be made in the troposphere the instruments should have detection limits on the order of $1x10^5$ molecules cm^{-3} for OH and $1x10^7$ molecules cm^{-3} for HO_2 and RO_2. Specific (for example spectroscopic) methods of detection are preferable. A few techniques already developed are approaching these specifications. These are Long-path Differential Optical Absorption (OH), Laser-induced Resonance Fluorescence (OH and HO_2 indirectly after conversion to OH), and Electron Spin Resonance (RO_2, HO_2 in cryogenically collected ice-matrices). Intercomparison of the existing techniques is highly desirable.

Indirect methods involving chemical amplifiers (RO_2, HO_2) are subject to substantial calibration problems. Techniques involving chemical conversion need to be carefully investigated for potential interferences and, when possible, cross-calibrated against spectroscopic measurements. Finally, chemical tracer experiments (e.g. field studies of the conversion rate of molecules controlled by OH) are useful for placing constraints on time and spatially averaged OH-mixing ratios.

5.3. Techniques to Measure NOy Compounds

Presently, techniques are emerging that allow the measurement of NOy. Available techniques rely on the reduction of NOy-compounds to NO at a surface. Preliminary tests with NOy-mixing ratios of 200 ppt and greater indicate that heated molybdenum oxide surfaces and gold-catalyzed CO-reduction followed by chemiluminescence detection of NO seem to offer promising possibilities. More tests are needed to determine if the techniques are capable of quantifying NOy in the clean troposphere ([NOy] < 100 ppt).

Techniques also are available to measure HNO_3 and NO_3^-. The techniques include filter collection, various types of denuder tubes and tunable diode laser spectroscopy for HNO_3. The measurement of ambient concentrations of nitric acid is critically dependent on the inlet system, as well as the detector. Because nitric acid is so easily collected by surfaces the inlet lines may become the critical factor limiting the capability of the instrument. For this reason intercomparisons must be carried out as closely as possible with the sampling configurations used (or to be used) during actual field studies.

The only technique for particulate nitrate currently available is filter collection. Further techniques should be developed to more directly determine the chemical composition of aerosols.

5.4. Techniques to Measure NOx Emissions

The principal techniques for the measurement of NOx fluxes from natural sources use chemiluminescence detection of NO with converters to transform NO_2 to NO. The present approaches use enclosures to measure NO emissions from the soil and eddy correlation techniques to measure NOx emissions from soils and from lightning. For eddy correlation measurements considerable development is required to improve the response time (including response smoothing by sampling) of NOx detectors. A carefully arranged intercomparison of techniques to measure soil emissions (chambers and eddy correlation) should be undertaken.

THE REGIONAL OZONE PROBLEM

R.Guicherit (chairman), D.Derwent (rapporteur),
P.I.Grennfeldt, J.Jerre, D.Kley, J.Logan, S.Penkett,
B.Prinz, P.Taalas.
(WORKING GROUP III)

1. INTRODUCTION

Ozone is an important secondary pollutant in the boundary layer and comes from three sources,the stratosphere, the free troposphere and from production in the boundary layer itself. Although the contributions from the first two sources are not well defined, there is no question that ozone concentrations in excess of 80 to over 200 ppb observed during photochemical episodes on a regional scale require a predominantly photochemical source in the boundary layer. The regional scale is defined here by the spatial scale of precursor source regions and by typical transport distances for ozone, 1000-2000 km.

Elevated ozone concentrations are of concern, because they have an impact on

a) human health

b) vegetation

c) certain materials e.g.(rubber, plastic, paint)

d) the oxidizing capacity of the atmosphere

e) climate

In order to derive the most effective abatement strategy, research is urgently needed to fill gaps in understanding of the formation and removal mechanisms for ozone and its precursors, NOx and hydrocarbons.

It is the purpose of this chapter to address some of these questions.

2. FORMATION MECHANISMS

The oxidation of volatile organic compounds provides the peroxy radicals (RO_2 and HO_2), which through their O atom transfer reactions with NO give rise to O_3 formation in the coupled NOx - hydrocarbon system. The

I. S. A. Isaksen (ed.), Tropospheric Ozone, 403–411.

flux of peroxy radicals depends on the detailed kinetics and pathways in the degradation mechanisms of the volatile organics. These mechanisms are reasonably well known for simple aliphatic hydrocarbons (alkanes and alkenes) for typical summertime surface temperatures, but are less well understood at colder temperatures (e.g. winter conditions or higher altitudes).

Mechanisms for degradation of large hydrocarbon molecules, including aromatic hydrocarbons, are much less well known. There are departures from the generalised behaviour of hydrocarbon oxidation in the case of aromatics, because the RO_2 radical formed by O_2 addition is unstable and R and RO_2 radicals can become equilibrated, opening up possible competitive pathways for the radical species, R. Other gaps in knowledge lie in the degradation of oxygenated compounds (esters, alcohols, ethers and carbonyl compounds) either emitted directly (e.g. solvent use) or formed in atmospheric oxidation of other hydrocarbons. The presence of substituent groups in the carbon chain can alter the rates and pathways in the subsequent atmospheric degradation, particularly for the primary attack by OH(and by NO_3 at night) and the competing processes in the fate of oxy-radicals, RO.

Relatively high concentrations of peroxy radicals can build up during the degradation of organics at low NO concentrations. In this regime competition between reaction with NO and reaction with other peroxy radicals including HO_2 becomes significant. The kinetics and, in particular, the products of the self and cross-reactions of peroxy-radicals are in many cases unknown and it is important to establish a reactivity pattern for these processes. This is a major limitation in our current ability to assess the impact of NOx-control on regional scale pollution levels.

Observational data for NO_3 indicate that nighttime chemistry may provide a significant sink for NOx through processes involving heterogeneous reactions of N_2O_5 under conditions of high humidity. The chemistry of NO_3 and N_2O_5 (including heterogeneous processes) requires further laboratory study. Reactions on aerosols and droplets may also be important for odd-hydrogen species; recent evidence suggests a high sticking coefficient for the HO_2 radical.

3. TRANSPORT MECHANISMS

The time scale for hydrocarbon oxidation during photochemical episodes and for ozone deposition and removal is on the order of days. Atmospheric transport therefore plays a significant role in determining the spatial distribution of elevated ozone concentrations. Current regional scale models appear well able to handle the main features of the synoptic scale transport. Mesoscale processes, however, on a spatial scale of 5-50 km are difficult to treat in detail. They include:

- land-sea breezes,
- mountain-valley flows,
- low-level nocturnal jets
- effects of surface roughness changes.

These processes appear to have an important influence on diurnal variations in ozone concentrations measured close to the ground in remote rural areas.

Difficulties in simulating these processes may contribute to discrepancies between model results and ambient concentration data.

Improvements in the representation of transport mechanisms are anticipated following the use of tracers, the application of advanced mesoscale observation methods involving lidar or sodar, and through the development of dynamic mesoscale, fine mesh numerical weather prediction models. Transport processes are discussed further in working group I.

4. EMISSION INVENTORIES

Detailed and accurate emission inventories are essential for accurate simulation of the fate of atmospheric pollutants. The rate of ozone formation depends on the mix of hydrocarbon species and on the ratio of HC/NOx in a non-linear manner. Consequently inventories must provide detailed information on hydrocarbon speciation and on spatial and temporatal patterns. For the regional ozone problem, emission rates are required with spatial resolution of a few kilometers and temporal resolution of one to a few hours. Inventories are required for both anthropogenic and natural sources of hydrocarbons and NOx. The accuracy of present data bases, as far as speciation and release rates are concerned, is inadequate for accurate formulation of ozone generation. Considerable improvement in the methodology used for derivation of inventories is required, particularly for Europe, where consistency among emissions for different countries is essential.
There is considerable uncertainty in:

a) exhaust and evaporation emissions from motor vehicles,

b) fugitive emissions and emissions from solvents and

c) emission rates of natural HC's from vegetation. Information on detailed hydrocarbon speciation is essential.

The quality of inventories may be improved by using the following approaches:

a) Emission factors: direct measurements should be made for individual types of sources under real world operating conditions, including definition of the temperature dependence of emission rates.

b) Improvement is required in the methods used for collection and evaluation of statistical information needed for the inventories (e.g. distances travelled by vehicles, numbers of minor point sources).

c) Emission inventories need to be tested using atmospheric measurements (e.g. the budget over an urban area).

d) Emissions from vegetation: techniques used to determine fluxes must be chosen so that they do not perturb the environment of the plant.

Emission rates should be derived retrospectively and techniques must be formulated for prognostic (future) emissions. Methodologies may differ for retrospective and prognostic emissions.
Trends in emissions should be compared with trends in deposition fluxes and with ambient concentration data.

5.SURFACE DEPOSITION

The transfer of ozone as well as precursors and reaction products from the atmosphere to the surface can represent a sink for these species as well as an input to the surface systems.Vegetation is the primary receptor for ozone and the mechanisms for ozone deposition are fairly well understood. For many other species of importance in regional photochemical pollution, knowledge of deposition pathways and mechanisms is very limited. Methods for measuring the flux of some of these compounds have yet to be developed.

Most studies of deposition are conducted for only a limited range of situations. Field studies are not usually performed over complex terrain and inhomogenous land use categories.Thus, there is a strong need to:

a) develop techniques for measuring the dry deposition of ozone and related compounds, particularly in complex situations.

b) generate experimental data including deposition velocities for specific land use categories,

c) develop models for interpretation and generalization of field data,

d) develop methods for characterizing different land use categories as input to a regional scale data base.

6. MEASUREMENT STRATEGIES

A detailed understanding of the mechanisms responsible for elevated ozone concentrations requires a comprehensive set of field measurements, which may be used to develop and test concepts and models of the important chemical and transport processes. The types of measurements required fall into two general categories, 1) long term measurements using operational instrumentation, and 2) intensive field and aircraft campaigns, using research instrumentation. Long term measurements are required to determine the spatial and temporal characteristics of species concentrations and to establish trends. The data are also needed for model initialization, and for testing results of chemical-transport models.Intensive field campaigns provide much more detailed information for short time periods, and provide the only means for testing details of chemical mechanisms under conditions typical of rural air, where smog chamber results are not appropriate. In addition, measurement programmes commonly provide evidence of new phenomena which are not predicted from current understanding of chemical processes occurring in the atmosphere.

The first type of experiment seeks to investigate longer term aspects of the behaviour, or chemical climatology, of major trace gases in the atmosphere. Such an experiment is clearly needed for ozone and has been planned as the TOR experiment within Europe. The main objectives of these longer term experiments are to investigate trends in concentrations of ozone and its precursors, seasonal behaviour, frequency of episodes and to establish the mechanism of ozone formation in the various regions.

A second type of experiment involves intensive field campaigns with the primary objective of comparing measured and calculated free radical concentrations, particularly the hydroxyl, OH, hydroperoxy, HO_2 and alkylperoxy, RO_2, radicals.To an extent, these experiments are limited currently by inadequacies in the instrumental techniques employed for the detection of the free radical species. Steady progress will ultimately solve this problem and allow comprehensive tests of chemical mechanisms.

Further discussion of measurement strategies is discussed in working group 2.

7. EFFECTS OF REGIONAL OZONE

Elevated ozone concentrations during regional scale photochemical episodes approach and exceed levels at which adverse environmental impacts can be anticipated. This is the essence of the regional ozone problem. Appreciation of the critical environmental effects has changed over the years; initially health effects and material damage provided the major focus, while more recently crop damage and forest decline have been studied extensively.

A number of issues are central to many of the environmental effects. Ozone is a secondary pollutant and hence it almost invariably occurs in combination with other pollutants causing major complications. Averaging time adds further complexity. It is important to know whether exposure over extended periods (e.g., the growing season) has the greatest significance. Interaction with other environmental stresses, biological and ecological diversity also have to be considered carefully in evaluating the environmental impacts of ozone.

7.1. Effects of Ozone on Man

The main effects which have to be taken into account include:

- irritation, mainly from by-products of atmospheric hydrocarbon degradation, including ketones and aldehydes,

- alteration of physiological functions including athletic performance,

- changes in upper airway resistance, which may lead to enhanced sensitivity for pathogenic infections.

Most experimental and epidemiological studies have been carried out for specific exposure conditions, such as for the Los Angeles basin. The extrapolation of these studies to other areas of regional ozone formation is required.

7.2. Effect of Ozone on Plants

Considerable effort has been focussed on the direct effects of ozone on annual crops, which lead to reduction in yield and quality of agricultural and horticultural products.

More recently indirect or subtle effects of ozone on forest trees and other long living plants have become a central focus of research activities.The following topics are recognized as important research priorities.

- Effects of ozone on agricultural crops. Crop loss studies similar to those in the NCLAN network in North America should be done in other areas with regional ozone pollution. Especially in central Europe the studies should take into account the specific conditions in the areas with respect to plants, soil, climate and ozone exposure.

- Investigation of more direct effects of ozone on deciduous and sensitive coniferous trees. These studies should be done with respect to discolouration, leaf and needle deformation, necrosis, premature senescence and alteration of growth habit.

- Impact of ozone on nutrient metabolism in connection with lowered buffer capacity and nutrient supply in soils, caused by acid deposition. The increased uptake of nitrogen to forest ecosystems and the resulting nutrient imbalance is of some importance.

- Enhanced leaching of nutrients by perturbation of the cell membrane system.

- Impact of ozone on carbohydrate metabolism in connection with disturbed translocation of assimilates. This is most important for coniferous trees because the single needle age classes have different source-sink functions for carbohydrates.

- Derivation of more complex models of the effects of ozone on trees which bring together all the different aspects of the interactions between stand-specific exposure conditions, age and prehistory of the tree, soil as a predisposing factor as well as climate as a triggering or synchronising factor, together with other pollutants, including acid deposition.

- Impact of ozone on natural vegetation, the importance of which has yet to be assessed.

7.3. Effects of Ozone on Materials

The effects of ozone on rubber, plastic, paint and other organic materials are well known. It is important, however, to consider ozone exposure in combination with a wide range of other pollutants in assessing materials damage in urban areas. Long term concentrations may well have greater significance than peak concentrations.

8. MODELS OF REGIONAL OZONE FORMATION

A variety of regional models are being developed in Europe and the U.S. with the goals of 1) improving understanding of the mechanisms leading to elevated ozone concentrations, and 2) designing effective control strategies. These models must account for transport processes that occur over a wide range of time scales, and must simulate a complex, non-linear chemical system. Eulerian and Lagrangian (trajectory) models have been used to simulate specific ozone episodes of a few days duration, with varying degrees of success. We outline here some of the major uncertainties and difficulties in modelling regional ozone.

8.1. Spatial Resolution

Emission rates of NOx and hydrocarbons may vary over several orders of magnitude in quite small geographic regions, and the dependence of ozone formation on precursor concentrations is quite different in an urban/-industrial plume compared with an adjacent rural area. The magnitude of errors introduced by averaging emissions over grid boxes with dimensions as large as 150 km is unclear at present. Further work is required on the treatment of emissions which are highly inhomogeneous on the chosen grid of the transport model.

8.2. Chemical Mechanisms

Most models employ simplified chemical mechanisms, derived from more complex schemes, because of limitations on computer time and storage. Those mechanisms may be tested against smog chamber data for the NOx regime found in urban areas, but there is no satisfactory way at present to test the mechanisms for the low concentrations of NOx and hydrocarbons found in rural air. This introduces considerable uncertainty into regional models. High quality atmospheric measurements are required in order to test the mechanisms for rural conditions, as discussed further in section 6 and working group 2.

8.3. Application of Models to Long Term Averages

The present generation of regional models have been used to simulate short term episodes when ozone values are high.
The ability of these models to simulate relatively clean conditions is unclear, as the results may be quite dependent on the choice of model initialization. The methodology for using an episodic model to predict long term average values of ozone has not been developed. Models are required which can simulate the climatology of ozone on regional scales, in order to predict longer term (e.g. seasonal) averages and trends.

8.4. Model Validation

Progress in testing regional ozone models is severely hampered at present by the lack of ambient data with which to compare model results. There are large areas of the populated continents for which ozone monitoring data are unavailable and systematic coverage of different land use categories, such as forests and agricultural areas has not been attempted. Where there are data particularly in Europe, no real attempt has been made to intercalibrate the methods used and to harmonize siting criteria. There is a need to test models against a wide range of secondary pollutants other than ozone, but data are usually lacking.

Model validation is clearly an important prerequisite for the development of regional ozone control strategies. Clearly, there is a requirement to verify that individual model components reproduce the presently known atmospheric behaviour in terms of chemistry mechanisms, transport, deposition and input from the free troposphere. Validation involves a further layer of testing and evaluation when these individual components have been incorporated into the complete model. It implies testing whether the complete model is able to reproduce the behaviour of ozone and its precursors observed in the atmosphere. This is currently the only way that model reliability can be assessed, and is necessary in order to guarantee the widespread acceptance of the control strategies produced by the models.

8.5 Source Receptor Relationship

The ozone source-receptor relationship is a powerful concept in the evaluation of control strategies for regional pollution problems. Attribution of sulphur deposition has been achieved in Europe using the EMEP trajectory model. However, there appears to be no basis on which unambiguous attribution is possible for regional ozone in the eastern regions of the USA and within Europe. This arises because of the complications of non-linear chemistry, and long range transport of secondary pollutants and their precursors.

9. ABATEMENT STRATEGIES FOR REGIONAL OZONE

The only approach currently available for the evaluation of abatement of control strategies for regional ozone involves theoretical modelling. Historically, control strategies for ozone have been based on relatively simple photochemical box models (e.g. the OZIPP approach in the United States). For the development of realistic control strategies there is a need for validated models containing comprehensive descriptions of emissions, transport, chemistry and deposition. These models should resolve the important sources contributing to a given receptor and indicate reliably the impact of controls applied to the upwind sources. Comprehensive models of this type are currently under development in Europe and the United States.

Two different approaches have been used to evaluate possible control strategies. In the first, hydrocarbon or NOx emissions or both are reduced by arbitrary percentages, and the impact on downwind secondary pollutants is determined. In the second, the emissions from specific sources are reduced, reflecting the impact of specific abatement technologies. The results of these model studies point to the following general conclusions:

- the efficiency of a hydrocarbon only strategy compared with a hydrocarbon and NOx strategy depends on the HC/NOx ratio.

- hydrocarbon emission reductions reduce ozone concentrations

- NOx emission reduction may increase ozone in urban areas and decrease ozone in some rural areas.

It should be pointed out that the ozone concentrations obtained for specific control strategies depend strongly on the details of the chemical mechanism used, emphasizing the need for validation of the details of the chemistry.

BACKGROUND OZONE

P.Crutzen (chairman), J.Fishman (rapporteur), R.Bojkov, R.Cicerone, A.Ghazi, A.Marenco, G.Megie, W.Seiler, W.C.Wang. (WORKING GROUP IV)

1. INTRODUCTION

Through the production of highly reactive hydroxyl radicals by photolysis of ozone in ultraviolet solar radiation, ozone is one of the most important gases in the troposphere.The hydroxyl radicals which are present only at very low concentrations in the troposphere (global average about $5x10^5$ molecules cm^{-3}) attack virtually all hydrogen containing gases that are emitted into the atmosphere by both natural processes and anthropogenic activities and in addition such important gases as CO, NO_2 and SO_2. It is clear that tropospheric ozone is an essential compound in the atmospheric environment.

By far most ozone ($\sim$90%) is located in the stratosphere, from where a downward transport of ozone into the troposphere definitely takes place. However, for large regions of the "background" troposphere, here defined as all troposphere outside the planetary boundary layer of the continental regions affected by industrial activities, the influx of stratospheric ozone most likely is not the dominant source of ozone. There are various processes in the background troposphere that can destroy or produce ozone. For instance, during the oxidation of CO to CO_2, ozone may be produced or destroyed depending on the background concentrations of NO. The potential for ozone production in the troposphere is very large, but not fully substantiated because the atmospheric residence time of NOx is only of the order of a day. However, the anthropogenic emissions of NOx have been steadily increasing due to human activities during this century through fossil fuel and biomass burning and now dominate the global NOx budget. Consequently, the potential for ozone production has increased substantially during this century, so that ozone increases also in the background troposphere at least at northern mid-latitudes (> 30°N) should have occurred. Long term ozone data records in the free troposphere from a number of stations are in agreement with this expected development.

Anthropogenic activities also affect the atmospheric concentrations of other trace gases that play important roles in the photochemistry of ozone and hydroxyl. The most important example of a gas with rapidly increasing ($\sim$1% per year) atmospheric concentrations is methane. Again, the photochemical net result of the oxidation of CH_4 to CO (and to CO_2) depends strongly on the atmospheric concentrations of NOx, such that in

I. S. A. Isaksen (ed.), Tropospheric Ozone, 413–422.

NO-rich atmospheric environments (mid-latitudes of the northern hemisphere, upper troposphere) one may expect an additional production of 3-4 ozone molecules for each additional methane molecule that is broken down in the atmosphere. On the other hand, in NO-poor environments (i.e. the clean troposphere not affected by lightning), the net result may be a loss of 1 - 2 ozone molecules. Most importantly, however, in these regions, there may be a loss of between 2 and 3.5 odd hydrogen (OH + HO_2) radicals for each additional CH_4 molecules that accumulates in the atmosphere. These clearly imply very important potential developments about which we must learn much more in the future through strong research efforts. Indeed it may be asked whether the background photochemical system is a stable one. For instance, in "clean" environments, more CH_4 might lead to less O_3 and OH, producing more CO. The lowering of ozone and the increase in carbon monoxide,in addition to the effect of more methane, in their turn strongly contribute to lower OH. The opposite could be occurring in NO-rich environments, implying the possibility of a drastic shift of atmospheric photochemical activity from the clean to the more polluted background atmosphere.

Unfortunately, it is not possible yet to estimate the extent and magnitude of these important developments. The following text will, therefore, especially emphasize the unknowns and give recommendations for future research efforts, which must be of international and global extent.

Ozone is not only a photochemically, but also a radiatively important gas. The ozone abundance in the troposphere, although accounting for only 10% of total column, plays a significant role in the radiation budget of the troposphere. Of special importance in this respect is the upper troposphere and lower stratosphere. Its changes may lead to a perturbation of the energy balance with subsequent effect on tropospheric climate. Presently, there are two main issues related to climate-chemistry interactions involving tropospheric ozone. The first are the latitudinal and seasonal changes in vertical O_3 distribution caused by increases of atmospheric trace gases (CH_4, CO_2, CFC-11, CFC-12, and N_2O) and the second involves the increases of atmospheric water vapor resulting from a warming climate to be expected from an increase of atmospheric carbon dioxide and other trace gases. The latter can feed back on atmospheric photochemistry through altering the OH distributions.

Because of the photochemical and climatic importance of improving the understanding of water vapor in the upper troposphere and lower stratosphere, future measurement programs should address this particular aspect.

Since the most important regions from a climatic viewpoint is the middle to upper troposphere, it is important to evaluate the existing data base for any noticeable changes in O_3 during the last few decades and design observational systems to detect such changes in the future.

2. EXISTING DATA BASE

2.1. Ozonesonde and Surface Data

There are hundreds of places around the globe where ozone concentrations near the ground are being measured. However, most of the stations are in locations strongly influenced by local pollution and therefore, the information is of limited or no use for studies of the background surface ozone levels and their variations. Mountain and WMO-BAPMoN stations are better suited and efforts to continue existing and establish new surface ozone measuring programs must be encouraged.

In accordance with the internationally accepted data exchange policy, data are currently submitted monthly to the World Ozone Data Center (WO3DC) and published monthly. Unfortunately, there are a number of stations with good quality, data-gathering capability which so far have not submitted their data for international exchange.(e.g. Arkona, Zugspitze, Fichtelberg) In any existing and future project, emphasis should be put on timely data exchange via WO3DC.

There are less than 20 stations worldwide at which ozonesoundings are being made, with various frequencies and sometimes during non-overlapping years. If one disregards stations which have made only sporadic soundings (with a frequency less than weekly) we are to realize that there exists no stations with regular vertical ozone soundings in the Southern Hemisphere.It is obvious that the coverage is far from being globally satisfactory.

A number of international comparisons have shown systematic differences in ozone determinations by various types of sondes. It is, therefore, very important to assure continuity of use of the same type of ozonsonde and it is mandatory for any precise trend analysis to normalize the available data before starting the trend analysis. It is feasible to introduce consistent corrections both to past stratospheric and tropospheric ozone records and thus to utilize the existing ozone-sonde records for trend analysis. The corrected data should replace those that are on file in the WO3DC.

2.2. Trends in Tropospheric Ozone

As already stated in this report, only data with sufficient continuity (one or two decades and a frequency not less than once, and preferably 2 to 3 times weekly) could provide meaningful information for study of the tropospheric trends.

Analysis of the available and normalized ozonesonde data confirm the continuous increase of ozone concentration by more than 1 % per year in the lower troposphere and by somewhat less than that rate for the upper troposphere. Although the data are very sparse and do not represent a global average, they reveal some seasonal differences also in the estimated trends over the Northern middle latitudes. It appears that the rate of tropospheric ozone increase during the winter months (Nov-Dec-Jan) is somewhat stronger than the increase during the summer months (May-September).

2.3. Trends in Other Trace Gases

There is evidence that the mixing ratio of several chemically and climatically important trace gases relevant to the distribution and abundance of background ozone in the troposphere have changed with time as a consequence of human activity. Information on long-term trends of these species have become available during recent years because of monitoring at different fixed measuring sites, measurements during ship and aircraft expeditions, and analyses of air samples trapped in ice cores.

The measurements performed at various measuring sites produce data that are generally of high quality, satisfying international standards. They are thus good for estimating possible temporal trends. Depending on the residence times of the measured species, the trend analysis obtained from these data may be representative of a global scale (e.g.,for methane) or on a regional scale (e.g. CO).

Chemical analyses of air samples trapped in ice cores provide important information on the secular trends of trace species covering time periods of hundreds to several thousands of years. This information, however, is restricted to those trace gases that are long-lived and which do not react during the time of storage in the ice sample. Information on secular trends of short-lived trace gases may be obtained from measurements of secondary photochemical products such as nitrate and sulphate, hydrogen peroxide, which are dissolved in the ice. Because of their short residence times, the concentration of these species may only reflect the secular trend at the higher latitudes.

An important species, which has been routinely measured in tropospheric air for more than 20 years is methane (CH_4). Measurements from both hemispheres indicate a regular increase of the tropospheric CH_4 abundance by about 1 % per year during the 10 last years to present values of 1.7-1.8 ppmv. Data on tropospheric CH_4 mixing ratios measured in the 1960's display considerable scatter so that conclusions on CH_4 trends during this longer time period are less certain. Analyses of air trapped in ice cores suggest that CH_4 mixing ratios were 0.6-0.7 ppmv 200-300 years ago, indicating that CH_4 concentrations have increased by more than a factor of two over the past 2-3 centuries. Interestingly, the pattern of CH_4 increase has followed the increase in world human population.

Long-term measurements of CO have so far been carried out only at a few stations (e.g., Mauna Loa Observatory, 20°N, Cape Meares, 30°N, Cape Point, 30°S, and Cape Grim, 40°S). Additional data on the global CO distribution are available through some time-limited aircraft and ship measurements, as well as ground-based spectroscopic measurements of total vertical CO column. Most data have been obtained in marine air masses; measurements in clean continental air are lacking.

Because of its relatively short residence time, CO shows strong latitudinal and longitudinal gradients as well as a pronounced seasonal cycle which are similar to those for methane. The seasonal CO variability is on the order of a factor of two. A similar difference has been found between the two hemispheres, with higher values in the northern hemisphere.

The budget and thus the distribution and abundance of CO in the troposphere is strongly influenced by anthropogenic activities (e.g., fossil fuel combustion and biomass burning) which may contribute more than 50% of the total CO source. As a consequence, a temporal increase of the tropospheric CO burden similar to CH_4 is possible. However,because of the high variability of tropospheric CO mixing ratios, long-term trends are difficult to assess and requires continuous measurements over a long time period. Long-term records of CO by Seiler and coworkers (not yet published) for Mauna Loa (8 years) and Cape Point (9 years) have not shown a significant trend. In contrast, for the station at Cape Meares, a CO increase of a few percent per year has been reported. This conclusion was ,however,based on a relatively short observing period and may therefore have been biased by regional shifts in seasonal effects. The strong increase of tropospheric CO as reported from a few older spectroscopic data on the Jungfraujoch through comparison with recent in-situ measurements may likewise have been influenced by seasonal and spatial effects and therefore not be representative of large scale and global conditions.

Clearly, therefore, additional continuous CO measurements are urgently needed to allow estimates of a possible trend of the tropospheric CO burden. These measurements need to be performed at several latitudes to account for the latitudinal CO gradients which also may change in response to changes to individual CO sources. Also needed are stations in clean continental atmospheres, in the tropics, in particular, to investigate the effects of natural sources. Essential is intercalibration of CO standard used by different investigators.

2.4. Survey Flights and Integrated Airborne Studies.

Airborne missions have in the past contributed substantially to knowledge about the distribution of ozone and other important trace gases in the troposphere. There is no doubt that in the future they will contribute even more to increased knowledge about trace gas distributions in space and time, chemical budgets and atmospheric photochemistry in various parts of the world. Increasingly, emphasis will hereby shift from simple survey flights to well integrated, comprehensive experiments in which the essential trace gases that determine the photochemistry of the atmosphere, together with meteorological parameters are measured with well tested, intercompared and intercalibrated instrumentation. These efforts will increasingly require international collaboration and coordination.

3. EXPANSION OF THE DATA BASE

3.1. Requirements for measurements

Because of the large relative error in the tropospheric part of past ozone soundings and because of the very few ozonesonde stations, knowledge about trends in tropospheric ozone should be considered as preliminary and by no means satisfactory. Major measurement efforts are therefore necessary to improve the situation.

The tropospheric and surface ozone concentrations both exhibit very pronounced seasonal and day-to-day variability; the surface ozone also has a distinct diurnal behaviour. Therefore, formulation of data requirements should incorporate provisions for providing at least the mean (average) distribution and for the daily, monthly and annual variability. If one wishes to address the trend issue, the main additional requirement is continuity for a period longer than 1-2 decades.

The frequency of the observations is very important especially for determination of trends. Surface ozone is usually registered continuously from which 1-hour mean values are being deduced and that is more than sufficient for any evaluation.

When considering to apply data from very few stations to deduce global (hemispheric) behaviour, the location of the ozonesonde stations with respect to the climatological position of the planetary (Rossby) waves, with their year-to-year fluctuation could be a source of misleading tendencies. In such a case, analysis of observations made under similar synoptic conditions may aid in trend analysis. We strongly support WMO recommendations (1980, 1982) that the ozonesondes should be made not less than twice weekly and preferably three times weekly during the period of high variability (April - September in the Northern Hemisphere). Generally, one sounding per week is an absolute minimum. High quality of data is essential, including possibly necessary corrections for interfering gases, such as SO_2 and NOx.

With respect to surface ozone observations it should be noted that the intake tubes are located at various heights above the surface (2 - 20 m). Although it is desireable to suggest and strive for uniformity, this may be very difficult to achieve. Even in cases where a shift to a uniform level is logistically possible the question of discontinuity of the given station record due to the change of height of the intake must be carefully considered and corrected for.

For the analysis of both surface an lower tropospheric ozone information on the height of the atmospheric boundary layer (ABL) could be of importance. So far in submitting the data on a monthly basis for international exchange, there is no such requirement to the World Ozone Data Center (WO3DC). However, for long-term analyses, the monthly (climatological) average height of the ABL will be desirable. If such a requirement is suggested, a carefully considered uniform procedure for ABL determination should be formulated. Reporting of relative humidity and surface pressure could likewise be useful for analysis. It is suggested that the Ozone Commission of IAMAP consider and make relevant proposals for implementation by the WO3DC.

With reference to site selection of the surface ozone stations, they should not be affected by direct sources of contamination. It is also desirable that they be close (< 100 km) to a station making total ozone measurements. Combined analysis of total and surface ozone data could be of some help in consideration of sudden changes and in long-term variability analyses.

There is need of regular instrument calibrations (and intercomparisons between the stations) and the results of such should be deposited and published by the WO3DC along with the daily ozone concentrations.

3.2. Site Selection

Several factors influence how sites should be selected for the expansion of the ozone data base. In principle each scientific goal places different demands on site location and measurement frequency. In practice, financial and logistical limitations also enter the picture.

The main scientific goals that demand an expanded, improved ozone data base are:

i) detection of temporal trends - in regions influenced by certain identified sources, on hemispheric global scales,

ii) provision of adequate data to test photochemical theories.

Goal (i) is important so that we can evaluate the extent to which radiative forcing of climate change is occurring and to evaluate the rates and spatial patterns of photochemical change. The nature of the radiative greenhouse effect demands that we know the temporal trends of ozone as a function of altitude (especially in the middle and upper troposphere) and as a function of latitude.

The second goal includes such major enterprises as determining the mechanisms and rates of ozone production and how they affect the concentrations of other key photochemical oxidants such as OH, HO_2, H_2O_2 and NO_3. Indeed, a comprehensive goal of understanding tropospheric photo-chemistry is involved here. A goal such as this requires a carefully selected suite of accompanying measurements and critical companion model developments. Intense episodes of field investigations are thus required. Tests of the theories require e.g.data sets from regions and seasons with variable mixes of hydrocarbons and NOx, variable sunlight conditions, aerosol and moisture amounts, etc. Examples of such environments would span the range of (industrially unpolluted) air masses, tropical forests, boreal forests, the clean "background" troposphere and the marine boundary layer.

For a species as temporally and spatially variable as ozone, the most basic need is to develop an ozone climatology that specifies both the average behaviour and the extent and type of variability. Surface-level sampling is well underway at a number of global sites. However, since no sampling network or satellite technique exists that could provide global three-dimensional information even outlines of such a climatology in a few major regions, particularly outside of the northern midlatitudes, would be a significant step forward.However, the problem of optimum site selection is insufficiently explored and should be investigated, e.g. during the preparatory phase of the proposed International Global Atmospheric Chemistry IGAC Programme of the Commission of Atmospheric Chemistry and Global Pollution (CACGP) of IAMAP.

4. UTILIZATION AND DEVELOPMENT OF REMOTE SENSING TECHNIQUES FOR GLOBAL STUDIES

4.1. Ground based lidar network

The lidar technique for the measurement of tropospheric ozone profiles is presently reaching an operational status. It presents several advantages over conventional techniques (i.e. ozonesondes).

i) As an active system where the emitting laser source can be directly controlled, it provides an absolute measurement of the ozone vertical concentration profile with high vertical and temporal resolutions in the troposphere.

ii) Lidar measurements can be performed continuously (during both daytime and nighttime in the troposphere) to yield a knowledge of the spatial scale of ozone variability.

iii) The airborne capability of such systems can further extend its usefulness to provide information on long-range transport of ozone.

However, as lidar measurements at a given site will only be obtained in clear sky conditions, a bias related to meteorological observations can be introduced in the data base when used for statistical or trend analysis. Complementarily with ozonesonde, measurements should thus be emphasized to remove such bias. Due to the rapid development of new laser techniques, a simple system can be designed to make measurements of ozone over the altitude range of 0.1 to 15 km with the following performing characteristics: vertical resolution .1 to 1 km ; integration time 1 - 15 min ; overall accuracy better than 5%.

Implementation of a ground-based lidar network for establishing a climatology of ozone in the free troposphere should be given high priority.

4.2. Analysis of Existing Satellite Data

The use of satellite data, under certain circumstances e.g. in areas as heavily affected by biomass burning and photochemical smog, may provide useful data for tropospheric ozone observations that,although no measurement technique presently exists to measure tropospheric ozone directly.

Preliminary use of such a technique in tropical latitudes has produced integrated tropospheric ozone values that are consistent with amounts measured by in-situ techniques. This method should be expanded to determine whether or not realistic seasonal behaviour and spatial gradients can be obtained in the Tropics where the variability of the ozone distribution in the stratosphere is considerably less than at higher latitudes.

4.3. Future possibilities

Tropospheric studies presently suffer from the lack of global observations of ozone and relevant trace species concentrations and require thus the development of satellite-based instrumentation which are the unique way of obtaining such global pictures. However the classical techniques using passive sensors, in the limb-scanning mode (as presently applied for stratospheric monitoring) cannot be used in the troposphere due to the very high probability of cloud interferences over long horizontal distances. Nadir looking instruments should, therefore be developed which should in addition provide vertical and temporal resolution compatible with the relevant scale of trace constituents variations.

The existence of the ozone maximum at stratospheric altitude, prohibits direct measurements of tropospheric ozone using passive sensors. Active techniques can in principle be developed in the IR, where advantage could be taken of the variation of the ozone absorption profile with pressure - ie. altitude. The use of tunable laser sources in the 9 - 10 μm region (CO_2 laser) either in the cw or pulsed mode could provide measurements of the ozone profile by using frequencies located in the wings of the absorption lines, which are essentially influenced by the lower altitude ranges. Use of a coherent (heterodyne) detection scheme could further increase the detection limit and accuracy of such measurements. Long lived trace constituents (CO, CO_2, CH_4, N_2O, CFCs) also present strong absorption features in the IR part of the spectrum. Inter ferometric or coherent laser techniques with high spectral resolution will also make possible spaceborne observations of total tropospheric content or vertical profile of these species. Whereas airborne systems have already been developed, in particular for ozone measurements, only carbon monoxide has presently been measured from space using such spectrometric techniques. Measurements of chemically active short lived species are far beyond the capacity of present day technology.

In any case, such spaceborne systems will require highly sophisticated instrumentation which could be only implemented on the large space platforms with increased weight and power capacities as foreseen for the end of the 1990's (Polar Platform, space station, co-orbiting platform). Ozone and trace species observations in the troposphere will thus have to rely, at least for the forthcoming decade, on ground based and airborne monitoring.

5. MODEL DEVELOPMENT

Unlike most other trace gases in the troposphere, ozone is one of the few that is not emitted directly into the atmosphere from a source originating at the surface. Its presence is the result of the oxidation of precursors emitted at the surface such as nonmethane hydrocarbons (NMHC's), CO, and CH_4, in the presence of oxides of nitrogen (NOx). Because both the transport and chemical processes that influence the amount of tropospheric ozone act on a wide range of time scales, models that will be used to obtain an understanding of tropospheric ozone must also address a broad spectrum of physical and chemical processes. It is impractical to believe that a single methodology can be used to provide a true understanding of atmospheric chemistry. Thus, a hierarchy of models must be developed adapted to the particular scientific problem.

For instance, in the case of fossil fuel combustion, NMHC's are released in the presence on NOx and large build-ups of boundary-layer ozone concentrations are often observed in industrialized areas during the summer. In such a photochemical system, it is important to understand how much of this ozone is transported into the free troposphere by convection, planetary boundary layer growth and decay, and by frontal (airmass) forcing. For biomass burning, concurrent emissions of NMHC's and NOx take place and ozone is likewise produced rapidly via photochemistry in the boundary layer. Transport of boundary-layer air to the free troposphere in most regions of widespread biomass burning (i.e., the Tropics) is primarily accomplished by convective processes in the intertropical convergence zone. During transport to this meteorological regime fresh NMHC emissions from tropical forests continuously change the NMHC/NOx ratio. Most likely also the tropical forests act as an important sink for tropospheric ozone since natural hydrocarbons react with ozone to destroy it. In addition, the large leaf-area of the forests result in efficient deposition sites for the destruction term in the tropospheric ozone budget.

For a more thorough review of chemical transport modelling we refer to the report of working group I.

SUBJECT INDEX

GPSR Compliance

The European Union's (EU) General Product Safety Regulation (GPSR) is a set of rules that requires consumer products to be safe and our obligations to ensure this.

If you have any concerns about our products, you can contact us on ProductSafety@springernature.com

In case Publisher is established outside the EU, the EU authorized representative is:

Springer Nature Customer Service Center GmbH
Europaplatz 3
69115 Heidelberg, Germany

Batch number: 10815974

Printed by Printforce, the Netherlands